INTERFACIAL SUPRAMOLECULAR ASSEMBLIES

INTERFACIAL SUPRAMOLECULAR ASSEMBLIES

Johannes G. Vos
Robert J. Forster

Dublin University, Ireland

Tia E. Keyes

Dublin Institute of Technology, Ireland

John Wiley & Sons, Ltd apologises for the error made to the authors' affiliations.

Johannes G. Vos and Robert J. Forster are affiliated to Dublin City University, Ireland and not Dublin University, Ireland.

Forster, Keyes and Vos - Interfacial Supramolecular Assemblies
ISBN 0 471 49071 7

JOHN WILEY & SONS, LTD

Copyright © 2003   John Wiley & Sons Ltd, The Atrium, Southern Gate, Chichester,
West Sussex PO19 8SQ, England

Telephone (+44) 1243 779777

Email (for orders and customer service enquiries): cs-books@wiley.co.uk
Visit our Home Page on www.wileyeurope.com or www.wiley.com

This publication is designed to provide accurate and authoritative information in regard to the subject
matter covered. It is sold on the understanding that the Publisher is not engaged in rendering
professional services. If professional advice or other expert assistance is required, the services of a
competent professional should be sought.

Other Wiley Editorial Offices

John Wiley & Sons Inc., 111 River Street, Hoboken, NJ 07030, USA

Jossey-Bass, 989 Market Street, San Francisco, CA 94103-1741, USA

Wiley-VCH Verlag GmbH, Boschstr. 12, D-69469 Weinheim, Germany

John Wiley & Sons Australia Ltd, 33 Park Road, Milton, Queensland 4064, Australia

John Wiley & Sons (Asia) Pte Ltd, 2 Clementi Loop #02-01, Jin Xing Distripark, Singapore 129809

John Wiley & Sons Canada Ltd, 22 Worcester Road, Etobicoke, Ontario, Canada M9W 1L1

Wiley also publishes its books in a variety of electronic formats. Some content that appears
in print may not be available in electronic books.

British Library Cataloguing in Publication Data

A catalogue record for this book is available from the British Library

ISBN 0-471-49071-7

Typeset in 10.5/12.5pt Palatino by Laserwords Private Limited, Chennai, India
Printed and bound in Great Britain by Antony Rowe Ltd, Chippenham, Wiltshire
This book is printed on acid-free paper responsibly manufactured from sustainable forestry
in which at least two trees are planted for each one used for paper production.

ac

Contents

Acknowledgments

The authors gratefully acknowledge the ongoing financial support of Enterprise Ireland, the National Science and Technology Development Board of Ireland, the Higher Education Authority under the Programme for Research in Third Level Institutions, the Electricity Supply Board and the European Union under the Training and Mobility of Researchers Programme. Han Vos wishes to thank Dublin City University for receiving an Albert College Senior Fellowship, which enabled him to devote time to writing this book. The authors are very grateful to colleagues and students for their essential contributions over many years. Professor Anders Hagfeldt, Dr Kees Kleverlaan, Dr John Cassidy, Dr Noel Russell and Professor John Kelly are all thanked for critically reading the manuscript and for their helpful comments and suggestions.

1 Introduction

Interfacial Supramolecular Assemblies comprise an electrochemically addressable solid surface functionalized with a film which incorporates molecular components that can be addressed electrochemically or photochemically. In these assemblies, specific bonding interactions exist between the surface and film and they are generally in contact with a solution. Typical of a supramolecular assembly, the individual building blocks retain much of their molecular character, but the overall assembly exhibits new properties, or is capable of performing a specific function beyond that possible when using the individual components.

1.1 Introductory Remarks

The development of chemistry is continuously reinvigorated by discovery and innovation. Following the discovery of atoms in the 18th century, synthetic techniques were developed and, since the beginning of the 19th century, molecular chemistry involving covalent bonding has been dominant. These synthetic capabilities were complemented by theoretical understanding and structural characterization, e.g. the correct structure of benzene was proposed by Kekule in 1865. At the beginning of the 20th century, coordination chemistry was added to the armor of the chemist when Werner defined the coordination bond and this development greatly promoted the development of inorganic chemistry. At around this time, progress was also made in the development of spectroscopic techniques and the discovery of the spectroscopic lines for the various elements led to the development of quantum chemistry.

During the 1970s, the picture of chemistry was that of a sophisticated science built on a good understanding of bonding and of the physical properties and behavior of compounds. In addition, many synthetic methods had been developed and a number of powerful techniques for their characterization were available. Without techniques such as X-ray diffraction, nuclear magnetic resonance spectroscopy, infrared spectroscopy, mass spectrometry and UV–visible spectroscopy, much of today's chemistry would be unthinkable. Other important developments have been the emergence of separation science, electrochemistry and photophysics. As will be shown below, the latter two techniques are of prime importance for the development of interfacial supramolecular chemistry. With this powerful array of techniques and knowledge, chemists started to consider more and more complicated systems, and as a result, interest in the molecular aspects of biological systems developed rapidly.

This has led to a well-developed biochemistry, and has resulted in a much improved understanding of the properties of enzymes, natural photosynthesis, respiration, etc. These studies revealed that the structure of natural systems is controlled by intermolecular forces and the importance of organization and self-assembly was soon recognized.

It is against this background that the interest in intermolecular interactions has developed. The term *supramolecular chemistry* was introduced in 1978 by Jean-Marie Lehn in an article in the journal *Pure and Applied Chemistry* and was defined as *chemistry beyond the molecule*. This definition implies that supramolecular chemistry deals with intermolecular interactions and with molecular assemblies. The central concept of supramolecular chemistry is that of *organization*. In biological systems, molecular assemblies are able to carry out specific functions because they are arranged in an appropriate manner. For example, in natural photosynthesis, it is not just the spectroscopic and redox features of the components that allow for effective charge separation, but more importantly, their relative orientation and intersite separation. It is this idea of utilizing and understanding organization and interaction that has attracted so many scientists into the area of supramolecular chemistry.[1]

One of the fascinations of scientists has long been the ability of nature to use supramolecular forces to create molecular assemblies for carrying out particular functions. As a result, one of the ultimate aims of supramolecular chemistry is to create molecular devices.

1.2 Interfacial Supramolecular Chemistry

From high-speed molecular computers to optoelectronic switching, technological advances in speed and miniaturization drive the search for novel materials with enhanced electronic properties. Supramolecular chemistry has played a major role in progressing research in this area, leading to novel classes of materials which are capable of light or electrically stimulated chemistry and long-range electronic communication.

Interfacial supramolecular assemblies use well-characterized redox centers and chromophores as building blocks to create assemblies on surfaces that are purposefully structured on the molecular level, while at the same time extending over supramolecular distances. Figure 1.1 illustrates how a surface can play an important, often decisive, role in dictating the overall structure and function of a molecular component. In terms of structure, a high degree of molecular organization can be best achieved by developing supramolecular architectures at solid interfaces. The surface impacts the supramolecular system in three important ways. First, the surface provides a platform for extended two-dimensional organization of the supramolecular adsorbate. Secondly, the packing density of the molecular species on the surface allows the extent and strength of lateral interactions to be controlled so that intermolecular communication, which may be individually weak, can collectively drive the assembly of defect-free structures. Finally, since the surface itself becomes

---

[1] A selection of suitable texts for further reading on this subject are presented at the end of this chapter.

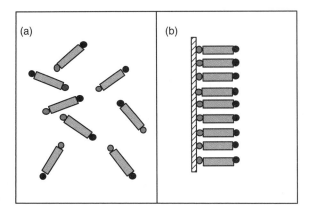

Figure 1.1 Schematic representations of (a) supramolecular and (b) interfacial supramolecular assemblies

a component of the supramolecular assembly, it participates in supramolecular function without eliminating the identity of each moiety in the structure.

A surface also provides a communicable interface, through which the adsorbate can be addressed directly. In doing so, it provides a powerful means of directing processes within the assembly. For example, if the surface is conducting it may be used to induce vectorial electron transfer.

Creating organized structures is an important goal in interfacial supramolecular chemistry. The Langmuir–Blodgett technique is important for the production of macroscopic materials that are organized on the molecular length scale. This approach allows amphillic molecules to be oriented at the air–water interface and then transferred sequentially onto a solid support. Despite the very elegant research conducted in this area, it seems unlikely that this approach will produce materials with the thermal, mechanical, and chemical stability required for practical applications. For this reason, Langmuir–Blodgett monolayers are not considered in this present book.

Synthetic flexibility is one of the most significant advantages of self-assembly allowing organic, inorganic and biological components to be used as building blocks. Organized molecular films deposited on solid surfaces are of great conceptual interest because their small thickness makes them 'quasi-ideal' two-dimensional systems. They constitute a novel 'bottom-up' approach to creating nanoscale structures. This approach contrasts with 'top-down' approaches that entail making existing devices so small that they eventually finish up as nanosized objects, with dimensions of no more than a few hundred nanometers. The top-down approach is typified by the manufacture of transistors on computer chips. Currently, such transistors are only ca. 200 nm in size and it is widely anticipated that they will break the 100 nm barrier in the near future. The bottom-up approach, in which interfacial supramolecular assemblies (ISAs) play an intimate part, involves constructing nanodevices from their constituent parts, i.e. atoms or small molecules. Fabrication is achieved by either physical relocation of the building blocks into their required locations, or by using molecular self-assembly. The former route involves techniques such as the use of laser tweezers or atomic force microscopy. However, the process is laborious and molecules occasionally stick to the substrate and break apart. Chemical manipulation of the kind described in this book is more elegant and vastly more subtle because it relies on instructions programmed into the system to determine the ultimate location of each building block.

The parallels with nature are obvious. Biotechnology is the only fully functional nanotechnology and life itself is intrinsically interfacial. Atomic-scale construction and information processing are mediated on the surfaces of protein and nucleic acid catalysts. Biological systems excel at atom-by-atom or molecule-by-molecule manipulation. Take, for example, the origins of life itself, namely a fertilized ovum, which is programmed to build, molecule-by-molecule, the most complex of self-assembled constructs, a living organism. In the laboratory, nothing even remotely as complex could be attempted in the foreseeable future, although it does not prevent scientists from deriving their inspiration from such complex functions. Simple instructions such as switchable lateral interactions, site-selective functionalization to create surface patterns, as well as self-healing and replication, can currently be encoded. These advances allow ISAs to be created that exist in the solid (ordered) regime, but are close to an order–chaos phase boundary, i.e. their structure is influenced by external factors. For example, permeation into an ISA can be switched on and off by the presence of key molecules in solution.

1.3 Objectives of this Book

The primary goal is to provide a molecular-level understanding of how ISAs are designed for specific functions, created, characterized and then used to address fundamental issues such as the distance dependence of energy and electron transfer, as well as applications such as molecular switching. This objective will be achieved by examining how the interplay of the physical and electronic structure, morphology and dynamical properties of an ISA influence its overall properties and functioning.

The intention is not to comprehensively review the literature that describes the multidisciplinary efforts of researchers to create interfacial supramolecular assemblies. The literature in this area is vast and involves research programs in chemistry, physics and biology, as well as analytical, materials and surface sciences. Rather, key examples of advances that have significantly influenced the field and will direct its future development are presented. In addition, some of the analytical methods, theoretical treatments and synthetic tools, which are being applied in the area of interfacial supramolecular chemistry and are driving its rapid development, will be highlighted.

1.4 Testing Contemporary Theory Using ISAs

Supramolecular chemistry has provided an experimental platform for testing many modern theories on bonding, molecular organization, photochemistry, and in particular, electron transfer theory. For example, in 1956, Marcus predicted that highly exoergic electron transfer reactions actually slow down with increasing driving force. Numerous bimolecular electron transfer reactions were studied in order to test this prediction, but slow diffusional mass transport inevitably limited the range of conditions under which rate measurements could be made. Supramolecular chemistry provided the first rigorous proofs of the validity of these contemporary theories by linking donor and acceptor species within a single electronically communicating entity. This approach allows the electron transfer rate-limiting reactions to be studied over a much wider range of driving forces.

Electron transfer remains one of the most important processes explored when using interfacial supramolecular assemblies and given the emerging area of molecular electronics, this trend is set to continue. Therefore, Chapter 2 outlines the fundamental theoretical principles behind the electrochemically and photochemically induced processes that are important for interfacial supramolecular assemblies. In that chapter, homogeneous and heterogeneous electron transfer, photoinduced proton transfer and photoisomerizations are considered.

1.5 Analysis of Structure and Properties

Modern surface analytical tools make it possible to probe the physical structure as well as the chemical composition and reactivity of interfacial supramolecular assemblies with unprecedented precision and sensitivity. Therefore, Chapter 3 discusses the modern instrumental techniques used to probe the structure and reactivity of interfacial supramolecular assemblies. The discussion here is focused on techniques traditionally applied to the interrogation of interfaces, such as electrochemistry and scanning electron microscopy, as well as various microprobe techniques. In addition, some less common techniques, which will make an increasing contribution to supramolecular interfacial chemistry over the coming years, are considered.

1.6 Formation and Characterization of Interfacial Supramolecular Assemblies

Chapter 4 discusses the formation and properties of spontaneously adsorbed monolayers, self-assembled monolayers and thin polymer films. This chapter considers how molecules can be immobilized on a surface in a controlled manner to create a useful ISA. The structural features of the layers are also considered. Self-assembled and spontaneously adsorbed monolayers offer a facile means of controlling the chemical composition and physical structure of a surface. These monolayers can exhibit low defect densities and high degrees of structural order over supramolecular distances. In contrast, polymeric ISAs tend to exhibit poorly defined primary structures, but their secondary structure is strongly influenced by external factors such as temperature, contacting solvent, ionic strength, etc. The possibility of controlling this secondary structure to achieve a specific function, e.g. exclusion of an interference in a chemical sensing application, is one of the most attractive features of these ISAs. Finally, Chapter 4 considers likely developments in the future, in particular, the role of molecular self-assembly in developing nanotechnology.

1.7 Electron and Energy Transfer Properties

Understanding those factors that control electron and energy transfer is not only of fundamental interest, but is vital for creating molecular electronic devices. Chapter 5 describes selected case studies which illustrate the key factors that control electron and energy transfer within interfacial supramolecular assemblies and especially across solid–film interfaces. In doing so, it seeks to identify those approaches that provide key fundamental insights and show the greatest promise for creating electrochemically and photochemically triggered molecular switches, sensors and biomimetic systems. It also considers the major challenges for the future and barriers to progress in the area.

Interfacial monolayer, multilayer and polymer species which exhibit interesting examples of light and electrically stimulated functions such as isomerization and proton transfer in ISAs are also presented in this chapter. Such materials may represent the precursors for electrooptic switches and addressable molecular-based machines.

1.8 Interfacial Electron Transfer Processes at Modified Semiconductor Surfaces

The development of functional supramolecular devices remains mainly conceptual. However, photovoltaic devices are one of the few exceptions. Dye-sensitized nanocrystalline semiconductor materials have received significant interest as a result of their application in solar energy conversion.

Chapter 6 takes the much studied supramolecular dye-sensitized TiO_2 as an example of an operational supramolecular interfacial device. The fundamental operation of these devices are discussed, including their mechanism of operation. The application of modified semiconductor surfaces as electrochromic devices is also considered.

In conclusion, this book is intended as an overview of the principles behind and state-of-the-art in interfacial supramolecular chemistry. The book is suitable for researchers and graduate students and focuses on assemblies that demonstrate at least the potential to produce useful devices such as solar cells, electrochromic devices, molecular wires, switches and sensors which are addressable by using electrochemical and optical stimuli. Molecular materials for nanoscale molecular devices remain an intriguing conceptual possibility.

Further Reading

Balzani, V. and Scandola, F. (1991). *Supramolecular Photochemistry*, Ellis Horwood, Chichester, UK.

Hamilton, A.D. (Ed.) (1996). *Supramolecular Control of Structure and Reactivity*, Wiley, Chichester, UK.

Kaifer, A.E. and Gomez-Kaifer, M. (1999). *Supramolecular Electrochemistry*, Wiley, Chichester, UK.

Kuhn, H. and Forsterling, H.-D. (2000). *Principles of Physical Chemistry: Understanding Molecules, Molecular Assemblies and Supramolecular Machines*, Wiley, Chichester, UK.

Lehn, J.-M. (1995). *Supramolecular Chemistry*, VCH, Weinheim, Germany.

Lindoy, L.F. and Atkinson, I.M. (2000). *Self-Assembly in Supramolecular Systems*, The Royal Society of Chemistry, Cambridge, UK.

Steed, J.W. and Atwood, J.L. (2000). *Supramolecular Chemistry, A Concise Introduction*, Wiley, Chichester, UK.

Vögtle, F. (1991). *Supramolecular Chemistry*, Wiley, Chichester, UK.

2 Theoretical Framework for Electrochemical and Optical Processes

Electrochemical and photochemical processes are the most convenient inputs and outputs for interfacial supramolecular assemblies in terms of flexibility, speed and ease of detection. This chapter provides the theoretical background for understanding electrochemical and optically driven processes, both within supramolecular assemblies and at the ISA interface. The most important theories of electron and energy transfer, including the Marcus, Förster and Dexter models, are described. Moreover, the distance dependence of electron and energy transfer are considered and proton transfer, as well as photoisomerization, are discussed.

2.1 Introduction

Electron, energy and proton transfer or molecular rearrangements are the most important events that occur in interfacial supramolecular assemblies. In this chapter, the general theories of electron transfer, both within ISAs and across the film/electrode interface, are described. Moreover, photoinduced electron, energy and proton transfer processes are discussed. As this book focuses on supramolecular species, the treatment is restricted to intramolecular or interfacial processes without the requirement for prior diffusion of reactants.

The objective of this present chapter is to provide a broad overview of the relevant theories to support understanding of subsequent chapters. For the interested reader, excellent and extensive reviews are available, which are recommended as a supplement to this chapter [1,2].

2.2 Electron Transfer

The Marcus theory is the most widely applied theory used to describe electron transfer reactions and is equally applicable to photoinduced, interfacial and thermally driven electron transfers. The fundamental difference between these processes lies primarily in the nature of the driving force. In the case of heterogeneous electron transfer, this may be controlled externally through an applied potential, whereas in homogenous and photoinduced electron transfer it is dictated by the electronic nature of the reactants.

2.2.1 Homogenous Electron Transfer

Marcus theory

Marcus theory remains the cornerstone on which many more sophisticated models of electron transfer have been based and is still broadly applied to both homogeneous and heterogeneous electron transfer reactions [3]. In the following section, the generic 'supermolecule', A–L–B, is used as a model to represent the system undergoing electron transfer, in which A and B are photo- or redox-active units that are connected through bridge L, which may be a physical linkage or a through space interaction. A or B may also represent a surface, for instance, an electrode or semiconductor in the case of interfacial electron transfer.

The homogeneous electron transfer process can then be represented as follows:

$$A–L–B \longrightarrow A^{\bullet+}–L–B^{\bullet-}$$

Figure 2.1 illustrates the classical model in which simple parabolas of equal curvature represent the free energies of the reactant state, A–L–B, prior to electron transfer and the product state, $A^{\bullet+}–L–B^{\bullet-}$, after electron transfer, as a function of the reaction coordinate. Parabolas are employed since the Marcus theory assumes that the reaction coordinates conform to a simple harmonic oscillator model. Two factors influence the rate at which an electron moves from the reactant to product curves. The first of these is the Franck–Condon Principle, which states that electron transfer is an instantaneous process. Therefore, no change in nuclear configuration, either inner sphere or outer sphere, can occur during the electron transfer. The second is the First Law of Thermodynamics, which states that energy must be conserved, and therefore electron transfer must be an isoenergetic process. The only point where both requirements are fulfilled is if crossover occurs at point * on the intersecting parabolas of Figure 2.1(a).

Under these circumstances, the rate of electron transfer is described by the following equation:

$$k_{ET} = \nu \exp(-\Delta G^{\ddagger}/RT) \tag{2.1}$$

where R is the gas constant, T the absolute temperature, ν the frequency factor, which describes the rate of reactive crossings of the transition state, and $\Delta G^{\ddagger}$ is the Marcus free energy of activation, as depicted on Figure 2.1. This parameter exhibits a quadratic dependence on ΔG^0, the standard free energy of the reaction, according to the following equation:

$$\Delta G^{\ddagger} = \frac{(\Delta G^0 + \lambda)^2}{4\lambda} \tag{2.2}$$

where λ is the total reorganization energy, i.e. the energy required to distort the reactant geometry and its surrounding media to attain the equilibrium configuration of the product state. The reorganization energy is comprised of two contributions, i.e. an outer-sphere component, λ_{out}, reflecting the contribution from reorganization

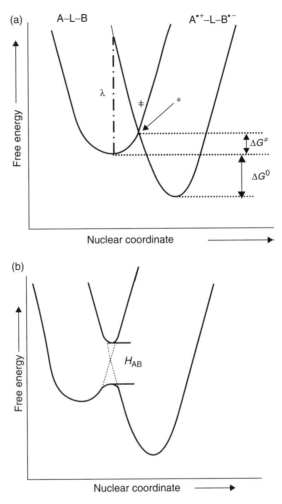

Figure 2.1 Parabolas representing (a) diabatic and (b) adiabatic electron transfer reactions

of solvent and surrounding media, and an inner-sphere component, λ_{in}, associated with changes in the molecular geometry of the reactant as it reaches the product state. These reorganization energies are defined in Equations (2.3) and (2.4) respectively, whereby $\lambda = \lambda_{in} + \lambda_{out}$.

$$\lambda_{in} = \sum_j \frac{f_j^r f_j^P}{f_j^r + f_j^P} (\Delta q_j)^2 \tag{2.3}$$

where f_j^r is the jth normal mode force constant in the reactant species, f_j^P is the jth normal mode force constant in the product species, and Δq_j is the equilibrium displacement of the jth normal coordinate.

The inner-sphere component of the reorganization energy represents the minimum energy required to change the internal structure of the redox center to its nuclear transition state configuration. Equation (2.3) is derived from the classical harmonic oscillator model and is an expression of the free energy associated with bond length changes accompanying electron transfer. It requires knowledge of the force constants associated with all molecular vibrations from both the reactant and product states. Because of the large number of parameters involved it is difficult to calculate. However, resonance Raman spectroscopy has been employed to provide estimates of oscillator frequencies and distortions for electron transfer reactions [4]. Nonetheless, many of the redox-active species typically employed in interfacial supramolecular assemblies, such as ferrocene, fullerene and $[Ru(bpy)_3]^{2+}$, undergo small or negligible bond length changes on oxidation or reduction and the most important contribution to λ is frequently from the outer-sphere reorganization energy. The latter contribution, λ_{out}, is given by the following:

$$\lambda_{out} = \frac{(\Delta e^2)}{4\pi \,\epsilon_0} \left(\frac{1}{2R_D} + \frac{1}{2R_A} - \frac{1}{r_{DA}} \right) \left(\frac{1}{\varepsilon_{op}} - \frac{1}{\varepsilon_s} \right) \tag{2.4}$$

where e is the electronic charge, ϵ_0 is the permittivity of free space, R_D and R_A are the radii of the donor and acceptor moieties, respectively, r_{DA} is the intramolecular distance between the donor and acceptor, and ε_{op} and ε_s are, respectively, the optic and static dielectric constants of the medium. This expression for the outer-sphere reorganization energy derives from the dielectric continuum theory and reflects changes in the polarization of the solvent molecules following electron transfer. The outer-sphere, or solvent, component of the free energy of activation arises because the charge on the redox center typically changes significantly during the electron transfer event. As indicated by Equation (2.4), the reorganization energy, and therefore the free energy of activation for outer-sphere reorganization, depends on both the static and optical dielectric constants of the solvent.

Equation (2.2) predicts a parabolic relationship between the free energy of activation and the overall free energy of electron transfer. As illustrated in Figure 2.2(a), over the range of ΔG^0 values known as the normal free energy region, the electron transfer rate is predicted to increase as ΔG^0 increases. As shown in Figure 2.2(b), when $\lambda = \Delta G^0$ and $\Delta G^{\ddagger} = 0$, the electron transfer becomes 'activationless', and the electron transfer rate, k_{ET}, reaches a maximum. As illustrated in Figure 2.2(c), a further increase in driving force leads to ΔG^0 becoming less than $\Delta G^{\ddagger}$, thus causing

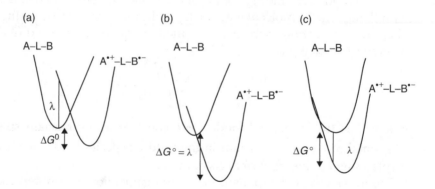

Figure 2.2 Reaction coordinate parabolas for diabatic electron transfer, illustrating the relationship between ΔG^0 and λ for: (a) the normal region, where $\Delta G^0 < \lambda$; (b) the activationless electron transfer region, where $\Delta G^0 = \lambda$; (c) the inverted region, where $\Delta G^0 > \lambda$

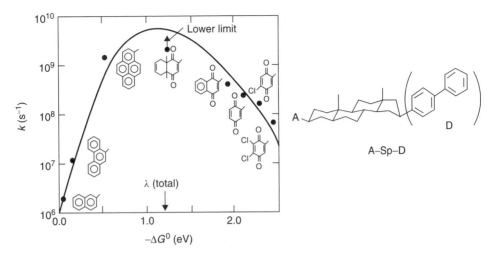

Figure 2.3 Evidence for the Marcus inverted region from intramolecular electron rate constants as a function of ΔG^0 in methyltetrahydrofuran solution at 206 K. Reprinted with permission from G.L. Closs, L.T. Calcaterra, H.J. Green, K.W. Penfield and J.R. Miller, *J. Phys. Chem.*, **90**, 3673 (1986). Copyright (1986) American Chemical Society

a decrease in the rate of electron transfer with any further increase in reaction driving force. The regime in which this behavior is observed is known as the *Marcus inverted region*.

Although the experiment involves a solution-phase supramolecular assembly, Figure 2.3 shows a plot of the intramolecular electron transfer rates versus the driving force for a biphenyl donor (D) and a series of acceptors (A) bridged across a cyclic alkyl bridge. This work, reported by Closs and co-workers [5], illustrates experimentally the trends predicted by the Marcus theory, i.e. $\log k$ vs. ΔG^0 produces a Gaussian shaped curve as electron transfer rates decrease at high driving forces. Confirmation of the apparently counterintuitive 'inverted region' was initially elusive, primarily because the early investigations focused on bimolecular redox reactions, where at high driving forces, diffusion of reagents rather than the electron transfer became rate determining. This obscured any evidence for the inverted region. However, through supramolecular synthesis, the inverted region has now been confirmed across a significant number of intramolecular photoinduced electron transfer reactions in solution.

The frequency factor, ν, from Equation (2.1), is the product of ν_n, the critical vibrational frequency associated with promotion of the transferring electron onto the product surface and the transmission coefficient, κ_{el}, which describes the probability of crossing over from the reactant to the product hypersurfaces once the transition state has been reached, according to the following:

$$\nu = \nu_n \kappa_{el} \tag{2.5}$$

For reactions in which electron transfer significantly distorts the bond lengths and angles of the molecule, the frequency factor is typically in the range 10^{13} to 10^{14} s^{-1}. In contrast, if the molecular structure is largely unperturbed by redox switching, then v_n is dictated by the dynamics of solvent reorganization, being typically in the range of 10^{11} to $10^{12}\,s^{-1}$.

As shown in Figure 2.1, electron transfer is basically a tunneling process, which may occur diabatically or adiabatically. The fundamental distinction between these forms of electron transfer lies in the degree of electronic coupling between the donor and acceptor orbitals. The degree of coupling is reflected in the magnitude of κ_{el}, which typically varies between zero and unity. In the case of strong electronic coupling, e.g. where reactants are linked by short conjugated bridges, then, as shown in Figure 2.1(b), there is significant flattening of the reaction hypersurface close to the transition state. Under these circumstances, the rate of crossing the barrier region is reduced, although the probability of electron transfer actually occurring once the transition state has been achieved is close to unity and the electron transfers across a single energy surface in passing from reactant to product. For this adiabatic process, the electronic transmission factor is approximately unity and Equation (2.1) then simplifies to Equation (2.6) below. Therefore, the maximum electron transfer rate of an adiabatic process is determined by v_n and hence is sensitive to Franck–Condon factors.

$$k_{ET} = v_n \exp(-\Delta G^{\ddagger}/RT) \qquad (2.6)$$

Figure 2.1(a) above illustrates the potential energy surface for a diabatic electron transfer process. In a diabatic (or non-adiabatic) reaction, the electronic coupling between donor and acceptor is weak and, consequently, the probability of crossover between the product and reactant surfaces will be small, i.e. for diabatic electron transfer κ_{el}, the electronic transmission factor, is $\ll 1$. In this instance, the transition state appears as a sharp cusp and the system must cross over the transition state onto a new potential energy surface in order for electron transfer to occur. Long-distance electron transfers tend to be diabatic because of the reduced coupling between donor and acceptor components; this is discussed in more detail below in Section 2.2.2.

As discussed above, this classical Marcus theory, introduced the concept that the rate of electron transfer is governed separately by electronic and nuclear (Franck–Condon) factors. The nuclear terms are responsible for the dependence of the electron transfer rate on solvent, temperature and exergonicity $(-\Delta G^0)$. The electronic factors account for the distance dependence and the dependence on the nature of the bridge. Marcus theory provides a good description of empirical data in the normal regime. However, this theory is not suitable for electron transfer reactions at low and intermediate temperatures and in particular for processes in which high-frequency intramolecular modes dominate λ. In such instances, poor fits in the Marcus inverted region are observed. In the following sections, the most widely used extensions of the Marcus theory, based on semi-classical and quantum mechanical models, developed to overcome these limitations, are outlined.

Semi-classical model of electron transfer

The semi-classical extension of electron transfer theory evolved from models developed by Landau, Zener, Marcus, and Hush [3,6]. The semi-classical Marcus–Hush model of electron transfer, is represented by the following equation:

$$k_{ET} = \nu_n \kappa_{el} \exp\left[\frac{-(\Delta G^0 + \lambda)^2}{4\lambda k_B T}\right] \tag{2.7}$$

where k_B is the Boltzmann constant. The difference between this expression and Equation (2.1) lies in the interpretation of ν_n and κ_{el}, i.e. the critical vibrational frequency and electronic transmission coefficient, respectively. According to semi-classical theory, for a diabatic electron transfer this quantity is defined by the following expression:

$$\nu_n \kappa_{el} = \frac{4\pi^2 |\mathbf{H}_{AB}|^2}{h}\left(\frac{1}{4\pi\lambda k_B T}\right)^{1/2} \tag{2.8}$$

Therefore, according to this approach, k_{ET} is a function of the free energy of the electron transfer, ΔG^0, the electronic coupling matrix between donor and acceptor, $|\mathbf{H}_{AB}|$, and the total reorganization energy, λ. The matrix coupling element is proportional to the overlap of the electronic wavefunctions of the donor and acceptor. According to this treatment, the electronic transmission coefficient is proportional to $|\mathbf{H}_{AB}|^2$. Under adiabatic conditions, $\kappa = 1$ and $|\mathbf{H}_{AB}| \geq 50$ cm^{-1}, while in contrast, for a diabatic electron transfer process, $\kappa \ll 1$ and $|\mathbf{H}_{AB}| \ll 50$ cm^{-1}. A common approach to using this model [7] involves fitting plots of k_{ET} versus the driving force, ΔG^0, to Equation (2.7) to yield estimates of λ and $|\mathbf{H}_{AB}|$.

Alternatively, rearrangement of Equations (2.2), (2.7) and (2.8) yields a linearized form of the semi-classical expression, as follows:

$$\ln\left[k_{ET}(\lambda T)^{1/2}\right] = \ln\left[\frac{2\pi}{\hbar}\frac{|\mathbf{H}_{AB}|^2}{(4\pi k_B)^{1/2}}\right] - \frac{\Delta G^{\ddagger}}{k_B T} \tag{2.9}$$

where $\hbar = h/2\pi$.

Plots of $\ln[k_{ET}(\lambda T)^{1/2}]$ versus $\Delta G^{\ddagger}/k_B T$ may then be employed to yield estimates of $|\mathbf{H}_{AB}|$ from the intercept. In such an approach, $\Delta G^{\ddagger}$ is estimated from Equation (2.2), λ_{out} is obtained from Equation (2.4), and an approximation of λ_{in} is typically made on the basis of spectroscopic data.

Quantum mechanical model of electron transfer

A quantum mechanical extension of Marcus theory was developed by Jortner to deal with inadequacies in the classical theory for describing nuclear tunneling in the inverted region. In the simplest description, a single mean vibrational frequency, ω, characterizes the intramolecular reorganization modes [8]. This approach takes greater account of diabatic electron transfer in the low- and intermediate-temperature regimes, i.e. $k_B T < \hbar\omega_\sigma \ll \hbar\omega$ and $\hbar\omega_s < k_B T \ll \hbar\omega$, respectively, where quantum effects become significant for high-frequency modes. In the high-temperature region, i.e. $\hbar\omega_s \ll \hbar\omega < k_B T$, electron transfer behaves classically and the activation energy is described by the Marcus theory.

The quantum mechanical approach is based on time-dependent perturbation theory and is derived from Fermi's Golden Rule for non-radiative decay processes [1].

This states that the quantum mechanical model for diabatic electron transfer is given by the following expression:

$$k_{ET} = \frac{4\pi^2}{h} |\mathbf{H}_{AB}|^2 \langle FC \rangle \tag{2.10}$$

In the above, FC is the Franck–Condon term and is defined as follows:

$$\langle FC \rangle = \left(\frac{1}{4\pi \lambda_S k_B T} \right) \sum_{w=0}^{\infty} \left(\frac{e^{-S} S^w}{w!} \right) \exp \left\{ - \left[\frac{(\lambda_S + \Delta G^0 + wh\omega)^2}{4\lambda_S k_B T} \right] \right\} \tag{2.11}$$

where $S = \lambda_{in}/\hbar\omega$, i.e. the inner sphere reorganization energy in units of vibrational quanta, and λ_S is the classical reorganization free energy described earlier in Equations (2.3) and (2.4). The summation is made over the quantum levels w, of this averaged high-frequency mode. Calculations based on this expression are, in general, computationally modeled and fitted to experimental data.

Figure 2.4 [9] shows that the principal differences between the semi-classical and quantum approaches described above are evident in plots of the dependence of k_{ET} on ΔG^0. Both models predict similar behavior in the normal and activationless regions for diabatic electron transfer processes, i.e. that k_{ET} increases with increasing driving force and then levels off. The main disparity between the two approaches lies in the inverted region where the influence of ΔG^0 on k_{ET} is less pronounced in the quantum model. In general, the less computationally intensive semi-classical model is used to describe electron transfer occurring in the normal region. Elucidation of the best fit of experimental data to semi-classical and quantum models reveals important information on the role of inner- and outer-sphere reorganization in controlling the rate of electron transfer. The Gaussian behavior predicted by the semi-classical model indicates that nuclear tunneling plays a minor role in dictating

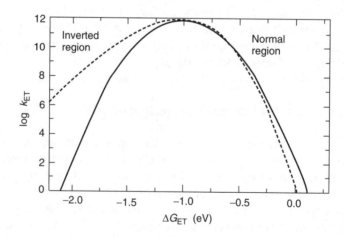

Figure 2.4 Computationally generated plots of log k_{ET} versus ΔG^0, where the continuous and dashed lines are calculated by using data obtained from semi-classical and quantum theories, respectively. Reprinted from K.S. Schanze and K.A. Walters, in *Molecular and Supramolecular Photochemistry*, Vol. 3, V. Ramamurthy and K.S. Schanze (Eds), Marcel Dekker, New York, 1998, Ch. 3, p. 80, Figure 2, by courtesy of Marcel Dekker Inc.

the electron transfer rate. In contrast, the lower slope of the inverted region in the quantum model indicates significant contributions to electron transfer from nuclear tunneling and the importance of intramolecular distortion in going from reactant to product surfaces. Quantum mechanical models of electron transfer are also available which are not limited to a single critical vibrational mode, as in the model described here. Naturally, application of such models requires significantly more computation, and the treatments presented here are the simplest and most popular models.

Influence of the bridge on the electron transfer rate

Supramolecular chemistry has proven to be a powerful tool for experimentally testing modern electron transfer theory. It has done so by providing materials for exploring the impact of parameters such as driving force, transfer distance and bridge electronic nature on intramolecular electron transfer rates.

Extensive solution-phase studies of supramolecular donor–acceptor systems have demonstrated that the bridge or spacer between interacting units is fundamentally important in dictating the extent and rate of intermolecular electron transfer. In both the semi-classical and quantum mechanical expressions for electron transfer dynamics, the primary origin of this distance dependence is in $|H_{AB}|$, the electronic coupling matrix element. This element is proportional to the extent of the overlap of the orbital wavefunctions of the donor and acceptor states. A useful empirical model to describe the decay of the electronic coupling matrix element with distance between the donor and acceptor states is given by Equation (2.12), which describes electron tunneling through a one-dimensional square barrier as follows:

$$H_{AB}(d) = H_{AB}(d_0) \exp\left[\frac{-\beta(d - d_0)}{2}\right] \tag{2.12}$$

where $H_{AB}(d_0)$ is the coupling element when donor and acceptor are separated by the van der Waals distance, d_0, $d(\geq d_0)$ is the center-to-center distance between donor and acceptor moieties, and β is the tunneling parameter, a constant corresponding to the rate of decay of coupling with distance. For example, β is 3 Å^{-1} for electron transfer across a vacuum, and typically lies between 0.7 and 1.2 Å^{-1} for donor–acceptor systems across most bridges including aromatic, aliphatic or protein linkers.

Studies on electron transfer rates through proteins, DNA and in supramolecular systems have shown that generally such rates decrease exponentially with increasing donor–acceptor distance according to the following equations [10]:

$$|H_{AB}| \propto \exp(-\tfrac{1}{2}\beta_{AB}r) \tag{2.13}$$

$$k_{ET} \propto \exp(\beta r) \tag{2.14}$$

$|H_{AB}|$ is also influenced by the electronic nature of the bridge between the active components as well as the relative orientation of these moieties. Consequently, these bridge attributes also impact on the electron transfer rate. Unsurprisingly, studies on donor–acceptor systems bridged through aromatic or conjugated covalent linkages reveal that π-bonds support fast electron transfer. Studies on donor–acceptor systems bridged through variable-length cyclo-alkyl chains have illustrated that

σ-bonds are equally effective in supporting efficient long-range electron transfer through a superexchange mechanism [11]. In a particularly revealing study, De Rege *et al.* linked identical porphyrin donor and acceptors moieties via π-conjugated σ-bonded bridges, and hydrogen (H)-bonded bridges [12], as shown in Figure 2.5. The driving forces and donor–acceptor distances across the complexes were essentially identical. Interestingly, the largest k_{ET}, at $8.1 \times 10^{-9}\,\text{s}^{-1}$, was observed for the H-bonded bridge. This study illustrated very effectively that covalency is not a prerequisite for efficient electron conduction and in particular that H-bonds are effective electron conductors. However, the mechanism by which the H-bond supports electron transfer remains poorly understood, and will presumably be an important focus for future electron transfer studies in supramolecular systems, not least because of the importance of H-bonding in mediating electron transfers in proteins.

Electron transfer has also been demonstrated to be mediated through solvents. Much recent work in this area has been conducted on supramolecular donor–acceptor systems with bridges that are sufficiently flexible to allow the centers to approach each other to within approximately 10 to 15 Å in a U-shaped configuration, while still minimizing through-bridge electron transfer [13,14]. In principle, any interaction between the active components in a flexible bridge may be mediated via the intervening solvent molecules and not necessarily the bridge.

Mechanisms of bridge-mediated electron transfer

A number of mechanisms for bridge-mediated electron transfer have been identified, all of which depend on the chemical nature of the bridge, the relative symmetry of the bridge and the reactants, and the distance between the active sites. Intramolecular electron transfer between donor and acceptor can be classified according to the properties of the intervening electronic orbitals and the distance between the donor and acceptor. Short-range processes, where the donor and acceptor are intimately linked, are simply electron and hole transfers. In long-range processes, the hole or electron transfers are bridge-mediated, and may occur via superexchange or hopping mechanisms.

Figure 2.6 illustrates the distinction between hole and electron transfers. Electron transfer involves interaction of the Lowest Unoccupied Molecular Orbital (LUMO) state of the donor (in the case of photoinduced electron transfer, with the LUMO state of the acceptor molecule). This operates over very short bridge lengths or through direct orbital overlap between donor and acceptor and is mediated through bridges with low-lying π or π^* LUMO orbitals capable of accepting and transferring an electron. Hole (an electron-deficient site) transfer is similarly observed for directly coupled donors and acceptors or through short bridging interactions. In this instance, the bridge LUMO is high in energy and unable to support electron transfer. The transfer of the positive hole is mediated by the Highest Occupied Molecular Orbital (HOMO) levels of donor and acceptor.

Electron transfer occurring over larger distances has generated significant interest as a result of its importance in understanding biological systems. Long-range electron transfers are also particularly important in photochemically active interfacial

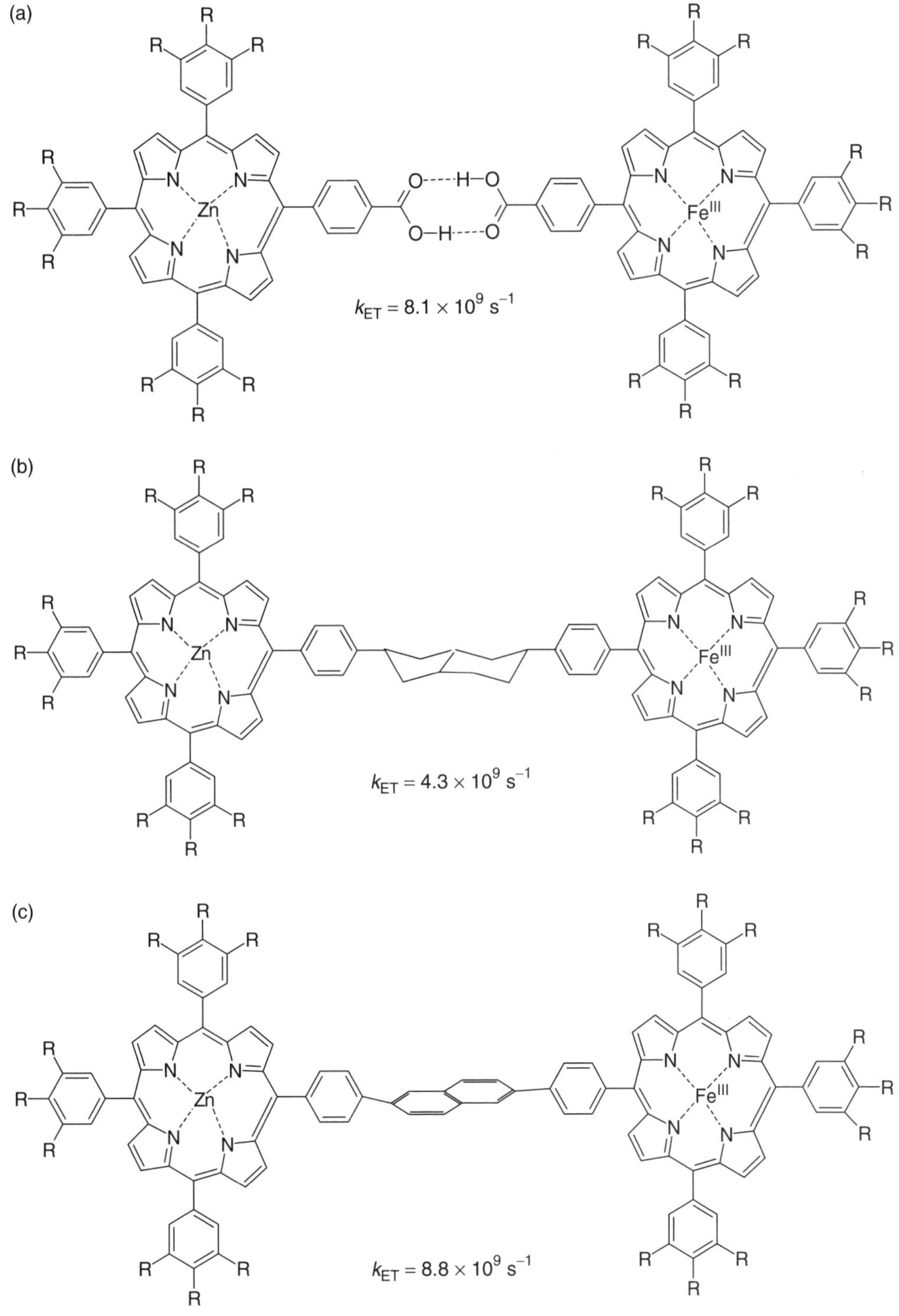

Figure 2.5 Molecular structures of zinc–iron porphyrin complexes across (a) hydrogen, (b) aliphatic and (c) aromatic bridges, and the corresponding rates of photoinduced electron transfer for each species, as reported by Rege *et al.* [12]

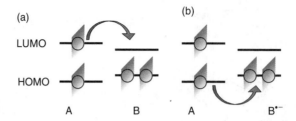

Figure 2.6 Schematic representations of the distinction between (a) electron and (b) hole transfer between the LUMO and HOMO orbitals of donor (A) and acceptor (B) species. Electron transfer is mediated through the LUMO levels, whereas hole transfer is achieved via the HOMO levels

systems since long bridges are frequently necessary to prevent surface quenching of excited states. Electronic coupling decreases exponentially with distance, and therefore long-range electron transfer is almost exclusively diabatic and is usually conveniently modeled by using the semi-classical Marcus–Hush theory or the diabatic quantum mechanical theories described above. The most important bridge-mediated mechanisms are superexchange and electron hopping. Observation of one or the other mechanisms depends on the extent of resonance between the donor and acceptor and the bridge orbitals. Figure 2.7 illustrates schematically the superexchange and electron hopping mechanisms.

Superexchange occurs when the bridge and donor energy levels are off-resonance, typically when the bridge levels lie in excess of 2 eV higher than the donor levels.

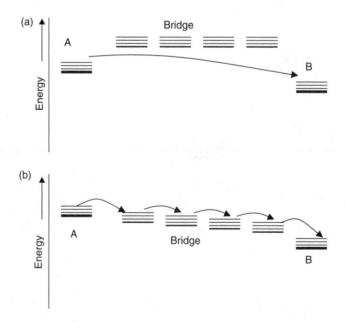

Figure 2.7 Schematic illustrations of the proposed mechanisms for electron transfer through a bridge: (a) superexchange, in which electron transfer is a concerted process mediated through a bridge whose energy levels lie well outside of resonance with the donor orbitals; (b) electron hopping, where the electron is transferred sequentially through a bridge whose energy levels lie in close resonance with the donor orbitals

Under such conditions, the transferring electron or hole does not populate the bridge levels directly. The role of the bridge in this instance is to spatially extend the wavefunctions of the donor and acceptor moieties, thus allowing them to couple. In doing so, high-energy virtual hole and electron transfer states arise within the bridge, and the electron will then transfer coherently between the donor and acceptor in a single-step process. In the superexchange mechanism, the electron transfer rate decreases exponentially with the distance between donor and acceptor.

Electron conduction, or resonant superexchange, occurs when the bridge virtual states are in close resonance with the donor. In this instance the transferring electron actually occupies the bridge electronic levels and transports in a sequential manner, from bridge state to bridge state, as a *polaron*. The rate of electron transport may be described as an additive function of each electron transfer step through the bridge. The electron conduction rate decreases inversely with distance between donor and acceptor. In a series of A–L–B complexes in which the length, but not the nature, of L varies, the dependence of electron rate on distance may be employed to distinguish between the two mechanisms. This conduction behavior is observed in 'molecular wires', which have low-lying, typically less than 1 $k_B T$, energy gaps between the donor and bridge orbitals. Examples of such systems include conducting polymers such as the polypyrroles and polyacetylene [15].

2.2.2 Heterogeneous Electron Transfer

Understanding the effects of distance, molecular structure, solvent reorganization and microenvironment on the dynamics of electron transfer has been revolutionized by investigations using interfacial supramolecular assemblies, especially self-assembled and spontaneously adsorbed monolayers. An advanced understanding of how these processes impact the dynamics of heterogeneous electron transfer across the electrode/assembly interface will inevitably dictate the success of interfacial supramolecular chemistry in the development of useful devices. With the advent of microelectrodes and high-performance instrumentation, heterogeneous electron transfer reactions can now be studied under unusual conditions and over short time-scales. The ability to make localized, even atomically, resolved measurements of electron transfer dynamics allows precise maps of surface reactivity to be constructed. In tandem with these experimental advances, sophisticated theoretical models now exist which allow general conclusions to be drawn from specific chemical systems. The advantages of electrochemistry over its traditional spectroscopic rivals are becoming increasingly widely recognized, in particular its ability to provide *direct* information and to probe specific interactions with tremendous sensitivity.

The elementary steps of heterogeneous electron transfer

Section 2.2.1 above introduced the concept of homogenous electron transfer in a generic donor acceptor supermolecule A–L–B. Here, the most commonly applied theoretical treatments of heterogeneous electron transfer are discussed. In this scenario, A is now an electrode characterized by continuous or semi-continuous

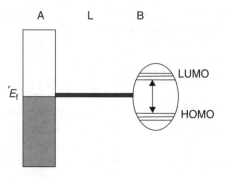

Figure 2.8   Schematic of a typical interfacial supramolecular assembly, A–L–B, when A is a metal electrode, illustrating the relative molecular band structures of A and B, respectively. E_f is the fermi level

density of states instead of the molecular orbitals which characterize a molecular moiety. Figure 2.8 presents a simple schematic illustrating the different orbital distributions of an adsorbed molecule on a conducting surface.

Figure 2.9 illustrates the elementary steps that are involved in the overall electron transfer process for a reactant that is adsorbed on an electrode surface, namely thermal activation, electronic coupling of the redox center with the electrode, and the instantaneous elementary electron transfer event itself.

Thermal activation

In many circumstances, it is convenient to think of the energy of the highest occupied molecular orbital (HOMO) donor state, or lowest unoccupied molecular orbital (LUMO) acceptor state, as being well defined and static in nature. However, the energies of these donor or acceptor states vary continuously. Random thermal fluctuations, as well as the ingress and egress of solvent dipoles from the solvation shell, cause the energy levels to fluctuate about some mean value. These fluctuations may make it easier or more difficult to oxidize or reduce the molecule at any particular instant. Electron transfer only occurs when the energies of the electrode and molecular states are identical, i.e. when resonance has been achieved. Typically, the percentage of oxidized adsorbate species that are sufficiently activated to achieve this condition is extremely small. This low percentage arises because the free energy required to create these activated states is typically many times higher than their average thermal energy.

Electronic coupling

Achieving resonance is a necessary but not sufficient condition for successful heterogeneous electron transfer. Once isoenergetic reactants have been created, the donor or acceptor orbitals and the electrode manifold states must couple electronically. The extent to which the redox centers and the electrode are electronically coupled typically increases exponentially with decreasing separation between the reactant and the electrode.

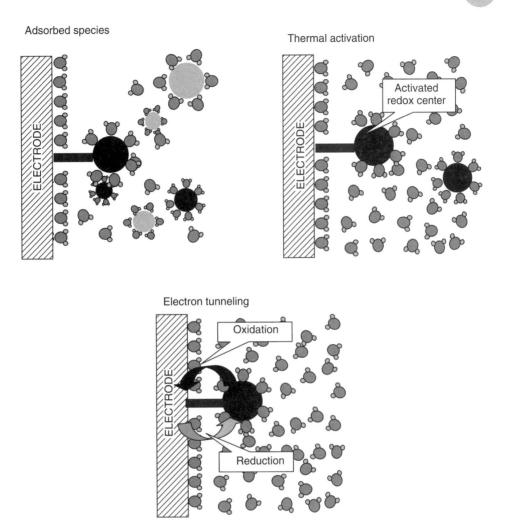

Figure 2.9 Schematics illustrating an adsorbate–electrode interface in aqueous solution, plus the corresponding thermal activation and electron tunneling steps associated with a heterogeneous electron transfer process

Elementary electron transfer

As described earlier in Section 2.2.1, according to the Franck–Condon principle, at the instant of electron transfer the redox center switches from an oxidized to a reduced form while still maintaining its internal structure and solvation shell. Given that electron transfer is instantaneous, energy cannot be transferred into, or out of, the surrounding medium. Therefore, as discussed above, the internal energies of the reactants and products must be identical. Electron transfer between the donor or acceptor moiety and the electrode, depending on the nature of the linkage, L, may occur by the electron hopping or superexchange mechanisms described above.

The two models most commonly applied to the heterogeneous electron transfer kinetics are the Butler–Volmer model, which is primarily a macroscopic approach and the Marcus theory, which was outlined for homogeneous electron transfer processes in Section 2.2.1 above, e.g. electron self-exchange between redox sites within an assembly. This latter approach permits prediction of the impact of microscopic reaction parameters such as electrode materials, solvent and molecular structure on the heterogeneous reaction dynamics.

Butler–Volmer model

The Butler–Volmer formulation of electrode kinetics [16,17] is the oldest and least complicated model constructed to describe heterogeneous electron transfer. However, this is a *macroscopic* model which does not explicitly consider the individual steps described above. Consider the following reaction in which an oxidized species, Ox, e.g. a ferricenium center bound to an alkanethiol tether, $[Fe(Cp)_2]^+$, is converted to the reduced form, Red, e.g. $[Fe(Cp)_2]$, by adding a single electron:

$$Ox + e^- \underset{k_f}{\overset{k_b}{\rightleftharpoons}} Red$$

The situation for a chemical, as opposed to an electrochemical, reaction is considered first. Simplified activated complex theory assumes an Arrhenius-type dependence of the forward rate constant, k_f, on the chemical free energy of activation, $\Delta G^{\ddagger}$, according to the following equation:

$$k_f = \frac{k_B T}{h} \exp\left(\frac{-\Delta G^{\ddagger}}{RT}\right) \tag{2.15}$$

Electrochemistry has the significant advantage that the driving force for the reaction can be controlled instrumentally. This capability contrasts sharply with homogeneous redox reactions where the temperature, or the chemical structure of a reactant, must be altered if the driving force is to be changed. Therefore, for a heterogeneous electron transfer reaction, the free energy of the reaction depends on the electrical driving force, i.e. the applied potential relative to the formal potential, $E^{0\prime}$, and $\Delta G^{\ddagger}$ must be replaced by the electrochemical free energy of activation, $\Delta \overline{G}^{\ddagger}$.

The electrochemical rate constant for the forward reaction, i.e. reduction, is given by the following equation:

$$k_f = \frac{k_B T}{h} \exp\left(\frac{-\Delta \overline{G}_f^{\ddagger}}{RT}\right) \tag{2.16}$$

As illustrated in Figure 2.10, both 'chemical' and 'electrical' components contribute to the electrochemical free energies of activation. The dashed line of Figure 2.10 shows that a shift in the potential of the electrode, to a value E, changes the energy of electrons within the electrode by $-nFE$. Under these circumstances, the barrier for the oxidation process, $\Delta \overline{G}_b^{\ddagger}$, is now less than $\Delta G_b^{\ddagger}$ by a fraction of the total energy change. This fraction is designated as $(1 - \alpha)$, where α is the transmission coefficient. This parameter takes on values between zero and unity depending on the shape of the free energy curves in the intersection region. Thus, the free energies

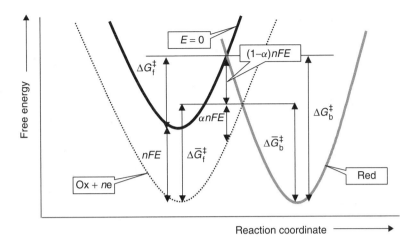

Figure 2.10 Parabolic free energy curves for heterogeneous electron transfer processes

of activation can be separated, as described by the following equations:

$$\Delta \overline{G}_f^{\ddagger} = \Delta G_f^{\ddagger} + \alpha nFE \tag{2.17}$$

$$\Delta \overline{G}_b^{\ddagger} = \Delta G_b^{\ddagger} - (1 - \alpha)nFE \tag{2.18}$$

Substitution into Equation (2.15) yields Equations (2.19) and (2.20), which describe, respectively, the potential dependence of the reduction and oxidation reactions as follows:

$$k_f = \frac{kT}{h} \exp\left(\frac{-\Delta G_f^{\ddagger}}{RT}\right) \exp\left(\frac{-\alpha nFE}{RT}\right) \tag{2.19}$$

$$k_b = \frac{kT}{h} \exp\left(\frac{-\Delta G_b^{\ddagger}}{RT}\right) \exp\left(\frac{(1 - \alpha)nFE}{RT}\right) \tag{2.20}$$

The first exponential term in both equations is independent of the applied potential and is designated as k_f^0 and k_b^0 for the forward and backward processes, respectively. These represent the rate constants for the reaction at equilibrium, e.g. for a monolayer containing equal concentrations of both oxidized and reduced forms. However, the system is at equilibrium at $E^{0\prime}$ and the products of the rate constant and the bulk concentration are equal for the forward and backward reactions, i.e. k_f^0 must equal k_b^0. Therefore, the standard heterogeneous electron transfer rate constant is designated simply as k^0. Substitution into Equations (2.19) and (2.20) then yields the Butler–Volmer equations as follows:

$$k_f = k^0 \exp\left[\frac{-\alpha nF(E - E^{0\prime})}{RT}\right] \tag{2.21}$$

$$k_b = k^0 \exp\left[\frac{(1 - \alpha)nF(E - E^{0\prime})}{RT}\right] \tag{2.22}$$

The dynamics of the system are described by k^0, with its units being s^{-1} for an adsorbed reactant. A redox couple with a large k^0 will establish the equilibrium concentrations given by the Nernst equation on a short timescale. Kinetically facile systems of this type require high-speed electrochemical techniques to successfully probe the electrode dynamics. The largest k^0 values that have been reliably measured are of the order of 10^6 s^{-1} and are associated with mechanistically simple reactions, i.e. there are no coupled chemical kinetics or significant structural differences between the oxidized and reduced forms.

The empirical Butler–Volmer formulation of electrode kinetics provides an experimentally accessible theoretical description of the kinetics of these systems. From a plot of $\ln(k)$ versus the overpotential, η ($\equiv E - E^{0'}$), α and k^0 can be obtained from the slope and intercept, respectively.

However, the Butler–Volmer formulation is deficient in a number of respects. First, the prediction that the rate constants for simple outer-sphere electron transfer reactions will increase exponentially with increasing electrical driving force agrees with experiment only over a limited range of overpotentials. The advent of microelectrodes and ideally responding redox-active monolayers allows heterogeneous electron transfer rates to be measured over very wide ranges of overpotentials or driving forces, η. Experimentally, one finds that k^0 initially depends exponentially on η, but then becomes independent of the driving force for sufficiently large values. Secondly, the Butler–Volmer formulation fails to account for the known distance-dependence of heterogeneous electron transfer rate constants. Finally, this model cannot predict how changes in the redox center's structure or the solvent affect k^0.

The Marcus theory, outlined above in Section 2.2.1 for homogeneous reactions, directly addresses these issues and is widely accepted as the most powerful and complete description of both heterogeneous and homogeneous electron transfer reactions. Its application to heterogeneous processes will be described in the following section.

Marcus theory of heterogeneous electron transfer

First, the standard heterogeneous electron transfer rate constant, k^0, i.e. when the electrochemical driving force is zero, is considered. The transition state theory [18] focuses on the intersection or crossover point of the free energy curves for the oxidized and reduced forms illustrated above in Figure 2.10. In this formulation, the rate of reaction depends on the product of the number of molecules with sufficient energy to reach the transition state at a particular instant in time and the probability at which they cross over the transition state. The number of molecules at the transition state depends on the free energy of activation, ΔG^0. The heterogeneous electron transfer rate constant is given by Equation (2.23), which is equivalent to Equation (2.1), describing homogeneous electron transfer:

$$k^0 = \nu\sigma \exp\left(\frac{-\Delta G^0}{RT}\right) \qquad (2.23)$$

The difference lies in the factor σ, which is introduced into the rate expression for heterogeneous electron transfer and represents an equivalent reaction-layer thickness (in cm).

As described earlier in Section 2.2.1, the Marcus theory assumes that the dependence of the free energy of activation on the reaction coordinate can be described as a simple parabola. Thus, the free energies of both outer-sphere solvent reorganization, ΔG_{os}, and inner-sphere vibrations, ΔG_{is}, contribute to the total free energy of activation, ΔG_{Total}, as described by the following equation:

$$\Delta G_{Total} = \Delta G_{is} + \Delta G_{os} \tag{2.24}$$

which is related to the reorganizational energy described in Section 2.2.1, since $\Delta G = \lambda/4$ in the Marcus normal region.

Therefore, unlike the empirical Butler–Volmer theory, in the Marcus formulation the heterogeneous electron transfer rate constant is sensitive to both the structure of the redox center and the solvent.

Heterogeneous electron transfers involving interfacial supramolecular assemblies are often assumed to be adiabatic because this simplifies the kinetic analysis. However, this assumption is often not justified and invalidates much of the subsequent analysis.

Potential-dependent heterogeneous electron transfer

The sensitivity of the heterogeneous electron transfer rate constant to the over-potential depends on the extent of electronic coupling between the reactant and the electrode [19]. For strongly coupled reactants, electron transfer occurs pre-dominantly through states near the Fermi level of the electrode and the *adiabatic* potential-dependent rate constant is given by the product of the frequency factor, v_n, and the density of acceptor states in the molecule, D_{Ox}, according to the following:

$$k(E) = v_n D_{Ox} (4\pi \lambda_{Ox} kT)^{-1/2} \exp\left\{-\left[\frac{(\lambda_{Ox} - E)^2}{4\lambda_{Ox} kT}\right]\right\} \tag{2.25}$$

where λ_{Ox} is the total reorganisation energy for the oxidation process. For diabatic systems, electrons with energies below the Fermi level may be transferred and one must then sum over all electron energies rather that just at the Fermi level, E_f. The Fermi function describes the distribution of occupied states within the metal and is described by the following equation:

$$n(E) = \left\{\frac{1}{1 + \exp[(E - E_F)/kT]}\right\} \tag{2.26}$$

The potential dependence of the diabatic electron transfer rate constant is given by the following:

$$k(E) = v_n N_{Ox} (4\pi \lambda_{Ox} kT)^{-1/2} \int_{-\infty}^{+\infty} \frac{e^{-[(\lambda_{Ox} - E)^2/4\lambda_{Ox} kT]}}{e^{(E - E_F)/kT} + 1} \, dE \tag{2.27}$$

where N_{Ox} is the number of donor states. The effects of the strength of the electronic coupling on $k(E)$ only become apparent at high driving forces, i.e. when the overpotential is more than half the reorganization energy. As discussed earlier in Section 2.1, perhaps the most significant consequences of Equations (2.25) and

(2.27) are that, unlike the Butler–Volmer theory, they both predict curvature in plots of ln(k) versus η. For extremely large driving forces, k no longer depends on the overpotential but reaches a maximum value when η is approximately equal to λ. This potential-independent-region electron transfer rate is equivalent to the *Marcus inverted region* described above for homogenous electron transfer reactions.

2.3   Photoinduced Processes

Photochemical and photophysical processes are the basis of many of the significant technological advances of the past 50 years. The advent of laser technology and subsequent development of optical data storage media have revolutionized the way in which information is stored and retrieved. This trend towards the development of optical devices is set to continue because they offer significant advantages in terms of speed and size of components. Miniaturization will require that well-defined light-addressable molecular systems are developed for integration into circuitry or devices. The escalating interest in developing efficient solar energy technologies, coupled with the ever-increasing insight into the mechanisms of natural photostimulated events such as photosynthesis and sight, will eventually lead to the generation of viable mimicking materials. An important element in these developments will be interfacial photochemistry. Photochemistry at the interface is implicit in biological systems in which membranes separate donor and acceptor sites. For example, in photosynthesis the photoexcitation of membrane-bound chlorophyll units results in a multi-step electron transfer, thus eventually leading to interfacial electron transfer to quinone acceptors. Photovoltaic technology is also inherently interfacial and the particular advantages of surface organization will likely lead to the development of photochemical molecular electronic devices which are interfacial in nature.

In order to describe the current status and likely future developments in this area, it is first important to outline the main photoinduced events that can occur in a supramolecular system and to understand how these photochemical processes are influenced by the presence of surfaces.

2.3.1   Photochemistry and Photophysics of Supramolecular Materials

In the following discussion on the fundamental photophysical properties of supramolecular systems, the simple model dyad A–L–B system is again employed for illustration. This consists of two components, A and B, which may be both molecular, or where one may be an electrode surface. At least one of these units is photoactive, and L is a linker or bridge which permits communication between the A and B components.

The most prevalent photoinduced processes in supramolecular and interfacial systems are electron transfer, energy transfer and nuclear motion, such as proton transfer and isomerization. Before discussing these processes, it is important to outline the fundamental properties of electronically excited states.

Excited-state properties

Photoexcitation of a ground-state molecule leads to the formation of an excited-state species whose lifetime will depend on the efficiency with which this species can dissipate its excess energy [20,21]. As illustrated in Figure 2.11, an excited supramolecular dyad, A^*-L-B, can return to the ground state by a number of physical and chemical pathways. The photophysical pathways available to an excited state species are illustrated in the modified Jablonski diagram shown in Figure 2.12. These photophysical processes may be radiative or non-radiative in nature. The most common radiative decay pathways are fluorescence (fluor) and phosphorescence (phos), which are distinguished on the basis of whether the transition involves a change in spin multiplicity. In fluorescence, the spin is conserved and this allowed process tends to be short-lived, with lifetimes typically between 10^{-12} and 10^{-6} s. In phosphorescence, spin is not conserved and this forbidden process yields relatively weak emissions that tend to be strongly Stokes-shifted with respect to the associated exciting absorbance. Phosphorescence, as a result of its forbidden nature, tends to be relatively long-lived, 10^{-6} to 1 s. Formal assignment of spin multiplicity of a given state is complicated by the presence of heavy atoms which cause spin–orbit coupling. Consequently, in many metal complexes emission is referred to simply as luminescence. For example, in the case of $[Ru(bpy)_3]^{2+}$ and its supramolecular analogues, the emission is formally

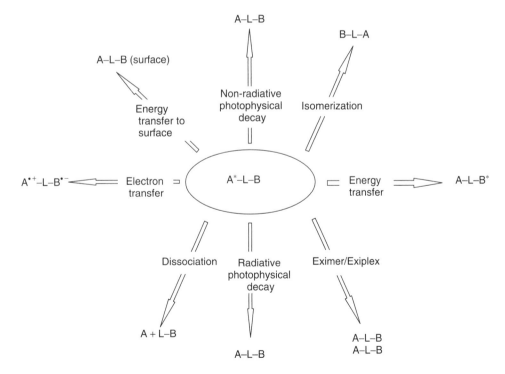

Figure 2.11 Possible photochemical and photophysical fates of the photoexcited dyad, A^*-L-B

phosphorescence, although the lifetimes and emission intensities are similar to those expected for fluorescent states. Processes that compete with emission, such as electron transfer, energy transfer and non-radiative photophysical pathways, will reduce the lifetime and intensity of the emission observed.

Non-radiative processes can also be distinguished on the basis of the spin multiplicity of the initial and final electronic states. Internal conversion (IC) is the non-radiative crossover between two states of identical multiplicity, while intersystem crossing (ISC) is a process in which spin is not conserved. In both instances, crossover between the states is isoenergetic, regardless of the multiplicity. Subsequent vibrational relaxation (VR) occurs to release excess vibrational energy (see Figure 2.12).

Whether a molecule is luminescent or not is dictated by the relative rates and efficiencies of non-radiative events and the possibility that other photochemical reactions can occur. Photochemically driven events of particular importance in a supramolecular device [21] are electron transfer, energy transfer and nuclear motion. In general, such processes will be competitive with luminescence and are typically detected by reduced luminescence in the luminophore unit in A–L–B. Figure 2.13 presents a simple energy diagram for photoinduced electron and energy transfer in the dyad, A*–L–B. The initial event is photoexcitation, which for the sake of this example is taken to occur at the electron or energy donor site, A. Photoexcitation provides the driving force for electron or energy transfer, which is not possible in the ground state.

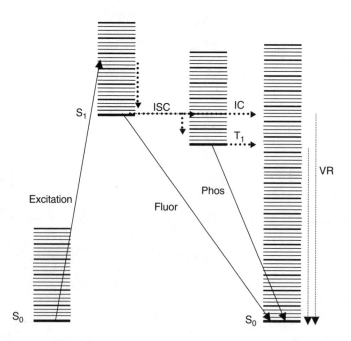

Figure 2.12 Modified Jablonski diagram illustrating the typical photophysical pathways for a molecule in a condensed medium: fluor, fluorescence; phos, phosphorescence; IC, internal conversion; ISC, intersystem crossing; VR, vibrational relaxation

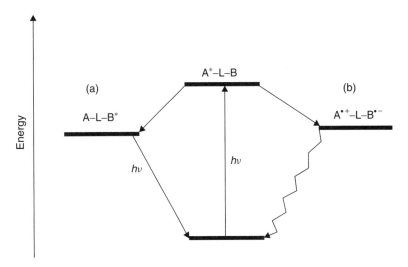

Figure 2.13 Schematic energy level diagrams illustrating the photoinduced (a) electron and (b) energy transfers in the supramolecular dyad A–L–B; straight lines indicate radiative transfers, while wavy lines represent non-radiative reactions

2.3.2 Photoinduced Electron Transfer

Photoinduced electron transfer occurs when an electron originating from an excited donor LUMO [22] is transferred to a ground-state molecule or electrode (as illustrated in Figure 2.14(a)). The initial step is photoexcitation of the donor of the dyad A–L–B, A + hv → A*, i.e. optical excitation stimulates an electron into the donor. Electron transfer proceeds between this donor state and an acceptor LUMO in A, provided that this process is thermodynamically feasible, and is mediated by the bridge electronic states. Such a process yields radical cations and anions of the donor and acceptor, respectively. The fate of these charge-separated species depends on the system, i.e. they may recombine in a ground-state redox process to produce A–L–B again, or may undergo further reaction.

The fundamental theories behind electron transfer were discussed above in Section 2.1. Indeed, some of the most important empirical proofs for these theories have originated from photoinduced electron transfer in supramolecular donor–acceptor complexes. The difference between thermally and photochemically induced electron transfer lies in both the orbitals participating in the reaction and in the additional thermodynamic driving force provided by the excited state. It is therefore important to consider the redox properties of excited-state species.

The excited state of a diamagnetic species with a closed-shell ground state is both a better donor and acceptor than its associated ground state. In addition to the ground-state redox potential, the excited state has the additional redox power of the absorbed photon, i.e. hv. Figure 2.15 illustrates this point, whereby ionization potentials (IPs) and electron affinities (EAs) for a ground- and excited-state molecule are compared. Excitation decreases the IP by $\Delta E_{\mathrm{HOMO-LUMO}}$, i.e.

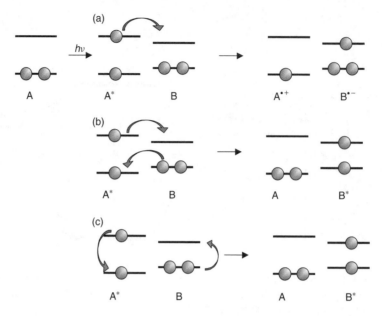

Figure 2.14 Schematic representations of the mechanisms of photoinduced (a) electron transfer, (b) Dexter (electron-exchange) energy transfer, and (c) Förster (dipole–dipole) energy transfer mechanism processes in the supramolecular dyad A–L–B; spheres represent electrons, while curved arrows indicate the directions of transfer

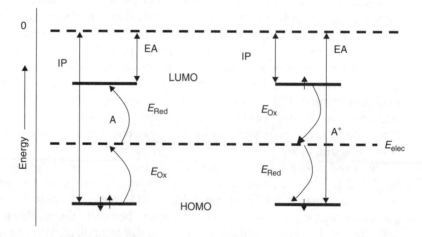

Figure 2.15 Schematic orbital diagram illustrating the relationship between ionization potential (IP) and electron affinity (EA) for the ground and excited states of a molecule and the corresponding ground- and excited-states redox potentials

the HOMO–LUMO energy gap, and increases EA by the same amount, thereby increasing both the oxidizing and reducing power of the excited state. This is reflected in the oxidation and reduction potentials of the excited-state species, which in this illustration become exergonic in the excited state.

If it is assumed that all excited-state energy is available as free energy for the excited-state processes, i.e. that there is no significant alteration in geometry between the ground and excited state, an estimate of the redox potentials of an excited state may be made according to the Rehm–Weller expressions, as follows:

$$E(A^+/A^*) = E^0(A^+/A) - E_{00} \tag{2.28}$$

$$E(B^-/B^*) = E^0(B^-/B) + E_{00} \tag{2.29}$$

where $E^0(A^+/A)$ and $E^0(B^-/B)$ are the standard ground-state oxidation and reduction potentials of donor A and acceptor B, respectively. E_{00} is the zero–zero spectroscopic energy for the acceptor and donor species, which corresponds to the energy gap between the lowest vibronic levels of the ground and excited electronic states. This parameter is frequently estimated from the fluorescence, λ_{max}, at 77 K.

For a photoinduced electron transfer and charge separation to be efficient in a supramolecular device, some structural and energetic prerequisites must be fulfilled. First, the electron transfer must be thermodynamically feasible, i.e. it must be exergonic. The free energy of a photoinduced electron transfer, ΔG^0_{pet}, may be calculated according to the following equation:

$$\Delta G^0_{pet} = [E^0(A/A^{\bullet+}) - E^0(B/B^{\bullet-})] - E_{00} - (e^2/\varepsilon r_{DA}) \tag{2.30}$$

The final term in the expression is referred to as the *Coulombic stabilization energy* of the charge-separated state, where e is the electronic charge, ε the solvent dielectric constant, and r the distance separating the donor and acceptor centers. In polar media, the Coulombic stabilization energy is typically less than 0.1 eV and is therefore usually neglected. Secondly, the electron transfer must be kinetically feasible, and the excited state must be sufficiently long-lived in order to allow electron transfer to occur. Once achieved, the quantum yield and lifetime of the charge-separated state must be as large as possible in order to provide an adequate population and time for the generated chemical potential to be put to use.

In developing supramolecular donor–acceptor complexes, the objective is typically to provide a sufficient driving force to promote photoinduced electron transfer, while also optimizing the lifetime of the resulting charge-separated product in order to drive further redox reaction. Charge recombination, where the electron in the acceptor LUMO is transferred back to the donor HOMO, is a possible, but undesirable, process. Strategies to slow or prevent 'back-reaction' are frequently thermodynamic, whereby 'back-electron transfer' is rendered endergonic, or so highly exergonic that it occurs in the Marcus inverted region.

Finally, as discussed before in Section 2.1, suitable orbital overlap must exist between the donor and acceptor, which will promote electron transfer and preferably prevent wasteful back reactions.

2.3.3 Photoinduced Energy Transfer

Apart from electron transfer, photoinduced energy transfer processes are widely observed within both natural and artificial systems and play a particularly important role at surfaces.

Photoinduced energy transfer occurs when the excitation energy from a donor species is transferred to an acceptor species, typically resulting in the generation of an excited acceptor state. There are two fundamentally different mechanisms of energy transfer, as illustrated above in Figure 2.14, which are most commonly referred to as the Dexter and Förster mechanisms [23].

Dexter energy transfer, also known as the electron exchange mechanism, involves the transfer of the excited electron from the LUMO of A, i.e. the donor, to the LUMO of the acceptor B, with the simultaneous transfer of the ground-state electron from the HOMO of B to the HOMO of A, (see Figure 2.14(b)). This 'energy'-transfer process is therefore essentially a double-electron transfer. Consequently, the same prerequisites broadly exist for a Dexter energy transfer as those for electron transfer. The process requires an overlap of the wavefunctions of the energy donor and the energy acceptor. In order for the process to be thermodynamically feasible, the excitation energy of the donor must exceed that of the acceptor species. In other words, the electron transfer between A* (LUMO) and B (HOMO), and B (LUMO) and A* (HOMO), must be exergonic. The overall spin quantum number of the species must be maintained, although two sites of different multiplicity may undergo electron exchange. Dexter energy transfer is the dominant mechanism in triplet–triplet energy-transfer processes. The rate of Dexter energy transfer, k_{den}, is expressed by the following equation:

$$k_{den} = \left(\frac{2|\mathbf{H}_{AB}|^2}{h} \right) \left(\frac{\pi^3}{\lambda RT} \right) \exp\left(-\frac{\Delta G^{\ddagger}}{RT} \right) \tag{2.31}$$

This expression clearly illustrates the importance of electronic communication in Dexter energy transfer, since the rate, like electron transfer, is proportional to the square of the electronic coupling matrix, $|\mathbf{H}_{AB}|$.

Förster energy transfer is a dipole–dipole mechanism whereby the deactivation of the donor species, $A^* \rightarrow A$, generates an electric field, or transition dipole, which stimulates the formation of B* (see Figure 2.14(c)). During the transition $B \rightarrow B^*$, a further transition dipole is generated, which in turn stimulates the deactivation of A*. The rate constant of Förster energy transfer, k_{fen}, is given by the following equation:

$$k_{fen} = \frac{c\phi_e K^2}{n^4 \tau d^6} \int_0^\infty f_d(v)\varepsilon_A(v)\frac{dv}{v^4} \tag{2.32}$$

where ϕ_e and τ are, respectively, the quantum yield and excited-state lifetime of the excited donor A*, ε_A is the extinction coefficient of B, K is the reorientation factor of the optical transition dipoles, taken to be 2/3 for random orientation, and d is the distance between donor and acceptor. The integral corresponds to the spectral overlap between donor emission and acceptor excitation. The Förster mechanism only occurs in multiplicity-conserving transitions since large transition dipoles are required if the mechanism is to be efficient.

Thermodynamic estimates of the driving force for energy transfer, by either the Förster or Dexter mechanism, may be made from the following:

$$\Delta G_{en} = E_{00}(A) - E_{00}(B) \tag{2.33}$$

This approach makes numerous assumptions, among them that the entropy difference between the ground and excited states is negligible and that the enthalpy difference between these states corresponds to E_{00}. However, it does provide a useful estimate and may provide insight into the nature of a given photoinduced reaction.

The efficiency of Förster energy transfer depends mainly on the oscillator strength of the $A^* \rightarrow A$ and $B \rightarrow B^*$ radiative transitions, whereas the efficiency of energy transfer by the electron exchange interaction cannot be directly related to an experimental quantity.

Distance dependence of energy transfer

Experimentally, one of the main methods of distinction between the Förster and Dexter mechanisms in an energy transfer is a study of the distance dependence of the observed process. From Equation (2.32) it is evident that the rate of dipole-induced energy transfer, k_{fen}, decreases as d^{-6}. This is typical of dipolar interactions and is reminiscent of the distance dependence of other such mechanisms, e.g. London dispersion forces. Therefore, the Förster mechanism can operate over large distances, whereas, in contrast, the rate of Dexter energy transfer, k_{den}, falls off exponentially with distance.

In the Förster mechanism, energy transfer occurs through dipolar interactions. This process is not coupled through bond interactions, and therefore orbital overlap and inter-component electronic coupling are unimportant. Dipole–dipole interactions may occur efficiently in systems where the donor and acceptor species are over 100 Å apart, whereas Dexter energy transfer is typically efficient only up to distances of approximately 10 Å.

Distinguishing between energy and electron transfer processes

A very significant distinction between electron transfer and energy transfer lies is the nature of the products of these processes. Energy transfer results in the formation of a new excited state species, e.g. $A-L-B^*$, whereas electron transfer usually results in a charge-separated state, i.e. ion radicals, e.g. $A^{\bullet+}-L-B^{\bullet-}$. This fact can be significant in experimental procedures aimed at distinguishing between electron and energy transfer. Laser flash photolysis, discussed later in Chapter 3, is probably the most useful method of elucidating the nature of a photochemical process, whereby the redox products resulting from electron transfer can be directly detected and compared with the absorbance of similar species generated through spectroelectrochemistry. However, it is important to note that energy transfer within a supramolecular system may lead to secondary reactions such as electron transfer, which can complicate the interpretation.

Electron transfer, Dexter energy transfer and Förster energy transfer may, in addition, be distinguished on the basis of the differences in their distance dependence. As described above, Förster energy transfer is a long-range process in comparison with electron transfer or indeed a Dexter-type energy transfer. In fact, the latter exhibits the strongest distance dependence of the three processes. This is because Dexter energy transfer is in fact a double-electron transfer and therefore has a tunneling decay parameter of $\beta_{\text{den}} \approx \beta_{\text{el}} + \beta_{\text{el}}$. In other words, such a parameter for a Dexter energy transfer constitutes, approximately, the sum of the tunneling parameters for each individual electron transfer step in that process. Therefore, although electron transfer may be observed across donor–acceptor distances of over tens of angstroms, depending on the nature of the bridge, Dexter energy transfer is rarely observed over distances greater than 10 Å. Thus, one approach to elucidating the mechanism behind a photoinduced process in an A–L–B dyad, albeit a synthetically demanding one, is to investigate the impact of lengthening L on the rate of quenching of the excited state moiety in this complex.

A final important distinction between electron and energy transfer processes lies in their temperature dependences. The rates of electron transfer and Dexter energy transfer are both temperature dependent, which is taken into account in Equations (2.1) and (2.31). In contrast, Förster energy transfer, which is not a thermally activated process, does not in general exhibit an intrinsic temperature dependence. Temperature-dependent studies of quenching rates of an excited chromophore by a donor or acceptor should therefore distinguish between Förster, and electron and Dexter transfers. However, this approach is not necessarily straightforward. For example, an apparent temperature sensitivity may be observed in a photoinduced energy transfer if a separate activated process occurs concurrently in the species which is slow at low temperatures but becomes competitive, and therefore observable, at high temperatures.

2.3.4   Photoinduced Molecular Rearrangements

Photostimulated molecular motion is an important photophysical phenomenon frequently exploited in molecular switches. The molecular electronic rearrangements accompanying optical excitation may stimulate nuclear rearrangement of the excited species. Like electron and energy transfer, such processes compete with radiative events and therefore reduce the measured lifetime and quantum yield of emission. The most important nuclear rearrangements in supramolecular species are proton transfer and photoisomerization.

Photoinduced proton transfer

Proton transfer is a particularly important transport process. Beyond acid–base reactions, proton transfer may be coupled to electron transfer in redox reactions and to excited-state chemistry. It is of enormous significance in biochemical processes where it is an essential step in hydrolytic enzyme processes and redox reactions spanning respiration, and photosynthesis where proton motion is responsible for sustaining redox gradients. In relatively recent times, proton transfer in the excited state has undergone significant study, primarily fueled by advances in ultrafast spectroscopy.

Under appropriate circumstances, the acidity of a labile proton in an excited molecule may alter by up to eight orders of magnitude when compared with the associated ground state. This difference in the apparent pK_a provides a significant driving force for proton transfer in the excited state. This additional driving force exists because the reorganization of electron density in the excited-state molecule responsible for changes in redox behavior also impacts on acidity. Such a

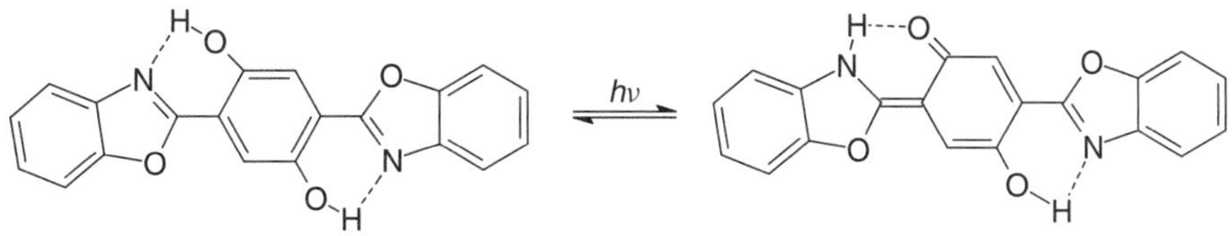

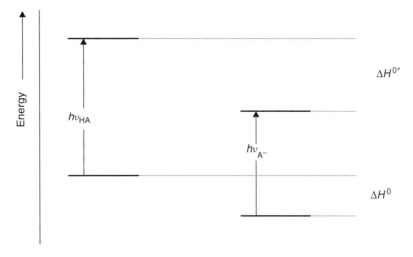

Figure 2.16 Intramolecular photoinduced proton transfer in 2,5-bis-(2-benzoxazolyl)hydroquinone, as reported by Grabowska *et al.* [24]

Figure 2.17 Potential energy diagram illustrating the Förster Cycle, in which the differences between the ground- and excited-state enthalpies, ΔH^0, of the protonated (HA) and deprotonated (A$^-$) species are shown

redistribution of electron density leads to spontaneous protonation or deprotonation of an active site on photoexcitation.

In instances where the same molecular species has a protonatable site, particularly one which is H-bonded in the ground state, this may lead to intramolecular proton transfer. Such behavior has been identified in many species, and depending on the nature and site of the excited state in relation to the labile proton, the acidity or basicity may increase. Such a reaction is illustrated in Figure 2.16 for 2,5-bis-(2-benzoxazolyl)hydroquinone, whose photoinduced proton transfer has undergone significant study [24]. Proton transfer is typically rapid, occurring on picosecond or sub-picosecond timescales, thus leading to excited-state equilibria between HA*, H$^+$ and A^{-*}, according to the following equation:

$$HA^* \rightleftharpoons A^{-*} + H^+$$

This equilibrium may be treated thermodynamically and the equilibrium constant may be estimated from the Förster Cycle, as shown in Figure 2.17. This cycle illustrates the enthalpy differences between the acid and base forms of the ground- and excited-state species. According to such a scheme, the difference between the ground- and excited-state pK_a values may be estimated from the ground-state parameters at 298 K, according to the following:

$$\Delta pK_a = 0.002\,09(\nu_{A^-} - \nu_{HA}) \tag{2.34}$$

where ν_{A^-} and ν_{HA} are the frequencies of the of lowest-energy absorbance bands for A^- and HA, respectively.

In instances where both HA* and A^{-*} are luminescent, the acidity of a proton in an excited-state species may be estimated from the following equation:

$$pK_a^* = pH_i + \log\left(\frac{\tau_{HA}}{\tau_{A^-}}\right) \tag{2.35}$$

where pH_i is the inflection point of a plot of emission intensity versus pH and τ_{A^-} and τ_{HA} are the emission lifetimes of the deprotonated and protonated forms of the species, respectively.

Although useful, the Förster Cycle is a theoretical estimate of pK_a^*, with the validity of its application being limited by the following assumptions:

- It assumes that there is no entropy change between the ground- and excited-state acid–base reactions. In order for this assumption to be valid, there should be little geometric distortion of the excited molecule and the acid–base reactions must originate from the same sites in both the ground- and excited-state species.

- That true acid–base equilibrium is achieved in the excited state. In short-lived excited states, where proton transfer is slow, e.g. diffusion controlled or intramolecular, this is unlikely to be the case.

The most common theoretical treatment of the kinetics of photoinduced proton transfer for strong H-bonds has employed the approach that proton transfer is an adiabatic process similar to electron transfer. However, like electrons, protons transfer can be diabatic and adiabatic. The theoretical approaches behind proton transfer, which are beyond the scope of this present text, are presented elsewhere [1a,25]. Although the theoretical treatment of proton and electron transfer may superficially be similar, there are a number of important differences between the two processes. The most significant difference is that although protons cannot tunnel over large distances, they can nonetheless *transport* over such distances, which is particularly evident in biochemical reactions. For example, during photosynthesis in the bacteria, *Halobacterium halobium*, the bacteriorhodpsin protein undergoes a cyclic photoinduced *trans–cis* isomerization during which one proton is actively translocated via the protein across the cell membrane.

The well-known Grötthus mechanism is thought to be an important mode of long-distance proton transfer processes, whereby the proton transfers along a H-bonded network, as shown in Figure 2.18. In this mechanism, successive protons are transferred by *hopping* along a hydrogen-bonded network, and the initial orientation of the hydrogen bonds is recovered via rotation of the H-bearing groups.

However, photoinduced proton transfer is still not well understood at the molecular level. In terms of its study, it has significant experimental advantages over electron transfer. It can be detected through vibrational spectroscopic techniques, can diffract X-rays and may undergo isotope exchange, thus permitting studies of kinetic isotope effects.

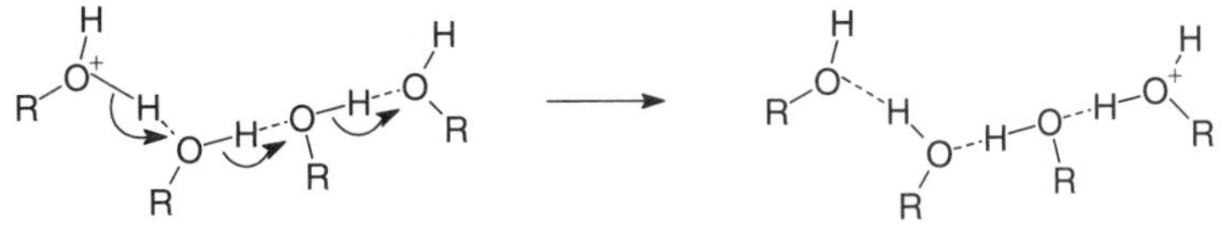

Figure 2.18 Schematic of the Grötthus mechanism of long-range proton transfer in water molecules. In this, hydrogen bonds and covalent bonds interchange, releasing a proton from one end of the chain as a new proton is introduced at the start of the chain

Photoisomerization

Like proton transfer, photoisomerization is a fundamentally important photochemical process. The two most important forms of photoisomerization are valence isomerization and stereoisomerization. The latter is probably the most common photoinduced isomerization in supramolecular chemistry. It may occur in systems in which the photoactive component has unsaturated bonds which can be excited, and this effect may be exploited for optical switching applications. A number of interfacial supramolecular complexes capable of undergoing *cis–trans* photoisomerization have been studied from this perspective – some examples are outlined in Chapter 5.

Stereophotoisomerization plays an important role in biology where, for example, the primary photochemical step in vision is the *cis–trans* isomerization of retinal, as shown in Figure 2.19. Indeed, an important approach to elucidation of the mechanism in the photochemistry of retinal proteins has been through interfacial studies [26]. However, for the purpose of illustration, a far simpler photoisomerization process is discussed here. Some of the most common photochemically induced *cis–trans* isomerizations occur across alkene-type bridges. Alkenes exhibit a high degree of conformational stability at ambient temperatures. Double-bond rotations on a ground-state potential energy surface possess high activation barriers, e.g. activation energies of 60 kcal mol^{-1} for thermal *cis–trans* isomerizations are relatively common. Photoexcitation of an electron from the π molecular orbital into a π^* orbital reduces the order of the olefinic bond, as a consequence the antibonding character of the excited orbital, creating what is, in essence, a temporary single bond. Consequently, the activation barriers to isomerization decreases significantly and rotation around this bond becomes facile. As illustrated for stilbene in Figure 2.20, once photoexcited, alkenes usually quickly depart from their planar geometries around the double bond to undergo large conformational changes. The reduction in bond order allows the release of any steric repulsion present between the vicinal

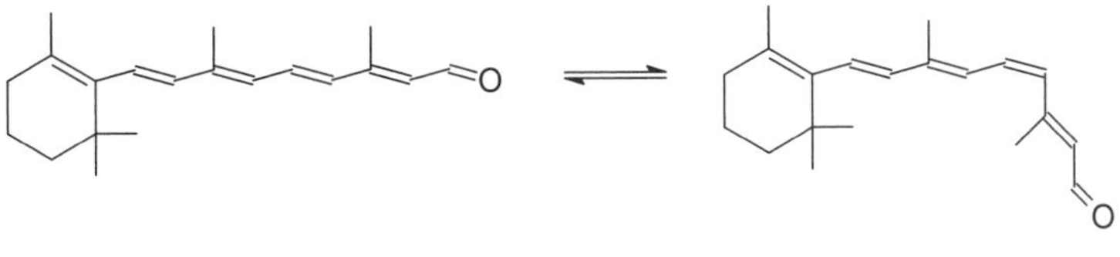

trans-Retinal *cis*-Retinal

Figure 2.19 Schematic illustrating the photoinduced *cis–trans* isomerization of retinal

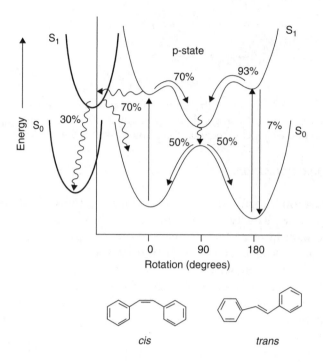

Figure 2.20   Potential energy diagram illustrating the room-temperature photoinduced isomerization of stilbene in solution; wavy lines represent non-radiative transitions, curved arrows isomerizations and straight lines radiative transitions

functional groups attached to the double bond at the planar ground-state geometries by rotation around this bond.

Like proton transfer, the study of photoisomerization reactions has also benefited from developments in ultrafast spectroscopy, as *cis–trans* isomerization reactions typically occur on or within a few picoseconds, i.e. of the order of the timescale of a molecular rotation.

A useful example of a photoisomerization is shown in Figure 2.20, which illustrates a potential energy diagram for the much-studied photoinduced *cis–trans* isomerization of stilbene. After $\pi-\pi^*$ photoexcitation, stilbene remains *trans* for approximately 70 ps, eventually rotating about the central bond to the lowest point on the excited-state surface, roughly midway between *trans* and *cis*, or about 90° degrees – this is known as the 'p-state'. This state is very short-lived, lasting for approximately 300 fs before internally converting to the ground state. Internal conversion is an isoenergetic process. The point on the potential energy surface at which stilbene relaxes to the ground state is near an energy maximum and the molecule can either rotate to the *trans-* or *cis-*conformations from this state. In fact, after photoexcitation, typically 50 % of each isomer is obtained. The central bond cannot achieve a 90° angle in the ground state because of steric repulsion between the phenyl groups.

Beyond its vital importance in the chemistry of vision, *cis–trans* isomerization offers potential advantages for organized assemblies capable of photoswitching. The latter is frequently a fast process which requires a strictly photochemical input. This results in a significant structural change that in an interfacial assembly may alter the ion transport by changing the density of a film or may allow for molecular recognition in one isomer which is not possible in another. Examples of how *cis−trans* isomerizations have been exploited for photoswitching in interfacial supramolecular assemblies are discussed further in Chapter 5.

2.4 Photoinduced Interfacial Electron Transfer

The discussion thus far has focused on photoinduced electron and energy transfer processes occurring intramolecularly between the molecular building blocks of the supermolecule. Indeed, many photochemically active supramolecular assemblies constructed on interfaces have been reported to undergo such homogeneous processes. When B is a solid interface, which may be metallic, semiconducting or insulating, the surface may perturb the photophysical properties of the adsorbates. For example, the surface may participate in energy or electron transfer following photochemistry within the supermolecule, rather than by merely acting as a passive platform. For the purposes of this discussion, we presume that B represents the solid, as indicated by | representing an interface in A−L−|B.

The most prevalent current applications in interfacial supramolecular chemistry are in photocatalysis and photovoltaics. For these processes, the adsorbed, photoactivated species interacts with the solid electrode, thus leading to quenching of the excited state. This process may occur directly, whereby the photoexcited species which plays the role of sensitizer, injects/accepts an electron directly to/from the electrode, as illustrated in Figure 2.21(a). Alternatively, photoinjection may be a secondary process following a prior intramolecular electron or energy transfer from another component in the supramolecular assembly (Figure 2.21(b)). For example, Meyer and co-workers have illustrated the lateral energy transfer from $[Ru(bpy)_2(dcp)]^{2+}$ to $[Os(bpy)_2(dcp)]^{2+}$ (bpy is 2,2-bipyridyl and dcb is 4,4-dicarboxy-2,2-bipyridyl) where these materials were co-adsorbed on a TiO_2 interface [27], which was competitive with interfacial electron transfer. Likewise,

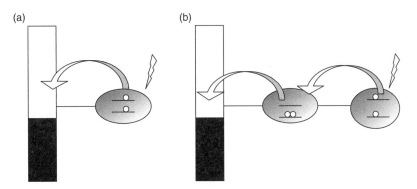

Figure 2.21 Schematics illustrating photosensitized heterogeneous electron transfer mechanisms: (a) a directly bound photosensitizer, which photoinjects directly to the electrode; (b) a remotely bound bridge sensitizer, which undergoes a prior photoinduced electron transfer, followed by ground-state heterogeneous electron transfer to the electrode

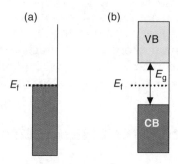

Figure 2.22 Schematics of the comparative energy levels of (a) a metal, and (b) an intrinsic semiconductor: E_f, Fermi level; E_g, energy gap; CB, conduction band; VB, valence band

the majority of recent studies on dye-sensitized photoinduced interfacial electron transfers have been conducted at semiconductor substrates, although conductors are receiving increasing interest.

The fundamental distinction between these two materials lies in their electronic structures, as illustrated in Figure 2.22. Conductors, such as metals, contain a continuum of states and the Fermi level, E_f, lies within these densely packed electronic levels and may be tuned by applying an external potential. Semiconductors possess discrete band structures rather than the continuum of levels characteristic of metals. The conduction and valence bands are separated by a forbidden energy gap, E_g. The magnitude of this gap is characteristic of a given semiconductor and when E_g is very large (>4 eV), the substance is regarded as an insulator. The Fermi level for intrinsic semiconductors lies approximately at the mid-point between the conduction and valence bands. Doping of a semiconductor material shifts the energy of E_f closer to the conduction band in the case of n-type semiconductors and closer to the valence band in the case of p-type semiconductors. A detailed treatment of the interfacial behavior of semiconductor materials is beyond the scope of this present work, but a number of monographs and reviews are available on this subject [28].

Interfacial photoinduced electron transfer from an electronically excited molecular state to an electrode is quite distinct from a molecule-to-molecule transfer process. In the former, electron transfer involves interactions between a very small number of discrete energy states. In dye-sensitized charge injection, the electron transfers from a molecular state into a wide continuum of acceptor levels. Therefore, in interfacial electron transfer numerous energetic pathways are open to the excited electron. Provided that the process is exergonic, the excited species will follow an activationless path where the strongest resonance between the molecular excited state and the electrode states exists to produce an optimal electron transfer rate. The simplest kinetic model of dye-sensitized interfacial electron transfer employed a quantum mechanical treatment of electron transfer, as introduced earlier in Section 2.2.1 (Equation (2.10)), where the kinetics of charge injection is described as a radiationless, diabatic process occurring from a single, photochemically equilibrated state of the adsorbed molecule.

For a dye-sensitized electrode with a large number of accessible acceptor levels, the summation over all of the terms of the Franck–Condon factor, $\langle FC \rangle$, reduces to the unweighted density of the final states [28]. The rate constant, k_{ci}, of the charge-injection process can therefore be given by the following:

$$k_{ci} = \frac{4\pi^2}{h}|\mathbf{H}_{AB}|^2 \frac{1}{h\omega} D_{Ox} \tag{2.36}$$

where $h\omega$ is the energy level spacing of the dye vibrational modes, and D_{Ox} reflects the density of acceptor states available within the solid ($0 < D_{Ox} < 1$).

For a semiconductor, the density of acceptor sites in the conduction band is given by the following relationship:

$$D_{Ox}(E) = 4\pi \left(\frac{2m^*_{de}}{h^2}\right)^{3/2} (E - E_{cb})^{1/2} \tag{2.37}$$

where E_{cb} is the energy of the conduction band edge, and m^*_{de} the density-of-state effective mass for electrons. The main limitation of this approach to interfacial electron transfer is that it assumes the electron transfers from a single excited state of the dye. Interfacial charge transfer in semiconductors has been shown to be ultrafast, and may, under certain circumstances, occur from multiple, 'vibrationally hot' excited states within the dye. Under these conditions, $|\mathbf{H}_{AB}|$ becomes time and wavelength dependent and Equation (2.36) is not an adequate description of the system.

A predictive mechanistic treatment of dye-sensitized photoinduced interfacial electron transfer has been described by Gerischer [29]. According to this treatment, the rate of dye-sensitized electron transfer, ρ_{dye}, can be described by the following:

$$\rho_{dye} = \int \kappa(E) D_{occ}(E) D_{unoc}(E) \, dE \tag{2.38}$$

where κ is the transfer frequency, and D_{occ} and D_{unoc} are the densities of the occupied and unoccupied states, respectively, which are integrated over all energies. However, this model does not specifically address the electronic behavior of the dye and the coupling between dye and conductor or semiconductor energy levels. This severely limits the application of this approach, which is, to date, the only predictive theory available for dye-sensitized electron transfer processes.

2.4.1 Dye-Sensitized Photoinduced Electron Transfer at Metal Surfaces

Interfacial electron transfer between a metal and an excited sensitizer, A*–L–|B where B represents a metal electrode, may be reductive, whereby the electron transfers from the conduction band of the metal to the singly occupied HOMO state of the excited adsorbed molecules, thus resulting in A⁻–L–B and a cathodic photocurrent at the electrode. Alternatively, it may be an oxidative process, wherein the electron is transferred from the adsorbate to the metal, so resulting in A⁺–L–B and an anodic photocurrent at the electrode.

Since the excited state of a given species is both a good donor and acceptor, both processes will be competitive. A steady-state photoelectrochemical current is generally only observed if the back electron transfer, between the ground-state product and the electrode, is prevented through the introduction of (typically solution-phase) sacrificial donor or acceptor species to re-oxidize or re-reduce the photochemical products.

A significant drawback of metals for photoelectrochemical applications lies in their ability to efficiently quench excited states via energy transfer processes, as discussed below. Direct detection of photosensitized electron transfer to or from a metal electrode surface has been observed [30]. However, unlike dye-sensitized semiconductor systems, little examination of the kinetics of such systems has yet been undertaken.

2.4.2 Dye-Sensitized Photoinduced Electron Transfer at Semiconductor Surfaces

The majority of surface-immobilized supramolecular systems studied for photovoltaics have employed large-band-gap, n-type semiconductors, which consequently absorb only in the UV region of the spectrum. The immobilized supramolecular species plays the role of photosensitizer, thus permitting photoinjection stimulated at visible wavelengths. Dye sensitization of semiconductors is dealt with in detail below in Chapter 6, and therefore only the basic principles are outlined here. Photoexcitation of the immobilized species generates the thermodynamic driving force for photoinjection into the conduction band of the semiconductor, as shown in Figure 2.23. Like metal photosensitization, two forms of dye sensitization tend to occur at the semiconductor surfaces. The first mechanism involves electron or hole injection from the electronically excited sensitizer molecule (A*) into the semiconductor (B) conduction band or valence band, respectively, as the primary step:

$$A–L–|B + h\nu \longrightarrow A^*–L–|B \longrightarrow A^+–L–|B(e_{cb}^-) \tag{2.39}$$

or:

$$A–L–|B + h\nu \longrightarrow A^*–L–|B \longrightarrow A^-–L–|B(h_{vb}^+) \tag{2.40}$$

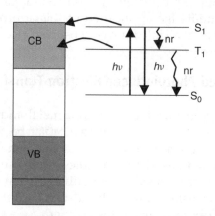

Figure 2.23 Schematic illustrating the dye sensitization of a semiconductor electrode via electron transfer; straight lines indicate radiative transitions, curved lines electron transfer, and wavy lines non-radiative (nr) transitions. Photoexcitation into the S_1 state of the dye may result in charge injection into the conduction band of the semiconductor or fluorescence and inter-system crossing, from where charge injection may occur from the triplet state or phosphorescence

Alternatively, charge injection into the semiconductor can involve the reductive or oxidative quenching of the dye excited state by a redox-active species (a supersensitizer (S)), followed by thermal interfacial electron transfer:

$$A^*-L-|B + S \longrightarrow A^- -L-|B + S^+ \longrightarrow A-L-|B(e^-_{cb}) + S^+ \qquad (2.41)$$

or:

$$A^*-L-|B + S \longrightarrow A^+ -L-|B + S^- \longrightarrow A-L-|B(h^+_{vb}) + S^- \qquad (2.42)$$

In order for injection of an electron from the excited state of the dye species into the conduction band of a semiconductor (as described by Equation (2.39)) to occur, the oxidation potential of the dye excited state (A^+/A^*) must be more negative than the conduction band potential of the semiconductor. Conversely, photoinduced hole injection from the excited dye into the semiconductor valence band (Equation (2.40)) requires the excited-state reduction potential of the sensitizer (A^*/A^-) to be more positive than the valence band potential.

After photoinduced electron injection, the strong interfacial electric field at the semiconductor solid–liquid junction draws the injected electron (or hole) into the semiconductor and towards the electrical contact. This process facilitates charge separation and reduces the chances of hole–electron recombination. The most important prerequisites for this process include, as described, good redox matching of the dye species excited state and the conduction band of the semiconductor, as well as strong orbital coupling between the immobilized dye and semiconductor.

The probability of an electron transfer, p_{et}, from an excited donor species to a semiconductor conduction band may be estimated from the following equation:

$$p_{et} \propto \exp\left[\frac{-(E_c - E^0_{D^*/D^{\bullet-}})^2}{4\lambda^* k_B T}\right] \qquad (2.43)$$

where E_c is the energy level of the conduction band, and λ^* is the reorganization energy of the excited state, which usually lies in the range 0.6 to 1.5 eV. The term $(E_c - E_{D^*/D^{\bullet-}})$ represents the thermodynamic driving force for charge injection. Photosensitized charge injection into a large-band-gap semiconductor is generally regarded as an ultrafast process. It therefore competes with luminescence, and in many instances, with other photophysical processes such as intersystem crossing.

One of the most convenient methods of studying the charge-injection process is by monitoring the radical species generated at the adsorbed layer. The rates of formation of these species then indicate the rate of charge injection, while their decay are generally indicative of back electron transfer. Ultrafast spectroscopic methods are described further in Chapter 3.

2.4.3 Photoinduced Interfacial Energy Transfer

It is well known that luminescent molecules near metal surfaces can suffer dramatically reduced luminescent quantum yields and lifetimes. This is attributed to non-radiative energy transfer to the metal surface [31]. This phenomenon is thought to commence when the luminescent species lies up to 200 Å from the metal surface. Therefore, this form of quenching is expected to be particularly prevalent in adsorbed monolayers. From the perspective of interfacial supramolecular species, it is likely to interfere with any light-induced function by providing an efficient competitive pathway for excited-state decay.

Most of the theoretical studies of this phenomenon were conducted in the 1970s and early 1980s, where energy transfer to a surface was observed to follow a Förster-type behavior. According to these models, the rate of energy transfer to the surface, k_{ien}, is approximated according to the following equation:

$$k_{ien} = \left(\frac{k_r}{4\check{D}^3} \right) \text{Im} \left[\frac{(\varepsilon_2 - \varepsilon_1)}{(\varepsilon_2 + \varepsilon_1)} \right] \tag{2.44}$$

where k_r is the radiative rate constant, ε_1 is the dielectric constant of the material separating the emitter and the metal, ε_2 is the complex dielectric constant of the metal, and $\check{D} = 2p\sqrt{(\varepsilon_1)}d/\lambda$, with d being the distance from the metal surface and λ the emission wavelength.

The quantum yield of the interfacial energy transfer, ϕ_{ien}, can be estimated from the following:

$$\phi_{ien} = 4d^3 \left\{ \text{Im} \left[\frac{(\varepsilon_2 - \varepsilon_1)}{(\varepsilon_2 + \varepsilon_1)} \right] \right\}^{-1} \tag{2.45}$$

The cubic distance dependence of the energy transfer rate ($\propto \sqrt{d^3}$, in Equation (2.44)) is attributed to an energy transfer mechanism. Dipolar interactions between the excited molecule as it relaxes to its ground state and the metal surface plasmon are thought to be responsible for interfacial energy transfer in metals. If the surface plasmon of the metal is in resonance with the energy of the emitted photon, the quenching is even more effective. Subsequent work has shown that interfacial energy transfer is considerably less effective at longer distances from the surfaces. Indeed, in the case of alkyl bridges, seven to ten carbons between a luminophore and the surface appear to be sufficient to significantly reduce the quantum yield of energy transfer.

2.5 Elucidation of Excited-State Mechanisms

Experimental methods for studying photoinduced electron transfer reactions are discussed in detail in Chapter 3. In an excited-state species, energy and electron transfer compete with other photophysical events, including emission. Consequently, energy and electron transfer are often detected by comparing the emission properties of the supramolecular species, using suitable model compounds as described below.

In terms of photophysics, electron transfer reactions create an additional non-radiative pathway, so reducing the observed emission lifetimes and quantum yields in A–L–B dyads in comparison with a model compound. However, there are other processes, such as molecular rearrangements, proton transfer and heavy-atom effects, which may decrease the radiative ability of a compound. One of the most important experimental methods for studying photoinduced processes is emission spectroscopy. Emission is relatively easy to detect and emission intensities and lifetimes are sensitive to competing processes. Studying parameters such as emission quantum yields and lifetimes for a given supramolecular species and associated model compounds can yield a wealth of information on the relative efficiencies of radiative and non-radiative processes, as well as the rates of competing photochemical reactions. In conjunction with time-resolved spectroscopic methods, photochemical products such as radical intermediate isomers or decomposition products may be detected and their lifetimes measured. This analysis allows accurate diagnosis of the photoinduced processes occurring within a supramolecular assembly.

For supramolecular assemblies, intramolecular processes may quench the emission of A^* to a degree which depends on the relative efficiency of the process when compared with emission. It is often useful to compare the photophysical and chemical behavior of the supramolecular species, e.g. A^*–L–B, with an appropriate model compound, for instance, AH, which contains the photochemically active component, A, in the absence of any units capable of interacting with A^*. For example, from luminescence lifetime measurements, the rate of electron transfer may be estimated by comparing the excited-state lifetime of the mononuclear model complex, τ_{Model}, with that of the supramolecular species, τ_{Supra}, by using the following equation:

$$k_{et} = \frac{1}{\tau_{Supra}} - \frac{1}{\tau_{Model}} \qquad (2.46)$$

In condensed media, emission occurs from the lowest vibrational state of the lowest electronically excited state [32]. For solution-phase species, this means that emission decays by typically following first-order kinetics, according to the following:

$$I = A \exp(t/\tau) \qquad (2.47)$$

where I is the emission intensity, A is the pre-exponential factor and τ is the lifetime of the emissive state corresponding to the time taken for I to decrease to $(1/e)$th of its initial value.

The measured radiative lifetime, τ, comprises contributions from all radiative and non-radiative processes that the excited molecule undergoes, according to the following equation:

$$\tau = \frac{1}{k_r + k_{nr}} \qquad (2.48)$$

where k_r and k_{nr} are the radiative and non-radiative decay rates, respectively. Since $k_r = \phi_r \tau^{-1}$, once the emission quantum yield, ϕ_r, is known, k_{nr}, i.e. the rate of non-radiative decay, can be determined.

When a fluorophore is encapsulated in heterogeneous media or immobilized on a surface, single exponential emission decays are rarely observed. Multi-exponential kinetics are attributed to the slow reorientation of the molecular environment after photoexcitation, and the heterogeneity of the microenvironment. Different species in the excited ensemble are oriented differently or exist in different microenvironments on the timescale of the emission which influences the excited-state lifetimes of the immobilized species. Studying the number and distribution of decays can provide information on the microenvironment of the immobilized fluorophore. When combined with fluorescence depolarization studies, detailed information on the motion of these species and their interaction with their environment can be obtained.

2.6 Conclusions

The two most important methods of both probing and stimulating supramolecular devices are photochemical and electrochemical techniques. The most prevalent events to occur in such devices are electron, energy and proton transfer, as well as molecular rearrangement. Any of these events can provide the basis for forms of transducable output on which molecular electronic devices may be based. Therefore, the theories most commonly applied to such electrochemically and photochemically triggered events are outlined in this chapter. In addition, an overview of the mechanisms by which such events occur is provided, identifying the molecular or physical parameters required to make such events feasible in a supramolecular structure.

Dye-sensitized semiconductors continue to be the focus of considerable research as a consequence of their importance in photovoltaic technologies, with the theories and mechanisms behind their operation also emerging. Considerably less is known about photoinduced interfacial processes in large photochemically active adsorbates on metals. ISAs on metals are becoming increasingly prevalent and are possible precursors for suitable molecular electronic devices. It would seem likely that over the coming years fundamental studies on the photophysics and chemistry of these materials will become more widespread.

References and Notes

[1] (a) Kuznetsov, A.M. and Ulstrup, J. (1999). *Electron Transfer in Chemistry and Biology. An Introduction to the Theory*, Wiley, Chichester, UK. (b) Kavernos, G.J. (1993). *Fundamentals of Photoinduced Electron Transfer*, Wiley, Chichester, UK.

[2] Balzani, V. (Ed.) (2001). *Electron Transfer in Chemistry*, Vol. 1, Wiley-VCH, Weinheim, Germany.

[3] Marcus, R.A. (1993). *Angew. Chem., Int. Ed. Engl.*, **32**, 1111.

[4] Myers Kelley, A. (1999). *J. Phys. Chem., A*, **103**, 6891.

[5] (a) Closs, G.L. and Miller, J.R. (1988). *Science*, **240**, 440. (b) Closs, G.L., Calcaterra, L.T., Green, H.J., Penfield, K.W. and Miller, J.R. (1986). *J. Phys. Chem.*, **90**, 3673.

[6] Hush, N.S. (1961). *Trans. Faraday Soc.*, **57**, 557.

[7] Gray, H.B., Winkler, J.R. and Wiedenfeld, D. (2000). *Coord. Chem. Rev.*, **200–202**, 857.

[8] Jortner, J. (1976). *J. Chem. Phys.*, **63**, 447.

[9] Schanze, K.S. and Walters, K.A. (1998). *Molecular and Supramolecular Photochemistry, in Organic and Inorganic Photochemistry*, Vol. 2, V. Ramamurthy and K. Schanze (Eds), Marcel Dekker, New York, Ch. 3, pp. 163–240.

[10] The parameters β_{AB} and β represent, strictly speaking, two different values, where the latter value, which incorporates reorganization energy as well as H_{AB} dependence, is anticipated to be larger. However, to simplify the argument, we assume here that they are the same.

[11] Paddon-Row, M.N. (2000). Electron and energy transfer, in *Stimulating Concepts in Chemistry*, F. Vögtle, J.F. Stoddart and M. Shibasaki (Eds), Wiley-VCH, Weinheim, Germany, pp. 267–291.

[12]   de Rege, P.J.F., Williams, S.A. and Therien, M.J. (1995). *Science*, **269**, 1409.

[13]   Han, H. and Zimmt, M.B. (1996). *J. Am. Chem. Soc.*, **118**, 2299.

[14]   Jolliffe, K.A., Bell, T.D.M., Ghiggino, K.P., Langford, S.J. and Paddon-Row, M.N. (1998). *Angew. Chem., Int. Ed. Engl.*, **37**, 916.

[15]   Skotheim, T.A., Elsenbaumer, R.L. and Reynolds, J.R. (Eds) (1998). *Handbook of Conducting Polymers*, Marcel Dekker, New York.

[16]   Butler, J.A.V. (1924). *Trans. Faraday Soc.*, **19**, 729.

[17]   Erdey-Gruz, T. and Volmer, M. (1930). *Z. Physik. Chem., A*, **150**, 203.

[18]   Chandler, D. (1986). *J. Stat. Phys.*, **42**, 49.

[19]   Miller, C.J. (1995). Heterogeneous electron transfer kinetics at metallic electrodes, in *Physical Electrochemistry: Principles, Methods and Applications*, I. Rubinstein (Ed.), Marcel Dekker, New York, pp. 27–80.

[20]   Suppan, P. (1994). *Chemistry and Light*, The Royal Society of Chemistry, Cambridge, UK.

[21]   Balzani, V. and Scandola, F. (1991). *Supramolecular Photochemistry*, Ellis Horwood, Chichester, UK.

[22]   Ground-state notation refers to the first excited state as being the lowest unoccupied molecular orbital (LUMO). Obviously, in the excited state the LUMO is half filled.

[23]   There is a third mechanism of energy transfer known as the *trivial* or *radiative* energy-transfer mechanism. Here, energy is transferred by radiative deactivation of a donor and reabsorption of this emitted light by an acceptor molecular entity. This mechanism is less important in supramolecular chemistry and is not dealt with in further detail here.

[24]   Grabowska, A., Mordziński, A., Kownacki, K., Gilabert, E. and Rullière, C. (1991). *Chem. Phys. Lett.*, **177**, 1.

[25]   (a) Bountis, T. (Ed.) (1992). *Proton Transfer in Hydrogen-Bonded Systems*, Plenum Press, New York. (b) Fleming, G.R. (1986). *Chemical Applications of Ultrafast Spectroscopy*, Clarendon Press, Oxford, UK.

[26]   Hong, F.T. (1999). *Prog. Surf. Sci.*, **62**, 1.

[27]   Farzad, F., Thompson, D.W., Kelly, C.A. and Meyer, G.J. (1999). *J. Am. Chem. Soc.*, **121**, 5577.

[28]   (a) Lewis, N.S. (1990). *Acc. Chem. Res.*, **23**, 176. (b) Miller, R.J.D., McLenden, G., Nozik, A.J., Schmickler, W. and Willig, F. (1995). *Surface Electron Transfer Processes*, VCH, New York. (c) Rajeshwar, K. (2001). Dye sensitization of electrodes, in *Electron Transfer in Chemistry*, Vol. 4, V. Balzani (Ed.), Wiley-VCH, Weinheim, Germany, pp. 279–343.

[29]   Gerischer, H. (1980). *Pure Appl. Chem.*, **52**, 2649.

[30]   Forster, R.J. and Keyes, T.E. (1998). *J. Phys. Chem.*, **102**, 10004.

[31]   Chance, R.R., Prock, A. and Sibley, R. (1975). *J. Chem. Phys.*, **62**, 2245.

[32]   Gilbert, A. and Baggott, J. (1991). *Essentials of Molecular Photochemistry*, Blackwell Scientific Publications, Oxford, UK.

3 Methods of Analysis

Characterizing the structure of interfacial supramolecular assemblies and their associated electrochemical and photochemical properties represents a significant challenge. The analytical tools for studying thin films have developed dramatically in recent years and it is now possible to probe their physical structure at the length scale of individual molecules and explore electron and energy transfer at femtosecond timescales. This chapter reviews the main techniques used in the field, identifies emerging approaches that are expected to be influential in the area over coming years and highlights the benefits from an interdisciplinary approach to the 'complete' characterization of interfacial assemblies.

3.1 Structural Characterization of Interfacial Supramolecular Assemblies

Interfacial supramolecular assemblies represent one of the most important approaches to creating architectures with both structural and functional control at the molecular level. The structural requirements for assemblies where READ, WRITE and ERASE functions are achieved through electron transfer reactions are particularly demanding. For example, very small positional, i.e. 1–2 Å, or orientational, 1–2°, changes in the interfacial arrangement of the interacting components can reduce electron transfer rate constants by orders of magnitude. Therefore, an essential ingredient in developing functional interfacial supramolecular assemblies is the ability to probe their physical structures and correlate these with their electronic and photonic properties.

The past decade has seen a dramatic improvement in the strategies and instrumentation available to characterize the structures of interfacial supramolecular assemblies. Current thrusts are towards *in situ* techniques that probe the structure of the interfacial supramolecular assembly with increasingly fine spatial and time resolution. The objective of this field is to assemble reaction centers around which the environment is purposefully structured at the molecular level, but extends over supramolecular domains. The properties of the assembly are controlled not only by the properties of the molecular building blocks but especially by the interface. Therefore, the focus is on both the interfacial and bulk properties of monolayers and thin films. Issues that need to be addressed include the film thickness, structural homogeneity and long-range order, as well as the electrochemical and photochemical properties of the modified surfaces. The intention in this present chapter is not to give a detailed physical background to all surface analytical techniques but to introduce the most important and contemporary characterization methods. Therefore, the basis of the measurements are outlined in this chapter, with illustrative examples of their application being presented in Chapter 4, *Formation and Characterization of Monolayers and Thin Films.*

3.1.1 Scanning Probe Microscopy

An important objective in this field is to create interfacial assemblies in which the position of the individual building blocks is controlled with molecular or atomic precision such that this order extends over macroscopic distances. Realizing this objective means that the physical structure and electronic properties of the interfacial supramolecular assembly must be probed with molecular resolution. Prior to the development of Scanning Tunneling Microscopy (STM) and related techniques, the ability to perform these atomic-scale measurements was extremely limited. As illustrated in Figure 3.1(a), STM provides topographic information by monitoring the electron tunneling current between a sharp microelectrode tip and a substrate, as the tip is scanned across the surface [1]. Since its development in the early 1980s and the recognition for its inventors with the 1986 Nobel prize [2], STM has found wide use in studies of interfacial supramolecular assemblies, dynamic processes and chemical reactions. The two essential technologies enabling microscopy by electron tunneling are the formation of a metallic tip that is ideally atomically sharp and precise positional control using high-resolution piezoelectrics. Typically, a tungsten wire is sharpened by electrochemical etching in an NaOH solution or a platinum/iridium wire is cut to form the probe. The few atoms closest to the scanned surface define the operational tip. Both vertical and lateral motion is controlled by piezoelectric transducers. For conventional experiments, an STM image is collected by scanning the tip in a plane less than a nanometer above the surface and the current is kept constant by continuously varying the distance between the tip and substrate through a feedback control loop. Thus, a

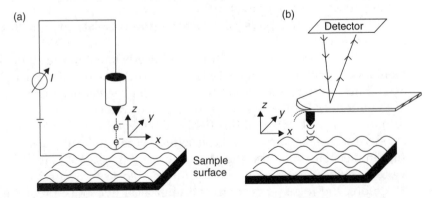

Figure 3.1 Schematic illustrations of (a) the scanning tunneling and (b) the atomic force microscopes

fixed tunneling current of pico- to nanoamperes is maintained by allowing the tip to move vertically; this motion reflects the electron density of the surface being scanned. In this way, an insight into the topography can be obtained at molecular length scales. While there is some uncertainty about the tunneling mechanism through insulators, organic molecules on conducting surfaces can be studied using STM. The most readily studied monolayers possess long-range order, e.g. self-assembled alkane thiol monolayers. One of the most novel uses of STM is to monitor dynamic surface processes including phase transitions, surface diffusion, epitaxial growth, and corrosion.

This technology has been extended to measurements in electrolyte solutions by independently controlling the potential of the substrate and the tip with respect to a reference electrode located in the solution. This electrochemical STM allows the progress of electrochemical reactions to be monitored *in situ* under potential control. The instrument uses a four-electrode configuration in which the potentials are controlled so that the current flowing between the substrate and tip is dominated by the tunneling current, while a predominantly Faradaic current flows between the substrate and the counter electrodes.

Although this application strictly belongs to nanofabrication technology, a scanning tunneling microscope can also be used to purposefully move atoms and molecules into desired positions on a surface. The objective here is to move the adsorbate along the surface without breaking the surface–adsorbate bond. Such a process requires that sufficient energy is input to cross the energy barrier to lateral diffusion, which is approximately 10–33 % of the adsorption energy. The desired lateral movement is accomplished by either field-assisted diffusion, caused by the action of the spatially inhomogeneous electric field near the probe tip on the dipole moment of the adsorbate, or by a sliding process where the tip–adsorbate attraction is used to reposition atoms. Another means to manipulate atoms is via perpendicular processes, whereby an atom is transferred to and from the tip by the application of a voltage pulse. While these 'molecule-by-molecule' approaches have the potential to create quite complex structures, they are generally too time and energy intensive to compete with directed synthesis and self-assembly approaches.

Atomic Force Microscopy (AFM), illustrated in Figure 3.1(b), measures deflections in the cantilever due to capillary, electrostatic, van der Waals and frictional forces between the tip and the surface. Mounting a fine tip on a piezoelectrically positioned cantilever spring allows surface forces of the order of 10^{-13} to 10^{-6} N to be measured. The principle advantage of AFM is that it can image non-conducting samples even when in contact with liquids and that it provides a more direct topographical image than the electron density map of a scanning tunneling microscope. AFM is particularly useful for imaging soft materials, e.g. interfacial assemblies incorporating biological materials. In this 'tapping mode' of AFM, a stiff cantilever is oscillated near its resonant frequency with an amplitude of about 0.5 nm. The interaction of the tip with the surface is detected as a shift to a new frequency.

3.1.2 Scanning Electrochemical Microscopy

Despite its ability to provide structural information with atomic resolution, STM does not provide any information about the chemical reactivity of the surface.

Bard and co-workers have developed the technique of Scanning Electrochemical Microscopy (SECM) [3], to provide information about the redox activity of a wide variety of assemblies. In common with STM, SECM uses high-resolution piezoelectric elements to scan a microelectrode tip across the interface of interest. However, in SECM the microelectrode acts as a working electrode in an electrochemical cell that contains a redox-active species. A redox reaction occurs at the microelectrode, e.g. $Ox + ne^- = Red$, and by monitoring the current generated at the tip, the surface can be mapped in terms of its redox activity.

As illustrated in Figure 3.2, when the microelectrode is distant from the surface by several electrode diameters, a steady-state current, $i_{T,\infty}$, is observed at the tip. The magnitude of the current is the same as that observed for a microdisk in a conventional experiment. When the tip is near a surface, the tip current, i_T, differs from $i_{T,\infty}$, and depends on both the distance between the surface and tip, and the chemical nature of the surface. If the interfacial assembly efficiently blocks electron transfer, i.e. it is an electronic insulator, the mediator will not be regenerated, thus causing $i_T/i_{T,\infty}$ to be less than unity. If the 'Red' species becomes re-oxidized at the surface, then the flux of the 'Ox' species to the tip is enhanced, and $i_T/i_{T,\infty}$ is larger than unity. The actual current–distance relationship depends on the tip shape, e.g. disk or cone, and on the heterogeneous electron transfer rate constant of the 'feedback' reaction. Currently, the best spatial resolution reported is of the order of 50 nm.

The SECM technique can image interfacial supramolecular assemblies so as to identify heterogeneities in their conductivity values. Information of this type is

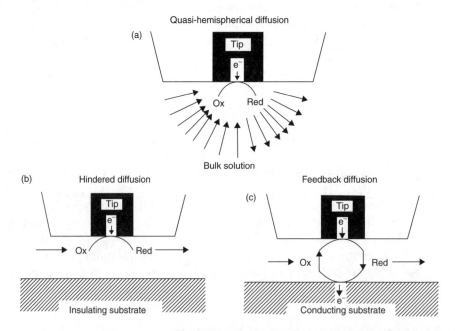

Figure 3.2 Operation of the scanning electrochemical microscope illustrating diffusion of a redox mediator to the microscope tip: (a) when the tip is far from any surface; (b) when the tip is near an insulator and the insulating surface blocks transport of the mediator to the tip; (c) when the tip is near a conducting surface that regenerates the mediator, this causing an enhanced current

essential if theoretical models that describe the structure of the electrochemical interface are to be developed. This microscopic technique can also be employed to distinguish regions of different reactivity, i.e. 'reaction rate imaging', on an electrode surface. Films on electrode surfaces can be probed with nanometer resolution by measuring the tip current as the tip passes through the film. Adsorption equilibria and kinetics on electrode and insulator surfaces can also be studied.

3.1.3 Contact Angle Measurements

Controlling the wetting of surfaces is of great technological importance and is a significant driving force in the development of interfacial supramolecular assemblies. Wetting of a solid by a liquid can be understood by placing a drop of pure solvent on the modified surface and measuring the contact angle between the surface and the drop as it advances or retreats across the surface [4]. Consider a drop of liquid resting on a solid surface. The drop rests in an equilibrium position because the three forces involved are balanced, namely, the interfacial tensions between solid and liquid, σ_{SL}, that between solid and vapor, σ_{SV}, and that between liquid and vapor, σ_{LV}. As illustrated in Figure 3.3, the contact angle or wetting angle, θ, is the angle between the tangent plane to the surface of the liquid and the tangent plane to the surface of the solid, at any point along their line of contact. The surface tension of the solid will favor spreading of the liquid, but this is opposed by the solid–liquid interfacial tension and the vector of the surface tension of the liquid in the plane of the solid surface.

The relationship between the cosine of the drop/surface contact angle and the three surface tensions is given by Young's equation, as follows:

$$\sigma_{SV} = \sigma_{SL} + \sigma_{LV} \cos \theta \qquad (3.1)$$

which can be interpreted as a mechanical force balance for the line of three-phase contact. Wetting is a direct consequence of Young's equation. If a droplet is present, the situation is described in terms of partial wetting, since the surface is only partially covered with the liquid. If the system is now perturbed in some way, e.g. by increasing the temperature or by switching the charge state of an interfacial assembly by changing its redox composition, the surface tensions will change. If at a certain point the sum of the solid–liquid and the liquid–vapor surface tensions equals the solid–vapor interfacial tension, the contact angle will be zero. Under these circumstances, a uniform film covers the whole solid surface and the liquid

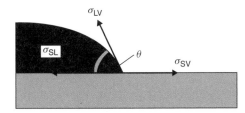

Figure 3.3 Forces acting on a liquid drop resting on a surface; the solid–liquid, solid–vapor and liquid–vapor forces are denoted by σ_{SL}, σ_{SV} and σ_{LV}, respectively

completely wets the solid. If the system is studied as a function of temperature, the transition temperature is called the wetting temperature (TW). In an approach that closely follows that used for bulk phase transitions, the order of the wetting transition is determined by the discontinuities in derivatives of the surface free energy. If a discontinuity occurs in the first derivative of the free energy, the transition is first-order, and will take place in a discontinuous manner. In contrast, if the first derivative of the free energy is continuous though a phase transition point, then that indicates that it is a higher-order or critical phase transition.

3.1.4   Mass-Sensitive Approaches

Many processes associated with interfacial assemblies, ranging from initial formation to selective binding and redox switching, are accompanied by a change in mass. These mass changes can be measured very accurately by using a piezoelectric crystal [5]. In this section, two types of mass-sensing phenomena are considered, namely bulk and surface acoustic wave devices. Both of these approaches employ piezoelectric crystals which are deformed mechanically by applying a potential or voltage in a controlled manner. By applying such a waveform, acoustic waves are generated which travel either through the bulk of the crystal, or along its surface. These devices are useful for monitoring the formation of an interfacial assembly, e.g. a self-assembled monolayer, because the velocity of the waves, and hence their frequency, depends on the mass of the crystal. Moreover, in redox-active systems, they provide useful information about mass changes that occur within the film when the oxidation state of an assembly, e.g. a metallopolymer layer, is switched.

Bulk acoustic wave devices

First, the underlying principles upon which bulk acoustic wave (BAW) devices operate are described. When a voltage is applied to a piezoelectric crystal, several fundamental wave modes are obtained, namely, longitudinal, lateral and torsional, as well as various harmonics. Depending on the way in which the crystal is cut, one of these principal modes will predominate. In practice, the high-frequency thickness shear mode is often chosen since it is the most sensitive to mass changes. Figure 3.4 schematically illustrates the structure of a bulk acoustic wave device, i.e. the quartz crystal microbalance.

For AT-cut quartz crystals operating in the shear mode, the oscillation frequency, f_0, is inversely proportional to the thickness d of the crystal, as described by the following equation:

$$f_0 = k_f/d \tag{3.2}$$

where k_f is the frequency constant, e.g. for an AT-cut quartz crystal at room temperature, k_f is 0.168 MHz cm. Since the mass of the crystal, m, is $\rho A d$, where ρ and A are the density and cross-sectional area, respectively, Equation (3.2) indicates that the oscillation frequency is inversely proportional to the mass of the crystal.

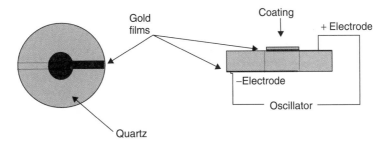

Figure 3.4 Configuration of a quartz crystal microbalance involving standing shear waves between facing gold electrode contacts

When a species binds to a crystal it increases its mass by Δm, thus causing the resonance frequency to shift by Δf, according to the Sauerbrey equation, as follows:

$$\Delta f = -\frac{1}{\rho_m k_f} f_0^2 \frac{\Delta m}{A} \tag{3.3}$$

where ρ_m is the density of the thin active coating on top of the piezoelectric substrate. It is essential to note that the Sauerbrey equation is only strictly applicable to rigid interfacial supramolecular assemblies. The quartz crystal microbalance is a very stable device that is capable of measuring mass changes as small as a fraction of a ng cm^{-2}.

From the perspective of electrochemically driven self-assembly or the characterization of redox-active interfacial films, the electrochemical quartz crystal microbalance (EQCM) has emerged as a powerful technique. With the EQCM, the very small mass changes that accompany electrochemical processes at electrodes can be detected. These processes include underpotential metal deposition, electrolyte adsorption, and mass changes accompanying ion and solvent movements in redox polymer films, as well as potential-controlled adsorption. The schematic layouts of two EQCM configurations are illustrated in Figure 3.5. The crystals are typically quite large, i.e. 5–10 mm in radius, with appropriately sized excitation electrodes. The crystal is mounted so that one of the exciting electrodes faces the solution of interest while the other typically faces air. The two excitation electrodes are electrically connected to an oscillator circuit which contains a broadband radiofrequency (RF) amplifier so that the electrode facing the solution is at 'hard ground'. The crystal is driven at the frequency at which the maximum current can be sustained in a zero-phase-angle condition by using a feedback loop. The output of the oscillator is then connected to a conventional frequency analyzer, thus allowing the mass change to be monitored. As illustrated in Figure 3.5, a critical feature of the EQCM is the potentiostat design. A dedicated Wenking potentiostat in which the working electrode is at hard ground is one option, although commercial potentiostats in which the working electrode is at 'virtual ground' can also be used. This objective can be achieved if the working electrode is *not* connected to the potentiostat and the potential difference between reference and hard ground is used to control the working electrode potential. The current is measured by the voltage drop across a resistor in series with the counter electrode connection. With this arrangement, the current and EQCM frequency can be measured simultaneously. The accessible

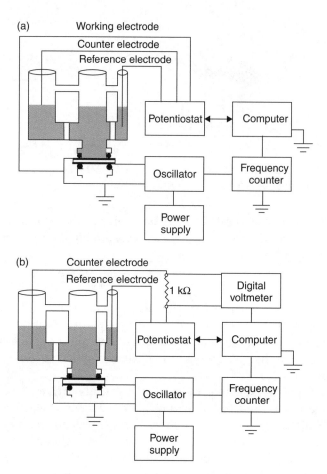

Figure 3.5 Schematic representations of typical electrochemical quartz crystal microbalance equipment: (a) using a Wenking potentiostat in which the working electrode is at hard ground; (b) using a conventional potentiostat in which the current is measured by the voltage drop across a 1 kΩ resistor in series with the counter electrode

timescale is typically in the milliseconds to tens of milliseconds domain, as dictated by the time constant of the QCM resonator and the relatively large area of the working electrode, which means that it responds relatively slowly to changes in the applied potential. However, this time domain is useful for monitoring interfacial supramolecular assemblies and the dynamics of electrochemical processes.

Surface acoustic wave devices

Unlike bulk acoustic wave devices, surface acoustic wave (SAW) detectors work by the interaction of a wave traveling down the surface of the device with species on or near the surface. An attractive mass-sensing device is the SAW oscillator which employs interdigitated electrodes. Figure 3.6 shows that the acoustic wave is originally created by an AC voltage signal applied to a set of interdigitated electrodes at one end of the device. The electrical field distorts the lattice of the piezoelectric

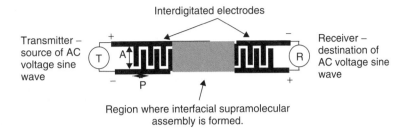

Figure 3.6 Schematic layout of a single-acoustic-aperture surface acoustic wave device

material beneath the electrode, thus causing a surface acoustic wave to propagate towards the other end through a region of the crystal known as the *acoustic aperture*. When the wave arrives at the other end, a duplicate set of interdigitated electrodes generate an AC signal as the acoustic wave passes underneath them. The signal can be monitored in terms of amplitude, frequency or phase shift. These devices operate at ultra-high frequencies (GHz), thus giving them the capability to sense as little as 1 pg of material.

3.1.5 Ellipsometry

Spectroscopic ellipsometry is a non-invasive optical technique for determining the index of refraction, extinction coefficient, film thickness, interface roughness, and composition of thin surface layers and multilayer structures [6]. This method measures the change in the state of polarization of light after reflection from the sample surface. Such measurements can be repeated for each wavelength across the spectral range of interest, thus determining the index of refraction and extinction coefficient as a function of the wavelength. These optical parameters are then matched to computer models to determine the structure, thickness and composition of the sample.

A typical experimental ellipsometer is illustrated in Figure 3.7. Monochromatic light, typically from a continuous wave laser, e.g. a He–Ne laser, is plane polarized (the angle of polarization is given by *p*) and impinges on a surface. A compensator is then used to convert the elliptically polarized reflected beam to a plane polarized beam (with *a* being the angle of polarization). The analyzer then determines the

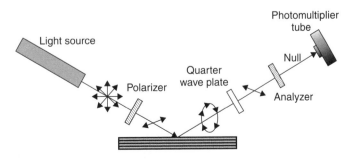

Figure 3.7 Components of a typical ellipsometer

angle by which the compensator has polarized the beam. These two angles, p and a, give the phase shift between the parallel and perpendicular components, Δ, and the change in the ratio of the amplitudes of the two components, $\tan \psi$, is given by the following:

$$e^{i\Delta} \tan \Psi = \left(\frac{E_{\text{reflected}(p)}}{E_{\text{reflected}(a)}} \right) \left(\frac{E_{\text{incident}(a)}}{E_{\text{incident}(p)}} \right) \tag{3.4}$$

where $\Delta = 2p + \pi/2$ and $\psi = a$. For a clean surface, Δ and ψ are directly related to the complex index of refraction of the surface, $\check{n}^s$ according to the following:

$$\check{n}^s = n^s(1 - ik^s) \tag{3.5}$$

where n^s is the ordinary refractive index and k^s is the extinction coefficient. When an interfacial supramolecular assembly is assembled on the surface, Δ and ψ are related to the complex indices of both the film and the substrate, as well as the film thickness. Therefore, in order to accurately determine the film thickness the refractive index must also be known.

3.1.6 Surface Plasmon Resonance

Surface plasmon resonance (SPR) represents a powerful way to monitor changes in the refractive index of a metal/film interface. In doing so, it can provide important information about the growth and structure of an interfacial supramolecular assembly, as well as the dynamic processes occurring within the interfacial region, e.g. host–guest binding. SPR is a phenomena that exists most commonly at the interface between a metal such as gold or silver and a dielectric material [7]. A surface plasmon is created by resonant excitation of the electrons at the surface of the metal, resulting in an associated surface-bound and evanescent electromagnetic wave of optical frequency. Excitation is commonly achieved by optical irradiation. The evanescent wave has a maximal intensity at the interface and decays exponentially away from the interface so that it is mostly contained within a distance corresponding to a fraction of a wavelength of light. Of central importance for studies of interfacial supramolecular assemblies is that the conditions required to resonantly excite surface plasmons depend on the refractive-index properties of the immobilized film. This sensitivity to interfacial phenomena allows small fractions of a monolayer coverage to be detected. Thus, SPR has the ability to measure, in real-time, the formation of an interfacial supramolecular assembly and the interactions of molecules with the functionalized interface due to changes in the interfacial refractive index. SPR is extremely useful for monitoring dynamic interactions within a fluid environment, which may be tailored to model conditions encountered in real-world applications, e.g. *in vivo* studies. More involved measurements allow the spatial distribution of material/refractive indices, interfacial roughness and interfacial spectroscopic properties to be determined.

The Kretchmann configuration illustrated in Figure 3.8 relies on the phenomenon of total internal reflection. This process occurs when light traveling through an optically dense medium, e.g. glass, reaches an interface with a material of lower optical density, e.g. air, and is reflected back into the dense medium. Although the

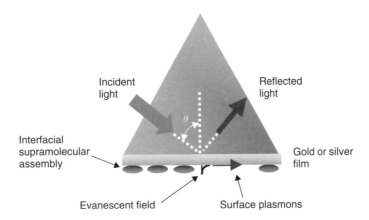

Figure 3.8 Kretchmann configuration of a metal-coated prism used for surface plasmon resonance investigations of interfacial supramolecular assemblies

incident light is totally internally reflected, a component of this light, the evanescent wave or field, penetrates the interface into the less dense medium to a distance of approximately one wavelength. In SPR, a monochromatic, p-polarized light source is used and the glass is coated with a thin metal film (of thickness less than one wavelength of light). The wave vector of the evanescent field, $\mathbf{K}_{ev}$, is given by the following:

$$\mathbf{K}_{ev} = \frac{\omega_0}{c} \eta_g \sin\theta \tag{3.6}$$

where ω_0 is the frequency of incident light, η_g the refractive index of the dense medium (glass), θ the angle of incidence of the light and c the speed of light in a vacuum. The wave vector of a surface plasmon, $\mathbf{K}_{sp}$, can be approximated to the following:

$$\mathbf{K}_{sp} = \frac{\omega_0}{c} \sqrt{\frac{\varepsilon_m \eta_s^2}{\varepsilon_m + \eta_s^2}} \tag{3.7}$$

where ε_m is the dielectric constant of the metal film and η_s the refractive index of the dielectric medium, i.e. the interfacial supramolecular assembly. The evanescent wave of the incoming light is able to couple with the free oscillating electrons, i.e. plasmons, in the metal film at a specific angle of incidence corresponding to the condition when $\mathbf{K}_{sp} = \mathbf{K}_{ev}$, and thus the surface plasmon is resonantly excited. This process causes energy from the incident light to be lost to the metal film, so resulting in a reduction in the intensity of reflected light which can be detected by a two-dimensional array of photodiodes or charge-coupled detectors. Equation (3.7) above shows that $\mathbf{K}_{sp}$ depends on the refractive index of the water/air medium above the metal film, which can be monitored up to a thickness of approximately 300 nm above the metal surface. Therefore, if the refractive index immediately above the metal surface changes, e.g. when triggered by the formation of an interfacial supramolecular assembly, a change in the angle of incidence required to excite a surface plasmon will occur. By monitoring the angle at which resonance occurs during an adsorption process with respect to time, an adsorption profile can be obtained.

A variety of other practical implementations of optical SPR have been demonstrated, including the use of metal-coated diffraction gratings, optical fibers, planar waveguides and even metal colloids.

3.1.7 Neutron Reflectivity

Neutron reflectivity is based on the interaction between a neutron beam and a surface [8]. In such an experiment, a collimated neutron beam, of wavelength λ and incident angle θ, is directed onto an interface and the reflected intensity, R, is measured as a function of the momentum transfer Q, as given by the following equation:

$$Q = 4\pi/\lambda \sin \theta \tag{3.8}$$

The measured reflectivity, $R(Q)$, depends upon the neutron refractive index profile perpendicular to the interface, defined as the z-direction. The neutron refractive index is a function of the scattering length density, Nb, which is the product of the number density N, in units of nuclei per cm^3, and the neutron scattering lengths, b, of the nuclei present. Since the neutron scattering length varies from nucleus to nucleus, chances in the nature and composition of the surface result in changes in reflectivity.

The reflectivity is determined by the Fresnel coefficients for the film and is composed of two terms. The first term, the Fresnel reflectivity, accounts for the reflection from a single interface, which causes the reflectivity profile to fall off as a function of Q^{-4}. The second term is the film factor, which accounts for the interference effects of neutrons reflected from the different interfaces of the film and depends on the scattering length density gradient. It is not possible to deconvolute the reflectivity profile into a scattering length density profile due to lack of phase information. However, after assuming a model for the Nb profile, a model reflectivity profile is fitted to the measured reflectivity. The scattering length density gradients in the system can be manipulated by isotopic substitution (usually deuterium substitution for hydrogen) to selectively increase the scattering from different regions of the film, solvent or substrate, while not drastically changing the chemical and physical properties of the system. There are some model-independent parameters, which can be derived from a reflectivity profile. Reflection of neutrons from the two interfaces of a particular layer leads to constructive and destructive interference. As a result, interference fringes appear in the reflectivity profile as a series of oscillations. The period of these oscillations, ΔQ, is related to the thickness of the film, d, as given by the following expression:

$$d = 2\pi/\Delta Q \tag{3.9}$$

The amplitude of the fringes also yields information about the sharpness of the interfaces in question, since large fringes correspond to a sharp interface, while small fringes are indicative of a diffuse interface.

Apart from yielding information about the film thickness and the nature of the interface, it is also possible to determine the volume fraction profiles of components present in a particular layer. For example, in the case of an adsorbed polymer film, the fractions of solvent, Φ_S, and polymer, Φ_P, can be extracted from the experimentally determined neutron scattering length Nb_E, as shown in the following equation:

$$Nb_E = \Phi_P Nb_P + \Phi_S Nb_S \qquad (3.10)$$

where Nb_P and Nb_S are the neutron scattering length densities of the polymer and solvent, respectively. In this way, detailed information about the structure of layers and their composition can be obtained, provided that the contrast at the interfaces is sufficiently large. In summary, the form of the reflectivity profile is dependent upon the film thickness and the scattering length densities. In some cases it can be used to directly obtain information on the film thickness and steps in Nb, although to obtain the complete Nb profile modeling must be used.

3.2 Voltammetric Properties of Interfacial Supramolecular Assemblies

Electrodes represent an unrivaled platform onto which interfacial supramolecular structures can be assembled. They can be fully characterized before assembly and offer a convenient means to both probe and control the properties of the film. The interest in this area has increased dramatically in recent years because adsorbed monolayers enable both the nature of the chemical functional groups and their topology to be controlled. This molecular-level control allows the effects of both chemical and geometric properties on electron transfer rates to be explored. Moreover, these assemblies underpin technologies ranging from electrocatalysis to redox-switchable non-linear optical materials.

3.2.1 Electrochemical Properties of an Ideal Redox-Active Assembly

Figure 3.9 illustrates the electrochemical and mass transport events that can occur at an electrode modified with a interfacial supramolecular assembly [9]. For monolayers in contact with a supporting electrolyte, the principal process is heterogeneous electron transfer across the electrode/monolayer interface. However, as discussed later in Chapter 5, thin films of polymers [10] represent an important class of interfacial supramolecular assembly (ISA) in which the properties of the redox center are affected by the physico-chemical properties of the polymer backbone. To address the properties of these thin films, mass transfer and reaction kinetics have to be considered. In this section, the properties of an ideally responding ISA are considered.

In this analysis, it is assumed that the redox couple is adsorbed on the electrode surface and is not present in solution, or its concentration is sufficiently low such that its contribution to the Faradaic current is negligible. Other assumptions are that all adsorption sites on the surface are assumed to be equivalent and the oxidized and reduced forms occupy equal areas on the surface. Moreover, adsorption and desorption are rapid and do not influence the kinetics of the electrochemical reaction. The free energy of adsorption and maximum or limiting surface coverage,

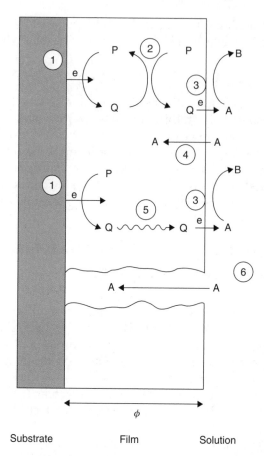

Figure 3.9 Schematic illustration of the processes that can occur at a modified electrode, where P represents a reducible substance in a film on the electrode surface and A a species in solution. The processes shown are as follows: (1) heterogeneous electron transfer to P to produce the reduced form Q; (2) electron transfer from Q to another P in the film (electron diffusion or electron hopping in the film); (3) electron transfer from Q to A at the film/solution interface; (4) penetration of A into the film (where it can also react with Q or at the substrate/film interface); (5) movement (mass transfer) of Q within the film; (6) movement of A through a pinhole or channel in the film to the substrate, where it can be reduced. From A.J. Bard and L.R. Faulkner, *Electrochemical Methods: Fundamentals and Applications*, 2nd Edition, © Wiley, 2001. Reprinted by permission of John Wiley & Sons, Inc

assumed to be equal to the surface activity, are considered to be independent of the applied potential. In addition, to observe an ideal response, the entire potential drop must occur at the electrode/ISA interface and the adsorbates must not interact laterally. Furthermore, it is assumed that the Faradaic and capacitive currents can be separated.

Voltammetry can provide a powerful insight into the thermodynamics and kinetics of electron transfer across the electrode/adsorbate interface and within the interfacial supramolecular assembly. Linear sweep voltammetry (LSV) and cyclic voltammetry (CV) are the most commonly used techniques to study the equilibrium behavior and kinetics of redox-active interfacial supramolecular

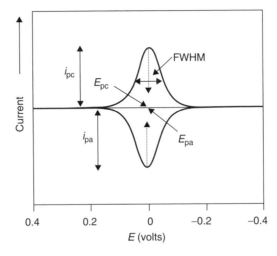

Figure 3.10 Current–potential curves obtained from cyclic voltammetry measurements for the reduction and oxidation of an adsorbed interfacial supramolecular assembly under finite diffusion conditions

assemblies. Consider the voltammetric response where an oxidized species, Ox, e.g. an osmium polypyridyl chloride center bound to a linear 4,4′-dipyridyl tether (p0p), [Os(bpy)$_2$Cl p0p]$^{2+}$, is converted to the reduced form, Red, e.g. [Os(bpy)$_2$Cl p0p]$^+$, by adding a single electron:

$$\mathrm{Ox} + e \underset{k_\mathrm{f}}{\overset{k_\mathrm{b}}{\rightleftharpoons}} \mathrm{Red}$$

Figure 3.10 illustrates the theoretical cyclic voltammogram expected for an adsorbed monolayer under finite diffusion conditions where the dynamics of heterogeneous electron transfer across the electrode/layer interface do not influence the observed response. Adsorption leads to changes in the shape of cyclic voltammograms when compared to solution-phase reactants since the redox-active material does not have to diffuse to or from the electrode surface [9]. In contrast to solution-phase species, the peaks for surface-confined species are Gaussian in shape and the peak-to-peak separation is zero. The peak current, i_p, for an ideally responding surface-confined reactant is given by the following:

$$i_\mathrm{p} = \frac{n^2 F^2}{4RT} \nu A \Gamma \tag{3.11}$$

where n is the number of electrons transferred, F is the Faraday constant, ν is the scan rate, R is the gas constant, T is the absolute temperature, A is the real or microscopic area of the electrode and Γ is the surface coverage or concentration of the redox-active adsorbate (in mol cm^{-2}). The surface coverage can be determined by measuring the Faradaic charge Q passed during exhaustive electrolysis of the assembly, e.g. using bulk electrolysis with chronocoulometry or slow-scan-rate voltammetry, according to the following equation:

$$\Gamma = \frac{Q}{nFA} \tag{3.12}$$

For an ideal Nernstian reaction where the adsorbates do not interact laterally, or at least where the interactions are independent of the surface coverage, a surface-confined species will follow the following relationships:

$$\text{FWHM} = 3.53\frac{RT}{nF} = \frac{90.6}{n} \text{ mV} \tag{3.13}$$

$$E_{pa} = E_{pc} \tag{3.14}$$

where FWHM is the *full-width at half-maximum* of either the cathodic or anodic wave.

3.2.2 The Formal Potential

The formal potential, $E^{0\prime}$, contains useful information about the ease of oxidation of the redox centers within the supramolecular assembly. For example, a shift in $E^{0\prime}$ towards more positive potentials upon surface confinement indicates that oxidation is thermodynamically more difficult, thus suggesting a lower electron density on the redox center. Typically, for redox centers located close to the film/solution interface, e.g. on the external surface of a monolayer, the $E^{0\prime}$ is within 100 mV of that found for the same molecule in solution. This observation is consistent with the local solvation and dielectric constant being similar to that found for the reactant freely diffusing in solution. The formal potential can shift markedly as the redox center is incorporated within a thicker layer. For example, $E^{0\prime}$ shifts in a positive potential direction when buried within the hydrocarbon domain of a alkane thiol self-assembled monolayer (SAM). The direction of the shift is consistent with destabilization of the more highly charged oxidation state.

For high coverages of redox centers with at least one charged oxidation state, $E^{0\prime}$ typically shifts in a negative potential direction with increasing electrolyte concentration, as described by the Nernst equation, i.e. $59/n$ mV per decade change in the electrolyte concentration. This shift is due to movement of charge compensating counterions into and out of the assembly as the redox composition is switched. The identity of the electrolyte ions can also influence the observed response through ion pairing. Therefore, it is important to distinguish between specific ion effects and simple changes in $E^{0\prime}$ brought about by changing the electrolyte concentration. This objective can be achieved by varying the concentration of the electrolyte ion in solutions of fixed ionic strength [11,12].

3.2.3 Effect of Lateral Interactions

The experimental response may deviate from that expected under ideal conditions for the following reasons. First, there may be interactions between the adsorbed molecules that cause the surface activity to differ from the surface concentration [13,14]. Alternatively, double-layer effects, ion-pairing, acid–base dissociation, and dispersion of the formal potentials can cause similar deviations. A non-zero peak splitting may indicate intermolecular interactions between the redox centers or that switching the redox composition triggers a structural change within the supramolecular assembly, e.g. adsorbate reorientation or the formation of an ion-pair. The full-width at half-maximum (FWHM) can either be greater or smaller than the theoretical value of 90.6/n mV. Beyond intermolecular interactions between redox centers and double-layer effects, large values of the FWHM may be caused by a distribution of formal potentials. Therefore, broad peaks may indicate a disorganized structure with the microenvironment or local electrostatic potential around each redox center being slightly different. Under these circumstances, the voltammetry becomes more ideal when the surface coverage of redox centers is reduced or ion-pairing reduces the change in charge density. Sharp voltammograms with small FWHM values are obtained for solid-like interfacial supramolecular assemblies such as densely packed electroactive SAMs.

3.2.4 Diffusional Charge Transport through Thin Films

For monolayers, at sufficiently short experimental timescales, i.e. high scan rates, the rate of heterogeneous electron transfer will influence the voltammetric response. Under these circumstances, ΔE_p will increase and the peak current will no longer increase in proportion to the scan rate. As discussed later in Chapter 5, these high-scan-rate voltammograms can be used to probe how certain factors, such as the structure and length of the bridging ligand, influence the electron transfer dynamics. For thicker films, e.g. metallopolymer coatings with thicknesses of the order of 1 μm, the layer is no longer exhaustively oxidized and the rate of charge transport through the layer influences the voltammetric response.

For a redox polymer film, under these conditions of semi-infinite linear diffusion control the peak current increases as $v^{1/2}$. In many films, electrochemical charge transport occurs by electron self-exchange reactions between neighboring oxidized and reduced sites. This electron hopping process is accompanied by the movement of charge-compensating counterions that are mobile within the layer. As illustrated in Figure 3.11, provided that the depletion layer remains well within the layer, and ohmic as well as migration effects are absent, the voltammetric

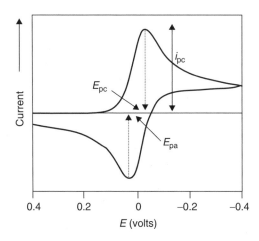

Figure 3.11 Reversible cyclic voltammogram obtained for an interfacial supramolecular assembly under semi-infinite linear diffusion conditions

response is reminiscent of that observed for a solution-phase reactant. The effective diffusion coefficient, D_{CT}, corresponding to 'diffusion' of either electrons or charge-compensating counterions, can be estimated from the well-known Randles–Sevçik equation, as follows:

$$i_p = 2.69 \times 10^5 n^{3/2} A D_{CT}^{1/2} C \nu^{1/2} \tag{3.15}$$

where C is the concentration of electroactive sites within the film. This equation is routinely used for the evaluation of D_{CT} when using cyclic voltammetry at relatively high scan rates, typically >50 mV s^{-1}.

The concentration profile of fixed oxidized and reduced sites within the film depends on the dimensionless parameter $D_{CT}\tau/d^2$, where τ is the experimental timescale, i.e. $RT/F\nu$ in cyclic voltammetry, and d is the polymer layer thickness. When $D_{CT}\tau/d^2 \gg 1$, all electroactive sites within the film are in equilibrium with the electrode potential, and the surface-type behavior described previously is observed. In contrast, $D_{CT}\tau/d^2 \ll 1$ when the oxidizing scan direction is switched before the reduced sites at the film's outer boundary are completely oxidized. The wave will exhibit distinctive diffusional tailing where these conditions prevail. At intermediate values of $D_{CT}\tau/d^2$, an intermediate i_p versus ν dependence occurs, and a less pronounced diffusional tail appears.

Under semi-infinite diffusion conditions, the peak-to-peak splitting is given by the following:

$$\Delta E_p = 57.0/n \text{ mV (at } 25\,^{\circ}C) \tag{3.16}$$

3.2.5 Rotating Disk Voltammetry

The majority of electrochemical measurements on interfacial supramolecular assemblies involve quiescent solutions in which mass transport e.g. of charge compensating counterions, occurs by diffusion. However, the hydrodynamic equations describing convective–diffusional transport have been solved for the rotating disk electrode. Rotating disk voltammetry is similar to cyclic voltammetry in that the working electrode potential is (slowly) swept back and forth across the formal potential of the analyte. However, as illustrated in Figure 3.12, this method differs significantly in that the working electrode itself is rotated. This rotational motion sets up a well-defined flow of solution towards the surface of the rotating disk electrode. The flow pattern is akin to a vortex that drives the solution, plus the analyte, towards the electrode. The resulting high rate of mass transport results in a sigmoidal dependence of the current on the applied potential and the height of this wave provides the analytical signal. The thin film of solution immediately adjacent to the surface of the electrode behaves as if it were stuck to the electrode. Thus, while the bulk of the solution is being stirred vigorously by the rotating electrode, this thin layer of solution is dragged along by the disk and, from the perspective of the rotating electrode, appears to be motionless. This layer is called the *stagnant* layer in order to distinguish it from the remaining bulk of the solution. Analyte is conveyed to the electrode surface by a combination of two types of transport. First, the vortex flow in the bulk solution continuously brings fresh analyte to the outer edge of the stagnant layer. Then, the analyte moves across the stagnant layer via simple

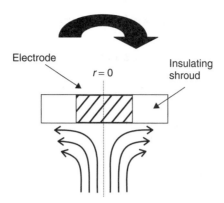

Figure 3.12 Schematic of the flows occurring at a rotating disk electrode

molecular diffusion. The thinner the stagnant layer, then the faster the analyte can diffuse across it and reach the electrode surface. Faster electrode rotation makes the stagnant layer thinner and so faster rotation rates permit the analyte to diffuse to the electrode more rapidly, thus resulting in a higher current being measured at the electrode.

The Levich equation predicts the current observed at a rotating disk electrode. This equation takes into account both the rate of diffusion across the stagnant layer and the complex solution flow pattern. In particular, the Levich equation predicts the height of the sigmoidal wave observed in rotated disk voltammetry. The sigmoidal wave height is often called the *Levich current*, i_L, and is directly proportional to the analyte concentration, Y. The Levich equation is given by the following:

$$i_L = 0.620nFAD_0^{2/3}\omega^{1/2}\upsilon^{-1/6}Y \tag{3.17}$$

where ω is the angular rotation rate of the electrode (rad s^{-1}) and υ is the kinematic viscosity of the solution (cm^2 s^{-1}). The kinematic viscosity is the ratio of the solution's viscosity to its density. For pure water, $\upsilon = 0.0100$ cm^2 s^{-1}, and adding supporting electrolyte does not dramatically affect this value, e.g. for 1.0 M KNO$_3$, υ is 0.009 16 cm^2 s^{-1}. Equation (3.17) indicates that the Levich or limiting current depends linearly on $\omega^{1/2}$ and a linear response is observed experimentally providing that the reaction under investigation is totally reversible, i.e. the kinetics of all associated chemical and electrochemical steps are fast when compared to the timescale of the experiment. In contrast, where kinetics influence the voltammetric response, e.g. when an interfacial supramolecular assembly mediates the oxidation of a solution-phase analyte at a slow rate, then a plot of i versus $\omega^{1/2}$ becomes non-linear. Under these circumstances, the current is described by the Koutecky–Levich equation, given as follows:

$$\frac{1}{i_t} = \frac{1}{i_k} + \frac{1}{0.620nFAY_0D_0^{2/3}\upsilon^{-1/6}\omega^{1/2}} \tag{3.18}$$

where i_k represents the kinetically controlled current. Under these circumstances, a plot of $1/i_t$ versus $\omega^{-1/2}$ yields a linear response and the intercept contains information about the rates of reaction that are occurring at the electrode surface.

3.2.6 Interfacial Capacitance and Resistance

Potential-step chronoamperometry is a sensitive approach to measuring the interfacial capacitance and resistance of an electrochemical cell containing an electrode onto which an ISA has been assembled. In this technique, a small-amplitude potential step is applied to the electrode in a potential region where the ISA is not electrochemically active. In a typical experiment, the potential is stepped by 50 mV and the resulting capacitive or double-layer charging current are then recorded as a function of time. This capacitive current versus time transient, $i_c(t)$, decays according a single exponential and is described by the following equation:

$$i_c(t) = (\Delta E/R_u) \exp(-t/R_u C_{dl}) \tag{3.19}$$

where ΔE is the pulse amplitude, R_u is the total cell resistance, and C_{dl} is the double-layer capacitance. The absolute slope of a semi-log current versus time plot represents the reciprocal of the cell time constant, $R_u C_{dl}$, while the intercept provides R_u. By comparing the R_u values obtained for the same electrode with and without modification, an insight into the permeability of the ISA towards ions can be obtained, while C_{dl} can provide information about the relative perfection of the monolayer, and in some cases, the film thickness.

3.3 Spectroscopic Properties of Interfacial Supramolecular Assemblies

Beyond the structural and electrochemical techniques discussed earlier, spectroscopic methods have been widely used for characterizing interfacial supramolecular assemblies. One of the principal advantages of spectroscopy is that it provides *in situ* information about the assemblies when in contact with a solution of interest. Moreover, spectroscopic techniques can *non-invasively* probe the structure of buried interfaces, e.g. within complex multilayer assemblies. Various techniques, such as luminescence spectroscopy, confocal microscopy and Raman spectroscopy, are now playing increasingly important roles in identifying the orientation of individual adsorbates and elucidating binding modes and lateral interactions. These sensitive techniques can also be used to probe dynamic processes and to characterize the assembly's structure as it relaxes following a perturbation, e.g. a potential step or sweep. Although technical and theoretical difficulties still exist, one of the most exciting developments is the *in situ* characterization of electrochemical interfaces using NMR spectroscopy [15].

Investigations into the absorption and emission properties are especially important since they can reveal if the photophysical properties of the molecular components are different when immobilized, compared to when they are dissolved in solution. These investigations typically involve both steady-state and time-resolved methods. Time-resolved or transient techniques yield information about the lifetime of the emitting state, while flash photolysis yields the absorption characteristics of the photochemically produced transient species. Information of this kind is essential for understanding the interactions between the molecular components and between the adsorbates and the solid surface. These investigations also play pivotal roles in the study of interfacial and intermolecular electron and energy transfer processes. This present section outlines the basic principles behind a wide range of these spectroscopic techniques. Moreover, it seeks to illustrate how spectroscopic methods can be used to achieve a deeper understanding of the physical structure and properties of interfacial supramolecular assemblies.

3.3.1 Luminescence Spectroscopy

Fluorimetry [16] represents one of the most important steady-state methods for detecting emitted light. Figure 3.13 illustrates the optical path for a typical steady-state fluorescence spectrometer. A light source, usually a continuum source, is directed via a monochromator onto a sample. The emission resulting from this excitation is directed through a second monochromator and detected at a right angle to the excitation source. Luminescence spectroscopy, and in particular, temperature- and solvent-dependence studies, yields important information on the nature and location of the lowest excited state in the building blocks used to create supramolecular assemblies. These techniques can also be used to identify processes such as eximer or exciplex formation, and energy and electron transfers. The luminescence quantum yield, ϕ_e, is a valuable measure of the efficiency of luminescence of a sample and is related to the rates of radiative, k_r, and non-radiative, k_{nr}, decays in a sample, according to the following equation:

$$\phi_e = \frac{k_r}{k_r + \sum k_{nr}} \tag{3.20}$$

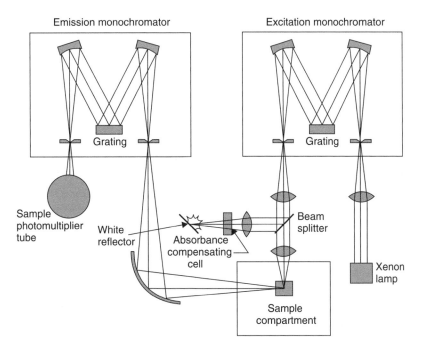

Figure 3.13 Schematic of the optical path within a typical spectrofluorimeter

The emission quantum yields are typically obtained by ratioing the areas under luminescence spectra for absorbance-matched samples of the species in question and a material of known quantum yield. This technique has been described in detail by Demas and Crosby [17] for ruthenium–polypyridyl systems.

The excitation and emission maxima may shift significantly on going from a dilute sample in solution to a solid or adsorbed species. Moreover, the magnitude of the fluorescence quantum yield depends on the matrix. For example, some compounds which do not luminesce in solution may exhibit a strong emission signal when immobilized on an inert matrix. Conversely, strong interaction of the luminophore with a conducting or semiconducting surface may result in photoinduced electron injection eliminating any observable emission in the bound species. These processes are discussed in greater detail in Chapters 5 and 6.

In studies on surface-modified electrodes, gold may be vapor deposited onto optically transparent glass slides, onto which the monolayer or thin film is then deposited. This approach produces films which have significant optical transparency, so allowing regular spectroscopic and photophysical techniques to be applied. When recording the fluorescence spectra of solid or surface-bound species, the technique may suffer from interfering scattered radiation and specular reflectance. To avoid this problem, the sample holders can be tilted at variable angles for front-face excitation and detection. Microscope and fiber optic attachments on conventional luminescence spectrometers offer even greater flexibility. Fluorescent detection techniques based on microscopy have greatly enhanced the sensitivity of the luminescent measurement. In conjunction with charge-coupled detector (CCD) devices, single photon counting detectors and, in particular, cooled intensified CCDs, fluorescence microscopy yields remarkably high sensitivity and the ability to obtain both images and spectral information.

3.3.2 Fluorescence Depolarization

Understanding the nature of the microenvironment around an immobilized molecule is frequently highly desirable as it impacts strongly on the reactivity of the adsorbate. If the immobilized molecule is fluorescent, the microenvironment can be studied by using fluorescence anisotropy [18].

When fluorescent species are illuminated by plane-polarized light, those molecules whose dipole moment is coincident with the plane of the polarized light are selectively excited. Owing to the initial random orientation of the molecules, the emission will be depolarized to a certain extent. Rotation of the fluorophore during its lifetime will result in further depolarization of the emitted light. Therefore, dynamic information about the motion of the fluorophore can be obtained by measuring the depolarization of the emitting species. Steady-state fluorescent data yield the relative emission intensities of the fluorophores parallel and perpendicular to the plane of polarization of the incident radiation, i.e. $I_{\parallel}$ and $I_{\perp}$, respectively. In this way, they reveal the degree of polarization of the fluorophores in the sample. As indicated by Equations (3.21) and (3.22), these measurements can be used to determine the polarization anisotropy, r, and the polarization, p:

$$r = \frac{(I_{\parallel} - I_{\perp})}{(I_{\parallel} + 2I_{\perp})} \tag{3.21}$$

$$p = \frac{(I_{\parallel} - I_{\perp})}{(I_{\parallel} + I_{\perp})} \tag{3.22}$$

These values can then be used in conjunction with the Perrin equation [19] to estimate the rate of rotation of a spherical fluorescent molecule, R_s, as follows:

$$r_0/r = \frac{(1/p - 1/3)}{(1/p_0 - 1/3)} = 1 + \frac{6R_s}{\lambda} \tag{3.23}$$

where r_0 is the fundamental anisotropy of the fluorophase, observed in the absence of other depolarizing processes. Time-resolved fluorescent anisotropy provides a powerful method for evaluating the orientational dynamics of an immobilized species. The time-dependent fluorescence anisotropy, $r(t)$, is given by the following:

$$r(t) = \frac{[I_{\parallel}(t) - I_{\perp}(t)]}{[I_{\parallel}(t) + 2I_{\perp}(t)]} \tag{3.24}$$

Experimentally, commercial steady-state fluorescence spectrometers can be equipped with polarizer attachments, either sheet or Glan–Thompson polarizers. Alternatively, sheet polarizers are usually easily incorporated into the sample cavity in the excitation and emission pathways. Likewise, for time-resolved spectroscopy, polarizers may simply be introduced into the excitation and detection paths. Frequently, the excitation source in time-resolved experiments is a laser which will be inherently polarized.

3.3.3 Epifluorescent and Confocal Microscopy

Fluorescence microscopy techniques are now available which are capable of studying supramolecular interfacial assemblies with excellent spatial and temporal resolution as well as exceptional sensitivity. These methods were initially developed for use in cellular biology, but are finding increasing application in interfacial supramolecular chemistry. This trend is set to continue as methods in single-molecule spectroscopy and time-resolved microscopy evolve.

As illustrated in Figure 3.14, in epifluorescence microscopy the sample is illuminated from above, and a single objective lens is employed with a dichromatic beam splitter that filters out any wavelengths of light that would make the image appear out of focus. The incident light is focused, through a dichroic mirror, by an objective lens into a cone-shaped beam so that the maximum intensity of the beam strikes one spot at a specific depth within the assembly. In a beam-scanned confocal microscope, the objective focuses the laser beam to a single diffraction-limited volume element which scans the x–y plane of the sample.

The fluorescence from this volume is focused, via the dichroic mirror, onto a pinhole aperture. Focusing the laser on a small area limits the total amount

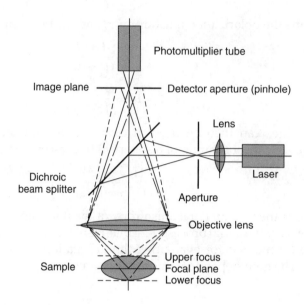

Figure 3.14   Schematic of the layout of a confocal microscope

of scatter and the pinhole eliminates light emitted from outside of the sample volume. Light emitted from regions above or below the focal point of the sample volume is focused elsewhere and is therefore almost totally restricted from reaching the light detector. Both axial and lateral resolutions are significantly improved when compared to non-confocal microscopes. The essential feature of confocal microscopy is the elimination of out-of-focus light, which yield very high sensitivity and a sub-micron spatial imaging resolution.

3.3.4   Near-Field Scanning Optical Microscopy

Near-Field Scanning Optical Microscopy (NSOM) is a technique which enables users to work with standard optical tools integrated with scanning probe microscopy (SPM). The integration of SPM and certain optical methods allows for the collection of optical information at resolutions well beyond the diffraction limit.

As illustrated in Figure 3.15, in NSOM the sample is excited by passing light through a sub-micron aperture formed at the end of a single-mode drawn optical fiber. Typically, the aperture is a few tens of nanometers in diameter. The fiber is coated with aluminum to prevent light loss, thus ensuring a focused beam from the tip. In a typical NSOM experiment, the probe is fixed while the sample is scanned using the same piezo-technology found in commercial scanning probe microscopes. The tip-to-sample distance may be regulated by employing a tuning-fork-based shear-force feedback, where this approach improves the force sensitivity and greatly facilitates the setup. The result is simultaneous, but independent, optical and topographic images of the sample. Only a very small portion of the sample is excited, which significantly reduces photobleaching. In addition to fluorescence imaging, chemical information can be obtained by using near-field spectroscopy

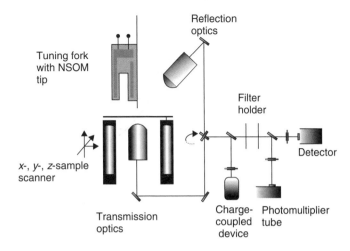

Figure 3.15 Schematic of the layout of an NSOM spectrometer

at resolutions better than 100 nm. UV–visible, IR and Raman spectroscopies have been successfully applied to the NSOM technique [20].

3.3.5 Raman Spectroscopy

Raman spectroscopy is the measurement of the wavelength and intensity of inelastically scattered light from molecules. Raman scattered light occurs at wavelengths that are shifted from the incident light by the energies of molecular vibrations, and is therefore a vibrational spectroscopy providing structural information comparable to infrared spectroscopy.

Figure 3.16 illustrates the mechanism of Raman scattering [21]. When a sample is illuminated with monochromatic light which is not coincident with an electronic transition, most of this light is scattered without a change in photon frequency, a process which is known as *Rayleigh* or *elastic* scattering. However, a tiny portion of the exciting wavelength will interact with the polarized field of the molecule as it vibrates – the molecule is then said to have been excited to a non-stationary or 'virtual state'. These photons will then be scattered either with additional energy (*anti-Stokes* lines), or less energy (*Stokes* lines), where the energy difference corresponds to the frequency of the molecular vibration with which the photons are interacting. This form of inelastic scattering is known as *Raman scatter*. Raman spectroscopy usually studies the Stokes lines as the anti-Stokes lines originate from molecules in vibrational levels other than the lowest; the latter are usually poorly populated at room temperature, thus leading to lower intensities.

Raman spectroscopy has become increasingly important as an analytical technique in recent years. This has largely been attributable to developments in laser, holographic notch filter and CCD technologies which have lent sensitivity to measurements of this inherently weak effect. The probability of Raman scatter occurring is small under most circumstances; however, two important complications of the Raman effect lead to strong enhancements of the Raman signal. The first is the *resonance Raman effect*. Here, the excitation wavelength is coincident with an electronic

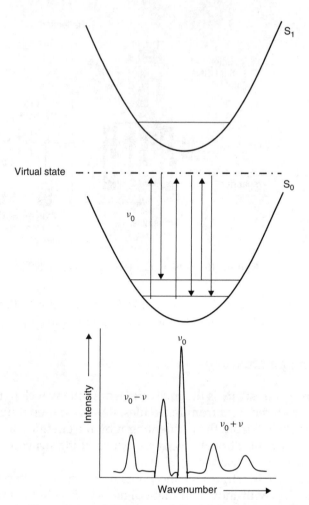

Figure 3.16 Schematic illustrations of Raman and Rayleigh scattering. Excitation to a non-absorbing 'virtual state' leads to Rayleigh (ν_0), Stokes ($\nu_0 - \nu$) and anti-Stokes ($\nu_0 + \nu$) Raman scattering

transition, which leads to signal enhancements of up to six orders of magnitude. Resonance excitation also leads to selection of the vibrational modes observed, whereby only the Franck–Condon modes, i.e. those distorted by the optical transition, are enhanced. This frequently provides useful information regarding the origin of the electronic transition with which the exciting laser is coincident. The second effect, which has particularly important implications for interfacial analytical applications, is *surface-enhanced Raman spectroscopy* (SERS). Raman signals have shown enhancement by to up to 14 orders of magnitude when molecules are adsorbed on surfaces. This effect is observed at certain metal surfaces or colloids of high reflectivity and a suitable roughness. The coinage metals, i.e. gold, silver and copper, have proven to be the most suitable for SERS applications. The mechanism of SERS is complex and a number of theories have developed. Broadly described, enhancement is associated with two effects, namely the increase in electric field experienced by the molecule due to interaction with surface plasmon waves and charge-transfer enhancements whereby electronic transitions between metal and adsorbate are thought to occur. The involvement of surface plasmons means that the exciting wavelength must correspond to the wavelength of the metal's surface plasmon in order to induce enhancement. Submicroscopic surface roughness is critical in both mechanisms. The signal enhancement observed in SERS depends on the distance of the Raman-active species from the metal surface. While monolayers show superior surface enhancement, there is evidence that SERS may operate at distances of up to tens of nanometers. SERS spectra conducted at rough surfaces, with the exception of single nanoparticles, are completely depolarized. Furthermore, for adsorbed species the selection rules are relaxed and vibrational modes forbidden in the normal Raman spectrum of a species are frequently observed when that species is adsorbed. Importantly, from the perspective of Raman spectroscopy of surface-modified electrodes, both the frequencies and intensities of vibrational bands in SERS are influenced by the applied potential. The relationship between these parameters and the applied potential is variable, and depends inherently on the nature of the particular vibration.

The use of confocal Raman microscopy provides excellent three-dimensional spectral mapping. This technique increases the surface-scattered-light collection efficiency and minimizes the interference from the solution phase. When used in conjunction with *in situ* electrochemical techniques, Raman microscopy provides an immensely sensitive method of probing the structure, e.g. orientation, surface coverage, etc., as well as the dynamics, kinetics of assembly, redox switching effects, etc. of interfacial species. One of the many advantages of Raman over IR spectroscopy is that water is an extremely poor Raman scatterer. This property allows the interrogation of interfaces without any strongly interfering signals from aqueous solvent.

Figure 3.17 illustrates the layout of a typical Raman microscope. The excitation source is monochromatic, since the small signals are recorded as shifts, or Δv, from the excitation line. Lasers are universally used since as well as providing monochromatic light, they offer high power and collimation. The exciting laser

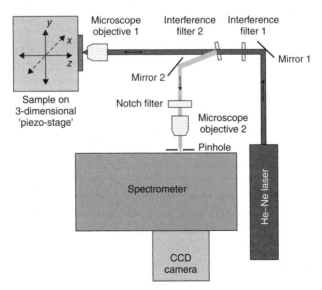

Figure 3.17 Block diagram of a typical confocal Raman microscope employing CCD detection

light is focused to a diffraction-limited spot on the sample with a microscope objective of high numerical aperture. Light scattered from the sample is directed backwards along the same optical pathway as the excitation source, collimated by the same objective and then focused via a holographic notch filter, which eliminates a significant portion of the Rayleigh scatter, through a pinhole onto a monochromator and CCD photodetector. Only light from the focus can penetrate through the pinhole; light from other depths in the sample is efficiently blocked. As a consequence, out-of-focus regions do not contribute to the overall signal. This focusing ability enables the user to obtain spectra at variable depth without interference from materials outside of that sample volume. This is vitally important for interfacial analysis as it allows the elimination of interference from species such as solution and electrolyte from the spectra.

3.3.6 Second Harmonic Generation

Second Harmonic Generation (SHG) is a versatile method for studying molecular orientation within ISAs. It is the most extensively investigated non-linear optical technique at electrode surfaces. SHG involves irradiating the surface with a pulsed laser, and collecting light at twice the frequency of the fundamental frequency of the laser near the reflected angle [22,23]. This technique is inherently surface-selective because the process which leads to frequency doubling is forbidden in the bulk of centrosymmetric crystals, but is allowed at the surface where this inversion symmetry is broken. The second harmonic response from a metal electrode is large when compared to that obtained from the adsorbed film, so this technique is primarily sensitive to the properties of the interfacial assembly. However, the intensity of the frequency-doubled light depends on the surface non-linear susceptibility, which is itself sensitive to both the charge on the surface and the surface morphology.

3.3.7 Single-Molecule Spectroscopy

Single-molecule fluorescence spectroscopy [24,25] is an emerging area which will have profound implications for studying supramolecular assemblies. In conventional optical measurements, a molecular ensemble is excited, thus leading to population averages and observation of multiple simultaneous photochemical or physical events. Such spectroscopy may obscure dynamic or conformational information which may be particularly important for molecules which are immobilized on a surface. In interfacial supramolecular assemblies, the molecular environment can be heterogeneous and slow to adjust, which can lead to significant inhomogeneous broadening of spectra. In contrast to conventional spectroscopy, single-molecule spectroscopy provides information on dipole moment, conformation, diffusion, fluorescence wavelength and quantum yield. Beyond this information, the observation of a single molecular species leads to some unexpected observations as a result of the quantum-confined nature of the measurement. For example, various dynamic processes, e.g. reversible photobleaching, are highly sensitive to the local chemical environment and often do not correlate well with the behavior observed for measurements on supramolecular assemblies.

The methodology for single-molecule spectroscopies is based primarily on the near-field microscopy or far-field confocal spectroscopy described above. For example, single dye molecules adsorbed on a surface can be imaged by using NSOM. Requirements for the single-molecule fluorescent technique include low concentrations, plus small probed spaces, where the latter is accomplished by using laser and high-resolution laser spectroscopy. Other limitations include potentially high photodecomposition rates, arising from the intense laser sources being used. The relatively long lifetimes of electronically excited states when compared to vibrational lifetimes restrict the maximum number of excitation emission cycles or photons which can be emitted by a molecule under saturation conditions to about 10^8 per second.

Perhaps the most exciting development in single-molecule spectroscopy in terms of interfacial supramolecular assemblies is the coupling of time-resolved spectroscopic methodologies. For example, confocal fluorescence microscopy has been coupled with single photon counting to provide both confocal images and excited-state lifetimes of single molecules. NSOM has been combined with a number of spectroscopic techniques to obtain spatial, spectral and temporal information on the target species. For example, porphyrin rings on glass and carbon have been identified and studied in this way [26].

3.3.8 Spectroelectrochemistry

Spectroelectrochemistry may be used to probe the effect of changing the redox state of components within an ISA on the spectral properties of the assembly. It may also provide information about the rates of electron transfer at the electrode/electrolyte interface. As illustrated in Figure 3.18, spectroelectrochemistry employs a two- or

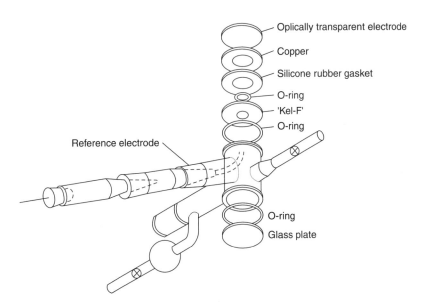

Figure 3.18 Illustration of a typical cell used for spectroelectrochemistry experiments; in this arrangement, the light beam passes along the vertical axis

three-electrode thin-layer cell in a configuration in which the working electrode can be irradiated. The working electrode is often optically transparent gold or indium tin oxide (low resistance). The excitation source is typically a continuous-wave broad-spectrum lamp. By using this setup, changes in the UV–visible spectrum arising from redox reactions occurring within the ISA deposited on the working electrode can be monitored to identify the redox moiety involved or to diagnose lateral interactions. As well as electronically conducting working electrodes, optically transparent TiO_2 (semiconductor) can be used onto which an ISA, e.g. a ruthenium-containing monolayer, is assembled (see Chapter 6). By employing a continuous lamp source, this apparatus can be used to measure the photocurrent as a function of wavelength for generating a photocurrent action spectrum.

3.3.9 Intensity-Modulated Photocurrent Spectroscopy

In Intensity-Modulated Photocurrent Spectroscopy (IMPS) the cell is perturbed by a sinusoidally modulated light source. As illustrated in Figure 3.19(a), the light source is typically a continuous wave (CW) source, such as a He/Ne laser which is modulated by using an acousto-optic modulator, after which the beam is split onto a photodiode and the sample. The observed photocurrent from the sample will exhibit a harmonic response that can be shifted in phase and magnitude with respect to the sinusoidally modulated source. The IMPS signal, i.e. the optoelectrical admittance, Y, is defined as the quotient of the harmonic current response and the harmonic element of the light intensity (Figure 3.19(b)). This parameter is a complex number, and can hence be plotted in the complex plane. The frequency dependence of the IMPS signal yields kinetic information about the processes leading to or the quenching of the photocurrent, such as hole and electron transfer across the supermolecule–electrode interface.

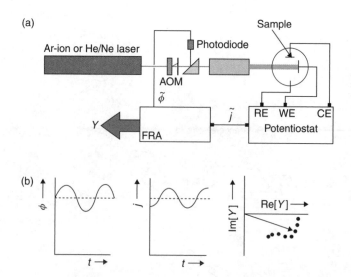

Figure 3.19 Intensity-modulated photocurrent spectroscopy, showing (a) the layout of a typical spectrometer, and (b) the response obtained: AOM, acousto-optic modulator; RE, reference electrode; WE, working electrode; CE, counter electrode; FRA, frequency response analyzer

3.4 Time-Resolved Spectroscopy of Interfacial Supramolecular Assemblies

On absorption of a photon, a single molecule, either in solution or at a surface, achieves an excited state from where it will usually equilibrate with its environment before re-establishing its ground-state configuration. In steady-state luminescence, excitation is usually continuous and the spectrum obtained represents an integration of all of the molecules in the sample undergoing various photophysical/chemical processes. Although steady-state techniques yield valuable parameters such as luminescence quantum yields, much of the kinetic data, such as relaxation dynamics, electron and energy transfer rates, are lost. In supramolecular systems capable of light-driven functions, dynamic information is of enormous value and is obtained by using time-resolved spectroscopic techniques. Currently, the timescale of these spectroscopic techniques ranges from seconds to femtoseconds, thus allowing the investigation of a vast range of dynamic photochemical processes.

3.4.1 Flash Photolysis

Laser flash photolysis provides a method by which short-lived chemical species, as well as charge and energy transfer reactions, may be studied with comparative ease. In these experiments, the formation or decay of a transient absorption signal is measured as a function of time after a high-energy pulse of light is delivered to the sample. To be able to study a reaction by flash photolysis, the lifetime of the transient species must significantly exceed the lifetime of the exciting photoflash. This technique provides one of the most effective methods for producing transient species such as radicals, i.e. electronically excited states, with concentrations high enough to permit characterization of their spectral properties and reactivities. The use of a laser for sample excitation gives the technique the possibility of single-wavelength excitation and micro- to femtosecond time resolution, so allowing the detection of events arising from phosphorescence to vibrational relaxations.

Figure 3.20 illustrates the typical layout of the optical components used in a flash photolysis instrument operating at nanosecond timescales. The output from a pulsed source, either a laser or a flashlamp, is directed onto a sample cuvette at right angles to or collinear with an analyzing beam. The latter is a continuum source such as a tungsten or arc lamp which may itself be pulsed for greater sensitivity. A carefully aligned optical system with good overlap of exciting and analyzing beams is required to produced optical irradiation of the sampling volume. The detector is usually maintained at 90° to the exciting source. Traditionally, streak cameras and photomultiplier tubes have been used for single-wavelength detection of transients. However, multichannel detectors such as photodiode arrays, and more recently, intensified charge-coupled devices, have allowed the collection of entire transient spectra with averaging over a few laser pulses.

An alternative to laser flash photolysis which is useful for studying opaque (but reflecting) samples, is diffuse reflectance spectroscopy [27]. This spectroscopic technique measures the ratio of the intensity of light reflected from the sample, I, to that reflected from a background or reference reflective surface, I_0. In time-resolved

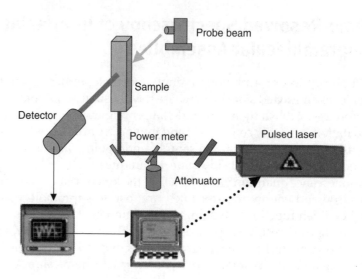

Figure 3.20   Schematic layout of a typical flash photolysis instrument operating at nanosecond timescales

experiments, *I* refers to the reflected light intensity of the incident monitoring beam when the sample has been excited, and I_0 the reflected light in the absence of an exciting pulse. Experimentally, the difference between this technique and conventional flash photolysis lies in the arrangement of the exciting and detection optics. Typically, the sample is held at a 45° angle to the monitoring beam, while the pulsed excitation source, normal to the monitoring beam, is arranged so that the two beams are coincident on the sample face. Diffuse reflected analyzing light may then be monitored normal to the cuvette face and focused through a monochromator onto a detector such as a PMT.

3.4.2   Time-Resolved Luminescence Techniques

The fluorescence lifetime of a fluorophore is highly sensitive to its molecular environment. Many macromolecular events, such as rotational diffusion, resonance-energy transfer, and dynamic quenching, occur on the same timescale as the fluorescence decay. Thus, time-resolved fluorescence spectroscopy can be used to investigate these processes and gain insight into the chemical surroundings of the fluorophore.

The instrumentation for time-resolved fluorescence spectroscopy is similar to that used for steady-state experiments. An excitation beam with a narrow wavelength range is directed at the sample, where it excites fluorescence. The fluorescence emission is collected at a 90° angle from the excitation to prevent light from the excitation source from interfering with the detection of the weaker fluorescence emission. The collected fluorescence emission enters a spectrograph and a detector then records the emission spectrum. The key differences between steady-state and time-resolved spectroscopy are the replacement of the continuous light source with a pulsed source and the use of gated detection of the fluorescence emission. A range of detection techniques are commonly employed in fluorescence lifetime techniques, with

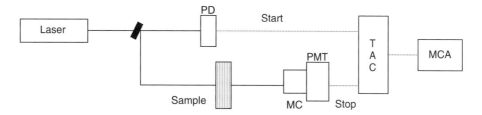

Figure 3.21 Schematic showing the typical components of a time-correlated single photon counting experiment: PD, photodiode; TAC, time-to-amplitude converter; MCA, multichannel analyzer; MC, PMT, photomultiplier tube

streak cameras, phase modulation and, time-gated systems with time-correlated single photon counting being among the most widely used approaches.

Time-correlated single photon counting (TCSPC) [28] is one of the most sensitive methods for studying time-resolved emission. In this technique, single photon events are detected after excitation and a statistical distribution of photons representing the decay of the excited state is built up over time.

Figure 3.21 illustrates the typical components of a time-correlated single photon counting experiment. A high-frequency excitation source, either a flashlamp or laser, is employed to excite the sample. Part of this optical pulse is focused on a photodiode, in order to generate a start voltage ramp at a time-to-amplitude converter (TAC). The stop pulse is generated by fluorescence from the sample detected at, for example, a PMT or a microchannel plate PMT (MCP-PMT). The TAC provides an output signal which is proportional to the time interval between the start and stop pulses. The time difference between the start and stop pulses from the TAC is processed by a multichannel analyzer (MCA). A significant advantage of TCSPC is its sensitivity, where in some instances less than one emitted photon per exciting pulse is sufficient to obtain a reliable signal-to-noise ratio. Front-face sample arrangements and the high sensitivity of this method make it applicable to both solid and surface measurements.

3.4.3 Femtochemistry

Ultrafast spectroscopy is usually distinguished from *fast* spectroscopy in that the former studies processes which occur in less than 10 ps. The commercial availability of solid-state femtosecond lasers, such as Ti–sapphire, makes femtosecond spectroscopy available for a wide range of applications [29].

Optical excitation of metals with intense femtosecond laser pulses can create extreme non-equilibrium conditions in the solid where the electronic system reaches several thousand degrees Kelvin on a sub-picosecond timescale, while the lattice (phonon) bath, stays fairly 'cold'. As illustrated in Figure 3.22, photoexcited hot electrons may transiently attach to unoccupied adsorbate levels and this change in the electronic structure may induce vibrational motions of the adsorbate–substrate bond. For high excitation densities with femtosecond pulses, multiple excitation/de-excitation cycles can occur and may eventually lead to desorption of adsorbate molecules or reactions with co-adsorbed species. After 1–2 ps, the hot electron

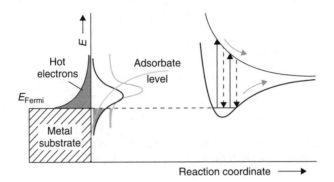

Figure 3.22 Effect of creating photoexcited hot electrons on the adsorbate–surface bonding interaction

distribution equilibrates with the lattice and the heated phonon bath may induce chemical reactions by thermal activation. Time-resolved experiments with a pair of two excitation pulses allow phonon- or electron-driven mechanisms to be distinguished and provide insights into the rates and pathways of energy flow in surface chemistry.

This approach has the potential to resolve the time evolution of reactions at the surface and to capture short-lived reaction intermediates. As illustrated in Figure 3.23, a typical pump–probe approach uses surface- and molecule-specific spectroscopies. An intense femtosecond laser pulse, the 'pump' pulse, starts a reaction of adsorbed molecules at a surface. The resulting changes in the electronic or vibrational properties of the adsorbate–substrate complex are monitored at later times by a second ultrashort 'probe' pulse. This probe beam can exploit a wide range of spectroscopic techniques, including IR spectroscopy, SHG and infrared reflection-adsorption spectroscopy (IRAS).

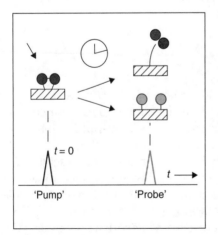

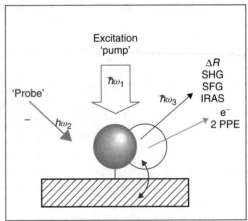

Figure 3.23 Schematic illustration of the femtosecond 'pump–probe' technique used to initiate and follow the reactions of adsorbed species: SHG, second harmonic generation; SFG, sum frequency generation; IRAS, infrared reflection–adsorption spectroscopy; PPE, photons per event; ΔR, change in reflectivity

3.5 Conclusions

The characterization of many interfacial supramolecular assemblies, especially those that are structurally disordered, e.g. those based on polymers or biomaterials, is challenging because of their complexity, and a 'complete' characterization requires no less than the location of every atom in a macroscopic sample! While the periodicity of crystalline materials, such as organized self-assembled monolayers, reduces this problem to locating every atom in a microscopically sized unit cell, no such symmetry is present in amorphous materials, and determining their structures at this level is impossible. However, as discussed in this present chapter, various diagnostic tools, such as surface plasmon resonance, neutron reflectivity and infrared spectroscopy, can be used to probe the structures of ISAs, to understand the interactions between the molecular building blocks and to probe dynamic processes triggered by electro- and photochemical stimuli. Although these approaches often provide information about *average* properties (or simple property distributions), their resolution continues to increase significantly. Appropriate structural characterization is an essential, but often neglected, prerequisite when seeking to create functional interfacial supramolecular assemblies.

References

[1]   Frommer, J.E. (1992). *Angew. Chem., Int. Ed. Engl.,* **31**, 1298.

[2]   Binnig, G. and Rohrer, H. (1987). *Angew. Chem., Int. Ed. Engl.,* **26**, 606.

[3]   Bard, A.J., Fan, F.-R. and Mirkin, M.V. (1994). Scanning electrochemical microscopy, in *Electrochemical Chemistry*, A.J. Bard (Ed.), Marcel Dekker, New York, pp. 244–270.

[4]   Mittal, K.L. (Ed.) (1993). *Contact Angle, Wettability and Adhesion*, VSP Publishing, Zeist, The Netherlands.

[5]   Hepel, M. (1999). Electrode–solution interface studied with the electrochemical quartz crystal nanobalance, in *Interfacial Electrochemistry*, A. Wieckowski (Ed.), Marcel Dekker, New York, pp. 599–630.

[6]   Tompkins, H.G. and McGahan, W.A. (1999). *Spectroscopic Ellipsometry and Reflectometry: A User's Guide*, Wiley, New York.

[7]   Raether, H. (1977). Surface plasma oscillations and their applications, in *Physics of Thin Films*, Vol. 9, G. Hass, M.H. Francombe and R.W. Hoffmann (Eds), Academic Press, New York, pp. 145–261.

[8]   Penfold, J. (1991). Data interpretation in specular neutron reflection, in *Neutron, X-Ray and Light Scattering*, P. Lindner and T. Zemb (Eds), Elsevier, New York, p. 223–235.

[9]   Bard, A.J. and Faulkner, L.R. (2001). *Electrochemical Methods: Fundamentals and Applications*, 2nd Edition, Wiley, New York.

[10] Forster, R.J. and Vos, J.G. (1992). Theory and applications of modified electrodes, in *Comprehensive Analytical Chemistry*, Vol. XXVII, G. Svehla (Ed.), Elsevier, Amsterdam, p. 465–485.

[11] Buttry, D. (1990). *Langmuir*, **6**, 1319.

[12] Forster, R.J. and Faulkner, L.R. (1994). *J. Am. Chem. Soc.,* **116**, 5444.

[13] Brown, A.P. and Anson, F.C. (1977). *Anal. Chem.*, **49**, 1589.

[14] Laviron, E. (1974). *J. Electroanal. Chem.*, **52**, 395.

[15] Tong, Y.Y, Oldfield, E. and Wieckowski, A. (1998). *Anal. Chem.*, **70**, 518A.

[16] Lakowicz, R. (1999). *Principles of Fluorescence Spectroscopy*, Plenum Press, New York.

[17] Demas, J.N. and Crosby, G.A. (1971). *J. Phys. Chem.*, **75**, 991.

[18] Kawski, A. (1993). *Crit. Rev. Anal. Chem.*, **23**, 459.

[19] Heyward, J.J. and Ghiggino, K.P. (1989). *Macromolecules*, **22**, 1159.

[20] Pohl, D.W. (1991). Scanning near-field optical microscopy, in *Advances in Optical and Electron Microscopy*, Vol. 12, C.J.R. Sheppard and T. Mulvey (Eds), Academic Press, London, pp. 243–312.

[21] Schrader, B. (Ed.) (1995). *Infrared and Raman Spectroscopy, Methods and Applications*, VCH, Weinheim, Germany.

[22] Prasad, P.N. and Williams, D.J. (1990). *An Introduction to Non-Linear Optics in Organic Systems*, Wiley, New York.

[23] Corn, R.M. and Higgins, D.A. (1994). *Chem. Rev.*, **94**, 107.

[24] Basche, T., Moerner, W.E., Orrit, M. and Wild, U.P. (Eds) (1997). *Single-Molecule Optical Detection, Imaging and Spectroscopy*, Wiley-VCH, Weinheim, Germany.

[25] Nie, S. and Zare, R.N. (1997). *Annu. Rev. Biophys. Biomol. Struct.*, **26**, 567.

[26] Schenning, A.P.H.J., Benneker, F.B.G., Guerts, H.P.M., Liu, X.Y. and Nolte, R.J.M. (1996). *J. Am. Chem. Soc.*, **118**, 8549.

[27] Wilkinson, F. (1986). *J. Chem. Soc., Faraday Trans.*, **82**, 2073.

[28] O'Connor, D.V. and Phillips, D. (1984). *Time-Correlated Single Photon Counting*, Academic Press, London.

[29] Baskin, J.S. and Zewail, A. (2001). *J. Chem. Edu.*, **78**, 737.

4 Formation and Characterization of Modified Surfaces

By their very nature, heterogeneous assemblies are difficult to characterize. Problems include the exact nature of the substrate surface and the structure of the modifying layer. In this chapter, typical examples are given of how surface assemblies can be prepared in a well-defined manner. This discussion includes the descriptions of various substrate treatment methods which lead to clean, reproducible surfaces. Typical methods for the preparation of thin films of self-assembled monolayers and of polymer films are considered. Methods available for the investigation of the three-dimensional structures of polymer films are also discussed. Finally, it will be shown that by a careful control of the synthetic procedures, polymer film structures can be obtained which have a significant amount of order. It will be illustrated that these structural parameters strongly influence the electrochemical and conducting behavior of such interfacial assemblies and that this behavior can be manipulated by control of the measurement conditions.

4.1 Introduction

This chapter focuses on the formation and properties of spontaneously adsorbed monolayers, self-assembled monolayers (SAMs) and thin polymer films. Spontaneously adsorbed and self-assembled monolayers are similar in that both involve strong binding of a surface active functional group to a surface but are distinct in that a self-assembled monolayer also involves stabilizing lateral interactions between the adsorbates. The objective is often to create a single layer of molecules on a substrate in which the molecules exhibit significant long-range order, achieved by close packing of the highly oriented adsorbates. As illustrated in Figure 4.1, long-chain thiols and disulfides spontaneously form remarkably well-packed and stable monolayers on gold and other coinage metals [1,2].

Self-assembled and spontaneously adsorbed monolayers offer a facile means of controlling the chemical composition and physical structure of a surface. As discussed later in Chapter 5, applications of these monolayers include modeling electron transfer reactions, biomimetic membranes, nano-scale photonic devices, solar energy conversion, catalysis, chemical sensing and nano-scale lithography.

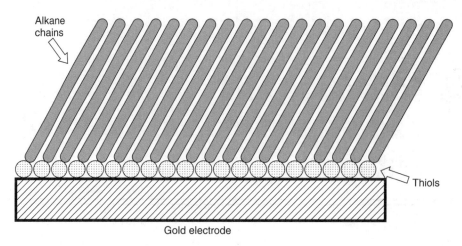

Figure 4.1 Representation of an organized monolayer on a substrate (electrode), which can be deposited either by the Langmuir–Blodgett (LB) method or by self-assembly. Reprinted from H.O. Finklea, in *Electroanalytical Chemistry*, Vol. 19, A.J. Bard and I. Rubinstein (Eds), Marcel Dekker, New York, 1996, p. 109, Figure 1, by courtesy of Marcel Dekker Inc

There is also a substantial interest in the properties of surfaces modified with thin polymer films and these studies possibly constitute the most extensively investigated area of supramolecular chemistry at surfaces. Such studies have ranged from evaluation of the electrochemical properties of polymer-modified electrodes to the fabrication of multilayers based on self-assembly of ionic polymers, the deposition of conducting polymers and nano-composites [1]. The popularity of polymers is related to the relative ease and the range of approaches available by which thin layers can be produced. Coupled with the tailorable properties of structural polymers, e.g. shapeability, formability, flexibility, low-cost, etc., these materials are ideal for a variety of important 'smart materials' applications. Surfaces can be modified with polymers by simple methods such as solvent removal, spin coating and dip coating, methods that do not require the presence of any specific molecular features in the polymers. In addition, the molecular composition of polymers can be modified by well-defined synthetic techniques. A wide range of additional methods, such as electropolymerization and self-assembly, is available. Although these methods require specific functional groups to be present, they offer important routes to ordered interfacial polymer layers. This synthetic flexibility make polymers ideal for fabricating patterned nano-structures and techniques such as embossing, imprinting and templating often rely on macromolecular components.

The formation of well-defined monolayers depends critically on the quality of the metal surface. For the modification of surfaces with thin polymer films, less stringent conditions are required. Since the coating is stabilized by adsorption, cleaning the substrate of organic contaminants is normally sufficient. It is generally observed that the nature of the substrate is not important. Among others, gold, platinum and carbon-type surfaces have shown to be suitable surfaces for producing stable polymer coatings.

4.2 **Substrate Choice and Preparation**

The platform onto which an interfacial supramolecular assembly is formed can influence its physical structure and these effects are particularly important for monolayers. For example, some monolayers may form a structure that is commensurate with the crystal lattice of a particular metal, thus leading to highly ordered structures. In the case of alkane–thiol monolayers, the identity of the metal, e.g. gold, silver, copper, etc., affects the structure and defect density of the monolayer. Thus, while it is possible to form close-packed monolayers that exhibit very significant long-range order on gold single crystals, less perfect structures are typical on silver and other coinage metals. In the case of thicker films, e.g. multilayers or polymer films, the underlying platform exerts less influence on the structure of the assembly, which is dominated by lateral interactions between the individual building blocks and the contacting solution.

The nature of the platform also impacts on the electrochemical and photophysical properties of the interfacial supramolecular assembly (ISA). For example, the density of states within gold, platinum and carbon electrodes are different, so causing subtle changes in the rate of electron transfer across the electrode/ISA interface. In addition, in terms of the photophysical properties, the nature of the platform can radically change the excited-state properties of a molecule upon adsorption. For example, if a adsorbate is located close to ($\leq$10 nm) a metal surface and is then 'pumped' into an electronically excited state, efficient energy or electron transfer is expected which will lead to quenching of the excited state. This process can dramatically increase the photostability of compounds that would ordinarily photodecompose in solution.

In this section, the preparation of electrodes for the assembly of ISAs is briefly considered. Gold is the most popular material on which thiol SAMs are formed. Gold has the advantage that it can be handled in air without the formation of an oxide surface layer, and can survive harsh chemical treatments which remove adventitious contaminants. Substrates that have been used include bulk gold in both polycrystalline and single-crystal forms, as well as thin films deposited on materials ranging from ceramics to quartz. The smoothness of the metal surface partially dictates the structure of the monolayer, especially the defect density. For the most highly oriented crystalline SAM, an atomically smooth substrate, i.e. a single crystal, and a high-purity electrode are desired. Controlling the defect density is especially important for applications where the SAM is used to block or control access of solution-phase reactants to the electrode surface. Evaporated or sputtered gold films, typically 50–200 nm thick, on glass, silicon or cleaved mica substrates represent particularly useful substrates on which to assemble monolayers. To improve the adhesion of the gold film to the oxide surface, a thin (1–5 nm) layer of chromium or titanium is often deposited first. After thermal annealing at high temperatures, the gold surface is largely composed of Au(111) domains and is frequently labeled as a Au(111) substrate. By controlling the thickness of the evaporated gold, films that are transparent in the visible region of the spectrum can be produced, thus facilitating spectroscopic investigation.

An essential part of forming high-quality SAMs is the production of clean gold substrates. Bulk gold can be conveniently cleaned by heating in a gas/air flame to produce a hydrophilic surface which is free of all organic contamination. Alternatively, electrochemical cycling in dilute acid so as to first oxidize and then reduce a gold oxide monolayer produces a clean surface. Moreover, the corresponding voltammogram allows the cleanliness and crystallinity of the gold, as well as its true or microscopic surface area, to be determined. Chemical etching can be achieved by using dilute aqua regia which reduces the damage caused by mechanical polishing. In the case of thermally evaporated gold films, monolayers are typically formed as soon as possible after their production so as to minimize contamination. However, it is important to note that monolayers will often successfully form on hydrophobic, i.e. gold surfaces coated with organic impurities, because of the limited 'self-cleaning' capacity of the gold–thiol bond formation. Alternatively, organic surface contaminants can be chemically oxidized and removed by treating the gold surface with a powerful oxidant such as 'piranha' solution, which consists of a 1:3 mixture of 30 % hydrogen peroxide and concentrated sulfuric acid at ca. 100 °C. *(Caution: this mixture reacts violently with organic material and can explode if stored in closed vessels!)* Oxygen plasma and UV production of ozone can also be used to remove contamination. The principal disadvantage of these chemical approaches is that they can create a layer of gold oxide onto which the SAM subsequently forms. This oxide layer can be removed by washing the electrode surface with ethanol.

Silver, copper and platinum have also been used as platforms onto which thiol SAMs have been assembled. Copper and silver tend to oxidize easily and are typically covered with an oxide layer. While the self-assembly process appears to efficiently remove the oxide layer on silver, the oxide film persists on copper to yield SAMs with a high defect density. Therefore, in seeking to form thiol SAMs on copper, the oxide should be removed using electrochemical reduction or by washing in mineral acid. Mercury has the distinct advantage that it is atomically smooth as well as structurally and chemically homogeneous. While it is easily oxidized by thiols, homogeneous, densely packed, defect-free SAMs can be prepared by using solution, but not vapor, phase deposition.

4.3 Formation of Self-Assembled Monolayers

Self-assembled monolayers (SAMs) have been formed using sulfur compounds, especially thiols, disulfides and sulfides, on a wide range of metals, including gold, silver, platinum and mercury. Significantly, this range of materials includes all of the common electrode materials except carbon, thus facilitating electrochemical investigations. Less common are SAMs based on adsorption of isonitriles. Chlorosilanes with long alkyl chains self-assemble on doped metal oxides such as SnO_2, silicon with a thin oxide coating, and even gold without any surface oxide. Significantly for the development of hybrid semiconductor molecular devices, densely packed SAMs can be formed on oxide-free silicon by generating alkane radicals.

Thiols and related molecules have the following attractive features. The monolayers can be easily formed on gold and other metals by exposing them to the thiol dissolved in solution. A particularly important feature of thiol-based SAMs is the strength of the thiol–metal bond. This high bond strength, typically of the order of

100 kJ mol^{-1}, means that the functionalized surfaces are stable when in contact with a wide range of solvents, and across a wide range of potentials and temperatures. While annealing to reduce the defect density can be a slow process, the monolayers can typically be deposited within a few minutes. In addition, monolayer formation does not require anaerobic, anhydrous or vacuum conditions, thus making this approach attractive for large-scale industrial exploitation. The availability of a wide range of ω-terminated thiols, e.g. halide, ether, alcohol, aldehyde, carboxylic acid, amide, ester, amine and nitrile, means that surfaces with a wide range of chemical reactivities and hydrophobicities can be prepared. Furthermore, by including heteroatoms, aromatic groups, conjugated unsaturated links and other rigid-rod structures such as sulfones and amides, the internal structure of the monolayer can be tailored for specific applications. For technological applications such as the formation of friction-free surfaces, corrosion protection, etc., the approach can easily accommodate curved or intricately shaped objects. Moreover, substrates can range in size from micro- to macroscopic and porous substrates can be functionalized. The use of thiols as molecular building blocks allows multicomponent monolayers to be formed either by simultaneous coadsorption or in two sequential steps. The mole fraction of each molecule in the SAM can be controlled by changing the mole ratio of the two molecules in the solution, the identity and temperature of the solvent, and the time of the deposition. The distribution of the two molecules may vary from intimately mixed to completely phase-separated.

4.3.1 Solution-Phase Deposition

SAMs can be formed by exposing a clean metal surface to a solution of the surface active molecule or complex at room temperature, followed by rinsing with a solvent. The bulk concentration has a significant influence over the quality of the monolayer formed. Very low, i.e. micromolar, concentrations result in slow self-assembly and favor the production of large crystalline domains. It is possible for the deposition solvent to become entrapped in the monolayer and this entrapped solvent may not be removed in the subsequent washing cycle.

The dynamics of monolayer assembly have been extensively investigated and these studies reveal that for deposition solutions containing millimolar concentrations of the thiol, monolayers form within a few seconds. However, these monolayers are highly defective and if left in contact with the deposition solution will slowly anneal over several days to give a low-defect-density surface. Detailed spectroscopic studies reveal that the alkane thiols initially adsorb horizontally onto the metal surface, followed by reorientation and infilling to give a dense monolayer. STM studies indicate that monolayer formation is accompanied by etching of the gold surface, with gold having been detected in the deposition solution during self-assembly.

The adsorption dynamics of species capable of forming spontaneously adsorbed monolayers have also been investigated [3]. For example, Figure 4.2 shows the change in the surface coverage with time when the concentration of $[\text{Os(bpy)}_2(\text{p3p})_2]^{2+}$ in the deposition solution is 10, 20 or 50 µM (bpy = 2,2'-bipyridine; p3p = trimethylene-4,4'-bipyridine). For each concentration, the surface coverage Γ initially increases relatively rapidly and then reaches a limiting value at times ranging from approximately 360 to 7500 s at bulk concentrations of 50 and 10 µM, respectively. The limiting

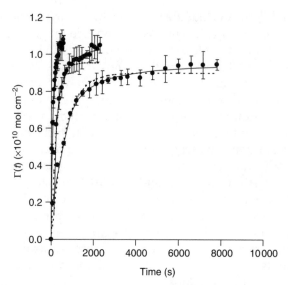

Figure 4.2 Temporal evolution in the surface coverage of $[Os(bpy)_2 (p3p)_2]^{2+}$ adsorbed onto a gold microelectrode (bpy, 2,2′-bipyridine; p3p, trimethylene-4,4′-bipyridine). From top to bottom on the left-hand side, the bulk concentrations of the complex dissolved in H_2O/DMF (2:1) are 50, 20 and 10 µM. The dashed and continuous lines represent the best fits of first- and second-order models, respectively, with the error bars representing data taken from at least three independent monolayers. From R.J. Forster and J.P. O'Kelly, 'Dynamics of $[Os(bpy)_2 (p3p)_2]^{2+}$ adsorption and desorption on platinum and gold microelectrodes', *J. Electrochem. Soc.*, **148**, E31–E37 (2001). Reproduced by permission of The Electrochemical Society, Inc

surface coverage is $1.1 \pm 0.1 \times 10^{-10}$ mol cm^{-2} for $20 \leq C_B \leq 100$ µM, i.e. for this range of bulk concentrations, dense monolayer coverage is obtained after long periods of time.

In seeking to understand those processes that contribute to the dynamics of monolayer formation, it is important to consider the role of mass transport to the electrode surface. Assuming linear diffusion conditions for micromolar concentrations in solution, a monolayer in which the surface coverage is 1.1×10^{-10} mol cm^{-2} will require a layer approximately 0.01 cm thick within the solution to be depleted of $[Os(bpy)_2 (p3p)_2]^{2+}$. The characteristic time, t, for this diffusion process is given by the following equation:

$$t = \delta^2/\pi D \qquad (4.1)$$

where δ is the film thickness and D is the diffusion coefficient of the complex in solution. Scan-rate-dependent cyclic voltammetry of the complex dissolved in solution indicates that D is $5.4 \pm 0.4 \times 10^{-6}$ cm^2 s^{-1}, and Equation (4.1) suggests that a dense monolayer can be assembled within less than 3 s. In contrast, Figure 4.2 indicates that dense monolayers take between approximately 400 and 7500 s to assemble, thus suggesting that the rate determining step is not diffusional mass transfer but the kinetics of surface binding.

The rate of thermally activated surface binding can be described in terms of the time-dependent fractional coverage, θ, by the following equation [4]:

$$\frac{d\theta}{dt} = kf(\theta)C_i \qquad (4.2)$$

where k is the rate constant, C_i is the bulk concentration and $f(\theta)$ describes the order of the surface binding reaction. For example, if adsorption takes place on a single site then $f(\theta)$ is $(1 - \theta)$, representing the fraction of vacant sites irrespective of whether the adsorbed layer is mobile or not.

Under these circumstances, Equation (4.2) can be integrated to yield the kinetic Langmuir equation, as follows:

$$\theta(t) = 1 - e^{-kC_0 t} \tag{4.3}$$

In contrast, if adsorption requires two sites and the adsorbed monolayer is mobile, then $f(\theta) = (1 - \theta)^2$, which is the probability of finding two free adjacent sites. Integrating Equation (4.2) with this probability factor yields the following second-order expression:

$$\theta(t) = 1 - \frac{1}{ktC_0} \tag{4.4}$$

Either approach can be modified to account for orientational effects if the molecule must be in a specific configuration for adsorption.

The dashed and thin solid lines of Figure 4.2 show the best fits obtained to models described by Equations (4.3) and (4.4), respectively. Both models reproduce the general shape of the experimental isotherm. This behavior contrasts with the behavior of other ionic and neutral surfactants [5]. Often, a two-step adsorption mechanism is observed in which initial rapid adsorption is followed by a second stage involving co-operative interactions between adsorbates [6,7]. However, the first-order model described by Equation (4.3) provides a poorer quality fit. The apparent rate constants obtained from the Langmuir approach increase from 1.4×10^2 to $13.5 \times 10^2 \ M^{-1} \ s^{-1}$ as the bulk concentration is increased from 10 to 100 μM. In contrast, the second-order approach yields a concentration-independent rate constant of $4.6 \pm 0.5 \times 10^2 \ M^{-1} \ s^{-1}$ over this concentration range. Studies of this kind suggest that an advanced understanding of the behavior of surfactants at the solid/solution interface depends critically on investigating not only the equilibrium structure of molecular assemblies, but also on probing the dynamics of their formation and destruction.

4.3.2 Electrochemical Stripping and Deposition

Interfacial supramolecular assemblies that involve *both* covalent binding to an electrode surface and strong lateral interactions, e.g. self-assembled monolayers of thiols, disulfides and sulfides, are among the most stable interfacial supramolecular assemblies known. Beyond their chemical and physical stability in a wide range of aqueous and organic electrolytes, they exhibit the widest potential range over which the assembly is stable. However, as illustrated in Figure 4.3, when a sufficiently negative potential is applied in a basic electrolyte, the gold–thiolate bond can be reduced and the monolayers are desorbed quantitatively [8]. Depending on the chain length of the thiol and the applied potential, desorption proceeds either from a few nucleation centers (long chain lengths and small driving force) or in a nearly homogeneous manner across the electrode (short chain lengths and large driving

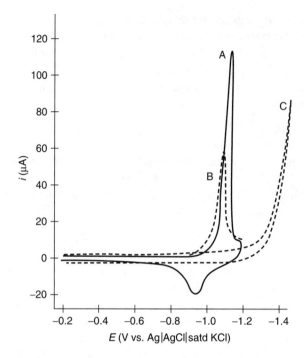

Figure 4.3 Electrochemical stripping and redeposition of a dodecanethiol SAM on Au(III) in 0.5 M KOH. The continuous (A) and dashed (B) lines represent, respectively, the first and second scans, while C indicates the cyclic voltammogram of bare Au(III). Reprinted with permission from M.M. Walcak, D.D. Popenoe, R.S. Deinhammer, B.D. Lamp, C. Chung and M.D. Porter, *Langmuir,* 7, 2687 (1991). Copyright (1991) American Chemical Society

forces). Given that the extent of desorption can be controlled through the time over which the reducing potential is applied and the magnitude of this potential, desorption represents a useful approach to controlling the surface coverage.

4.3.3   Thermodynamics of Adsorption

The Langmuir isotherm describes equilibrium adsorption when there are no lateral interactions between the adsorbed molecules, the limiting surface coverage is dictated simply by the size of the adsorbate, all adsorption sites on the surface are equivalent and adsorption is fully reversible. As discussed in Section 5.5.1 below, spontaneously adsorbed monolayers that fulfill these criteria will exhibit ideal voltammetric responses at slow scan rates, thus including a linear dependence of the peak current, i_p, on the scan rate, v, an FWHM of $90.6/n$ mV, where n is the number of electrons transferred, and a formal potential that is independent of the surface coverage.

A wide variety of quinones spontaneously adsorb onto various electrodes, including gold, platinum, carbon, and especially mercury. On mercury electrodes, these quinonoid monolayers often exhibit nearly ideal electrochemical responses in low-pH electrolytes, so making them attractive model systems for probing the thermodynamics of adsorption. In low-pH electrolytes, both the oxidized and reduced forms are neutral. Therefore, these supramolecular assemblies provide an opportunity to probe the effect of potential on the strength of adsorption without the added complication of any changing electrostatic effects.

The Langmuir isotherm is described by the following equation:

$$\Gamma_i/(\Gamma_s - \Gamma_i) = \beta_i C_i \tag{4.5}$$

where Γ_i is the surface excess of species i at equilibrium, Γ_s is the surface excess of species i at saturation, β_i is the adsorption coefficient of species i, and C_i is the concentration of species i in solution. Typically, the concentration of the surface active compound in solution is sufficiently low that activity effects are either negligible or are incorporated into the energy parameter.

For example, as illustrated in Figure 4.4, the cyclic voltammetric behavior of 2-hydroxyanthraquinone (2OH-AQ) monolayers on mercury microelectrodes is

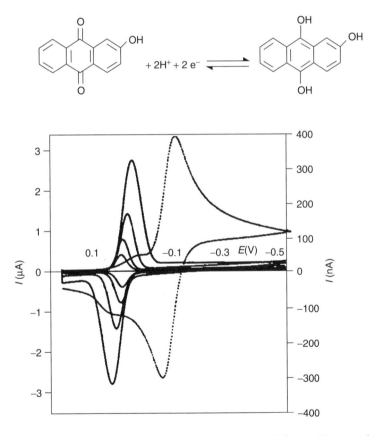

Figure 4.4 Cyclic voltammograms (continuous lines) obtained for a 20 μm radius mercury microelectrode immersed in a 10 μM solution of 2OH-AQ in 1.0 M HClO₄. The scan rates, from top to bottom, are 50, 20, 10 and 5 V s⁻¹; the current scale is on the right-hand side. The dotted line represents the cyclic voltammogram observed for the same electrode immersed in a 10 mM solution of 2OH-AQ; in this case, the scan rate is 50 V s⁻¹, with the current scale on the left-hand side. In all cases, cathodic currents are 'up', and anodic currents are 'down', with the initial potential being +0.250 V. Reprinted with permission from R.J. Forster, *Anal. Chem.*, **68**, 3143 (1996). Copyright (1996) American Chemical Society

nearly ideal in an $HClO_4$ supporting electrolyte [9]. The electrochemical response is consistent with reduction proceeding by a two-electron, two-proton, transfer mechanism. Figure 4.5 illustrates the dependence of the surface coverage on the solution-phase concentration of 2OH-AQ. The continuous line shown in this figure is the best fit to the experimental data provided by the Langmuir isotherm. The latter predicts that a plot of C_i/Γ_i versus C_i should be linear, and that the saturation surface coverage and the energy parameter can be obtained from the slope and intercept, respectively. As shown in the inset of Figure 4.5, plots of this type are linear (correlation coefficients > 0.995). The close agreement observed between experiment and theory suggests that the Langmuir adsorption isotherm provides a satisfactory description of adsorption in this system.

This analysis provides two key pieces of information—first, the surface coverage of the adsorbate within a dense monolayer, and secondly, the adsorption coefficient. For the 2OH-AQ monolayers, Γ_s is $1.0 \pm 0.08 \times 10^{-10}$ mol cm^{-2}. Once the surface coverage is known, the average area of occupation per molecule $\mathring{A}^2$, A_{Molec}, can be calculated according to the following:

$$A_{Molec} = \frac{10^{16}}{N_A \Gamma_s} \tag{4.6}$$

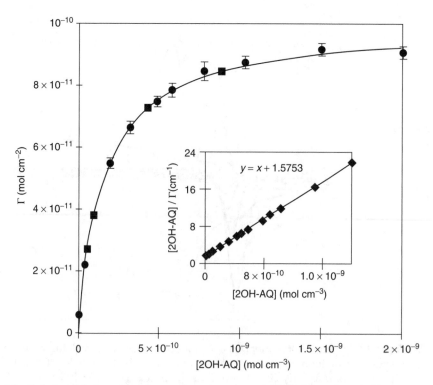

Figure 4.5 Relationship between the surface coverage of 2OH-AQ and its bulk concentration, where the continuous line represents the best fit to the Langmuir isotherm. The inset illustrates data plotted according to the linearized Langmuir isotherm. The supporting electrolyte is 1.0 M $HClO_4$. Reprinted with permission from R.J. Forster, *Anal. Chem.*, **68**, 3143 (1996). Copyright (1996) American Chemical Society

where A_{Molec} is measured in $Å^2$ and N_A is the Avogadro constant. For the 2OH-AQ monolayers, the area occupied per molecule is 166 ± 12 $Å^2$. By combining these experimental data with energy-minimized molecular modeling studies, a limited insight into the packing density, and perhaps the adsorbate orientation, can be obtained.

The adsorption coefficient can be used to determine the free energy of adsorption, $\Delta G_{ads}^{\ddagger}$, i.e. the difference in free energy between a surface active molecule in solution and in the adsorbed state, according to the following:

$$\beta_i = \exp(-\Delta G_{ads}^{\ddagger}/RT) \qquad (4.7)$$

The adsorption coefficient for the 2OH-AQ system in both the oxidized and reduced forms is $6.3 \pm 0.6 \times 10^6$ M^{-1}, corresponding to a free energy of adsorption of 38.7 ± 0.3 kJ mol^{-1}. The large $\Delta G_{ads}^{\ddagger}$ value found for this system confirms that, in common with many functionalized anthraquinones, 2OH-AQ is strongly adsorbed onto mercury.

Effect of lateral interactions

Voltammetric peaks that are broader or narrower than theoretically predicted, formal potentials that shift with changes in the surface coverages, and peak currents that do not depend directly on the scan rate, are all typically associated with adsorbates that interact laterally. Under these circumstances, the Langmuir adsorption isotherm does not describe the dependence of the surface coverage on the bulk concentration. The Frumkin adsorption isotherm accounts for lateral interactions by modeling the free energy of adsorption as an exponential function of the surface coverage, according to the following:

$$\beta_i C_i = \frac{\theta_i}{1 - \theta_i} \exp(g\theta_i) \qquad (4.8)$$

where $\theta_i = \Gamma_i/\Gamma_{sat}$, Γ_i is the coverage of the adsorbate in mol cm^{-2} at a bulk concentration C_i, Γ_{sat} is the saturation coverage obtained at high bulk concentrations, and β is the adsorption coefficient. Attractive interactions are indicated by $g < 0$, while repulsive interactions are indicated by $g > 0$.

In contrast to 2OH-AQ monolayers, 1-amino-2-sulfonic-4-hydroxyanthraquinone (1,2,4-AQASH) adsorbates exhibit significant lateral interactions, thus requiring the use of the Frumkin adsorption isotherm [10]. Figure 4.6 shows that the optimized Frumkin isotherm provides a satisfactory fit to the experimental surface coverages for monolayers assembled from both the reduced and oxidized forms.

Potential dependence of adsorption in reversibly bound systems

Voltammetry provides a powerful insight into the effect of the applied potential on the surface coverage, the free energy of adsorption, and the associated kinetics for electroactive films that form on electrode surfaces by *irreversible* adsorption.

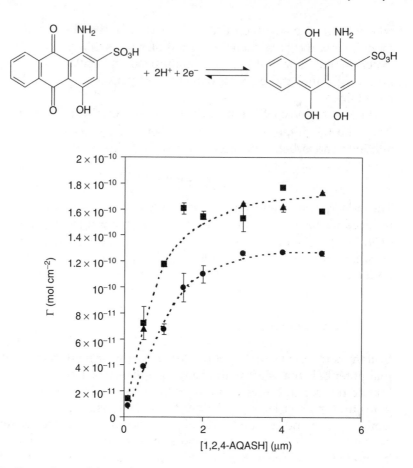

Figure 4.6 Dependence of the surface coverage on the bulk concentration of the quinone (where ■ and ▲ denote the areas under the anodic and cathodic peaks, respectively) and hydroquinone (where ● denotes both anodic and cathodic data) forms of 1,2,4-AQASH. The supporting electrolyte is 1.0 M HClO$_4$. The dashed lines represent the best fits to the Frumkin adsorption isotherm; where error bars are not shown, the errors determined from at least three independently formed monolayers are comparable to the sizes of the symbols. Reprinted with permission from R.J. Foster, T.E. Keyes, M. Farrell and D. O'Hanlon, *Langmuir*, **16**, 9871 (2000). Copyright (2000) American Chemical Society

However, when the adsorption process is *reversible*, one cannot use the traditional approach to measuring the effect of the deposition potential on the coverage, i.e. immersing a clean electrode in a deposition solution under potential control, followed by *ex situ* measurement of the surface coverage in a blank electrolyte solution.

In contrast, as discussed earlier in Section 3.2.1, studies of the interfacial capacitance allow the effect of the applied potential on the adsorption thermodynamics to be elucidated. For example, as discussed above, cyclic voltammetry reveals that the dependence of the surface coverage Γ on the bulk concentration of 2OH-AQ is accurately described by the Langmuir isotherm over the concentration range 20 nM to 2 μM. However, since adsorption is reversible in the anthraquinone system, the effect of changing the potential at which the monolayer is formed on the surface coverage, or the adsorption thermodynamics, cannot be investigated by *ex situ*

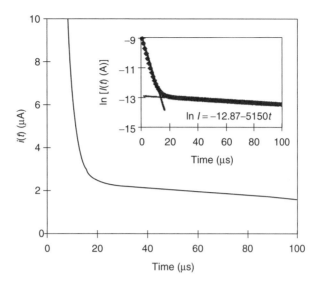

Figure 4.7 Current response for a 20 µm radius mercury microelectrode immersed in a 10 µM solution of 2OH-AQ following a potential step from −0.400 to +0.150 V. The supporting electrolyte is 0.2 M HClO₄. The inset shows the semi-log plot, where the time axis is referenced to the leading edge of the potential step. Reprinted with permission from R.J. Forster, *Anal. Chem.*, **68**, 3143 (1996). Copyright (1996) American Chemical Society

measurement of Γ in a blank electrolyte solution. This issue can be addressed by using µs timescale chronoamperometry to time resolve double-layer charging and heterogeneous electron transfer to the adsorbed anthraquinone moieties. As illustrated in Figure 4.7, this high-speed approach allows the double-layer capacitance, C_{dl}, to be measured even at potentials where the monolayer is redox-active, thus allowing the effect of potential and bulk 2OH-AQ concentration on C_{dl} to be probed. Significantly, when C_{dl} is measured at $E^{0'}$ as the concentration of 2OH-AQ in solution is systematically varied, the values of Γ_s and β_i obtained are identical to those measured voltammetrically. This observation suggests that C_{dl} depends linearly on Γ_i, at least at $E^{0'}$. As illustrated in Figure 4.8, capacitance data have been used to determine the adsorption isotherms as the deposition potential was systematically varied from −0.400 to +0.300 V, where the monolayer is fully reduced and oxidized, respectively. The lower capacity observed for the reduced form of the monolayer, trihydroxyanthracene, suggests that the reduced monolayer may be more compact than the oxidized anthraquinone films. Moreover, the magnitude of the free energy of adsorption, ΔG_{ads}, changes significantly, with values of -41.4 ± 2.7 and -26.5 ± 0.3 kJ mol^{-1} being observed for fully reduced and oxidized monolayers, respectively. These differences in the free energies of adsorption are thought to arise because of different extents of intermolecular hydrogen bonding in the oxidized and reduced films.

4.3.4 Double-Layer Structure

The electrode/electrolyte interface is intimately involved in many important processes, including voltammetric analysis, electrodeposition, metal corrosion, battery

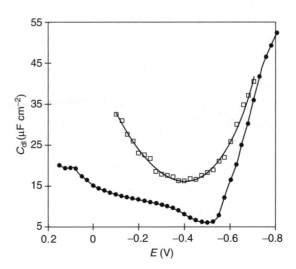

Figure 4.8 Dependence of the double-layer capacitance, C_{dl}, on the applied potential: ●, data obtained for a mercury microelectrode immersed in a solution containing 10 μM 2OH-AQ dissolved in 0.05 M NaClO$_4$ at a pH of approximately 5.8; □, data obtained for a clean mercury microelectrode in the same electrolyte solution but without any dissolved anthraquinone. Reprinted with permission from R.J. Forster, *Anal. Chem.*, **68**, 3143 (1996). Copyright (1996) American Chemical Society

charging and discharging, and electrochemical synthesis. One of the principal objectives in creating an interfacial supramolecular assembly is to control these processes. In order to achieve this objective, one must understand the structure of the electrochemical interface. However, a modified electrochemical interface is a complex system, and thus to understand its structure, aspects of solid-state and liquid physics must be combined with information about chemical reactivity. Despite this complication, many experimental techniques exist to probe the structures of these interfaces, and theoretical models describing such structures have been developed. Here, some of these models are described, the available techniques for probing the structure of the double layer are reviewed, and the effects of the interfacial structure on the kinetics of electrode reactions are explored.

The electrode/solution interface represents a distinctly different reaction environment to that encountered in homogeneous solution. Under equilibrium conditions, the time-averaged forces acting on a species in solution are isotropic and homogeneous, and there is no electric field. At the electrode/solution interface, the symmetrical force field breaks down, and the species experience an anisotropic electric field whose strength varies significantly with the distance from the electrode surface. Polarization causes solvent dipoles to become oriented, and a net excess ionic charge accumulates on the solution side of the interface near the phase boundary. Once the electrolyte side acquires a net charge, the electrode responds by acquiring on its surface an induced charge of the same quantity and opposite sign. Thus, a charge separation occurs at the electrode–solution interface, a potential gradient develops, and two planes of charge, the electrical 'double layer', are formed.

The actual interface is comprised of only the top layer of atoms on the electrode and the layer of adsorbed solvent and specifically adsorbed species. However, an interfacial region, which is different from the bulk phase, extends over considerably longer distances. This region encompasses the diffuse layer within the electrolyte, and if the interface has been synthetically modified, it also includes the adsorbed monolayer or polymer film. A space-charge region also exists within the electrode. This is very narrow, at most a few angstroms, within metallic electrodes, since a thick space-charge region would be associated with a potential difference that would cause a current to flow. The space-charge region can extend to several tens of angstroms within semiconductor electrodes where the density of charge carriers is much lower. The interface is electrically neutral overall, and the total charge integrated over the entire interface comprising electrode and associated electrolyte is zero.

First, we will review the equilibrium structure of an *unmodified* interface. This interfacial structure is partially dictated by the electrolyte ions. Most cations are strongly solvated because their radii are small and the free energy of solvation is high. This strong solvation means that even when the electrode is negatively charged there is usually little tendency for cations to lose their hydration sheath and adsorb directly on the metal surface. Thus, the distance of closest approach of the cations is determined by the radius of the inner solvent-coordination sphere. If the metal surface itself constitutes a plane, i.e. approaches atomic smoothness, then the distance of closest approach for the cation nuclei constitutes a plane termed the *Outer Helmholtz Plane* (OHP). In contrast, anions have larger radii and tend to be more weakly hydrated. Moreover, they are able to form relatively strong ionic or covalent bonds with the metal electrode surface, thus making it energetically feasible for them to shed their inner hydration sphere and become specifically adsorbed on the electrode surface. The plane formed by the nuclei of anions directly adsorbed on the metal surface is termed the *Inner Helmholtz Plane* (IHP). In the absence of specific adsorption, there is a well-defined potential at which the metal electrode has no net charge, and equal numbers of cations and anions will be present in the OHP. This potential is termed the *potential of zero charge* (PZC), and its determination is important in understanding the distribution of charge at the interface. Finally, the total charge on the metal, IHP and OHP will not generally cancel out, and thus a diffuse charged layer develops in the electrolyte which extends out into the bulk solution.

Capacitance of a modified interface

The electrode/electrolyte interface discussed above exhibits a capacitance whose magnitude depends on the distribution of ions on the solution side of the interface. In relatively concentrated electrolytes, the capacitance of the Helmholtz layer dominates the interfacial capacitance. For most metals, typical Helmholtz capacitances range from $20-60\,\mu\text{F cm}^{-2}$, and depend substantially on the applied potential, reaching a minimum at the potential of zero charge where there is no excess charge on either side of the interface.

Frequently, e.g. in the case of alkane thiol monolayers, the electrode is modified with a low-dielectric-constant layer. Film formation causes the Helmholtz layer to change from a mixture of ions and solvent with a high dielectric constant to an ion-free, often organic, layer with a low dielectric constant. The interfacial capacitance depends directly on the dielectric constant of the contacting medium. Therefore, film formation typically reduces the interfacial capacitance dramatically and causes it to become virtually independent of the applied potential.

Cyclic voltammetry in which the current response is measured as the applied potential is varied according to a triangular waveform, chronoamperometry, in which the exciting potential waveform consists of a sharp potential step, and frequency-based approaches, such as AC impedance, can be used to measure the interfacial capacitance. With regard to interfacial supramolecular assemblies, one significant feature is the ability to control the thickness of the film, e.g. by depositing a controlled number of electrostatically bound layers or by systematically varying the length of the adsorbate used to form a self-assembled monolayer. The reciprocal of the total interfacial capacitance C_T^{-1} can be represented by the sum of the reciprocal capacitances of the film, C_{film}, and the diffuse layer, C_{dif}, as follows:

$$C_T^{-1} = C_{film}^{-1} + C_{dif}^{-1} \tag{4.9}$$

$$C_{film} = \varepsilon_0 \varepsilon_{film}/d \tag{4.10}$$

$$C_{dif} = \varepsilon_0 \varepsilon_{SOLN} \kappa \cosh[ze(\phi_{PET} - \phi_{SOLN})/2k_B T] \tag{4.11}$$

where ε_0 is the permittivity of free space, ε_{film} and ε_{SOLN} are the film and solution dielectric constants, respectively, d is the monolayer thickness, z is the charge number of the electrolyte ion, e is the absolute electronic charge, k_B is the Boltzmann constant, T is the absolute temperature, and ϕ_{PET} and ϕ_{SOLN} are the potentials at the plane of electron transfer and in the bulk solution, respectively. The quantity κ is given by $(2n^0 z^2 e^2/\varepsilon_{SOLN}\varepsilon_0 k_B T)^{1/2}$, where n^0 is the number concentration of the ions in solution.

In the absence of specific adsorption of the electrolyte ions, only the diffuse layer capacitance is considered to depend on the potential or the concentration of the supporting electrolyte. Therefore, the diffuse-layer capacitance will become very large when the concentration of supporting electrolyte is high. The reciprocal relationship of Equation (4.9) means that the total capacitance then becomes dominated by the film capacitance. For example, as illustrated in Figure 4.9, plots of C_T^{-1} versus n, the number of CH_2 groups in the alkane chain, are linear for longer-chain-length alkane thiol monolayers [11].

The interfacial capacitance can also provide a significant insight into the permeability of interfacial supramolecular assemblies. While information of this kind complements studies using redox-active probes in solution, it also provides information on a significantly shorter length scale, i.e. that of electrolyte ions and solvent molecules. For example, for dense, defect-free monolayers, the limiting capacitance is very much lower (5–10 μF cm^{-2}) than that found for an unmodified interface (20–60 μF cm^{-2}).

Once the film capacitance is known, the dielectric constant of the layer can be determined by using Equation (4.10), provided that the thickness of the assembly is known. For self-assembled monolayers of long-chain alkane thiols, the experimentally determined dielectric constants fall in the range 2.3–2.6 and are consistent with a close-packed layer of alkane chains with essentially no penetration by electrolyte solvent or ions.

Forming a monolayer involves displacing specifically adsorbed ions and solvent molecules from the interface, which changes the double-layer capacitance from that

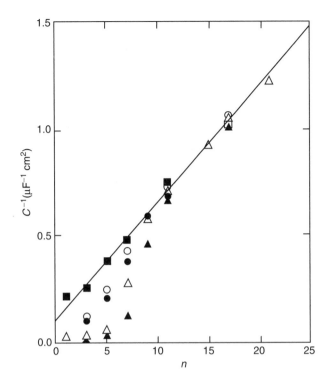

Figure 4.9 Dependence of the reciprocal of the interfacial capacitance on the number of methylene groups, n, within alkane thiol monolayers. The symbols represent capacitances obtained from cyclic voltammograms, with filled symbols indicating data obtained at 10 mV s^{-1}, and empty symbols those obtained at 100 mV s^{-1}: ■ □, 1 M KCl; ▲ △, 1 M HClO$_4$; ● ○, 1 M NaF. Reprinted with permission from M.D. Porter, T.B. Bright, D. Allara and C.E.D. Chidsey, *J. Am. Chem. Soc.*, **109**, 3559 (1987). Copyright (1987) American Chemical Society

observed at a clean unmodified interface. Determining the relationship between the surface coverage and the double-layer capacitance is an important objective since it allows the effect of potential on both the surface coverage and the adsorption thermodynamics to be investigated. For partial or sub-monolayer assemblies, it may be appropriate to represent the double-layer capacitance by an equivalent circuit containing two parallel capacitances describing the contributions from the modified and 'bare' regions of the electrode surface. For this model, the total double-layer capacitance, C_{dl}, is given by the following:

$$C_{dl} = C_{mono} + (C_{bare} - C_{mono})(1 - \theta) \qquad (4.12)$$

where C_{bare} and C_{mono} are the double-layer capacitances of a clean mercury electrode, and a mercury electrode modified with a dense monolayer, respectively, and θ is the fractional monolayer coverage. Equation (4.12) holds for all potentials where C_{bare} is greater than C_{mono}. This parallel capacitor model predicts that the linearized Langmuir isotherm can be expressed as follows:

$$\frac{C_i}{(C_{bare} - C_{dl})} = \frac{1}{C_{mono}}C_i + \frac{1}{C_{mono}\beta_i} \qquad (4.13)$$

where C_i is the bulk concentration of the adsorbate and β_i is the adsorption parameter. Hence, the interfacial capacitance corresponding to a dense monolayer on the electrode surface, together with the adsorption coefficient, can be determined from the slope and intercept of a plot of $C_i/(C_{bare} - C_{dl})$ versus C_i.

Interfacial supramolecular assemblies in which the molecular building blocks are capable of undergoing protonation reactions are important in catalysis, molecular electronics and biological systems. For example, thin films that undergo coupled electron and proton transfer reactions are attractive model systems for developing catalysts that function by hydrogen atom and hydride transfer mechanisms. In the field of molecular electronics, protonation provides the possibility that the energies of bridge states can be chemically modulated, thus giving rise to molecular switches. In biological systems, the kinetics and thermodynamics of redox reactions are often controlled by enzyme-mediated acid–base reactions. Measurements of the pH dependence of the interfacial capacitance provides a convenient means of determining the *in situ* pK_a of an acid/base moiety. In the model of Smith and White [12], the pH dependence of the total double-layer capacitance, C_{dl}, is described as follows:

$$1/C_{dl} = 1/C_F + 1/[C_S + C(f)] \qquad (4.14)$$

where $C(f)$ represents the variation of the capacitance with extent of monolayer protonation. According to this model, at pH values near the pK_a of the immobilized sulfonic acid groups, $C(f)$ reaches a maximum value, thus giving a local maximum in C_{dl}. Therefore, it is possible to determine the surface pK_a of a surface-confined acid or base by investigating the pH dependence of the interfacial capacitance.

4.3.5 Post-Deposition Modification

The diverse variety of functional groups that can be used to terminate SAM-forming thiols allows their properties to be tailored for specific applications after their formation, e.g. to promote protein binding. Significantly, this approach circumvents the difficulty of forming a low-defect-density monolayer with chemically and structurally complex molecules that tend to form rather disordered structures. It is important to note that the reactivity of these terminal groups may be significantly different from that expected on the basis of solution-phase reactions because of the low dielectric constant and long-range order of the monolayer. Covalent attachment of functional groups through the formation of amides and esters via acid chlorides and anhydrides, or by using carbodiimide coupling, are the dominant approaches to modifying SAMs. Such approaches produce stable modifications and are synthetically very flexible. For example, as discussed later in Section 4.9, proteins, enzymes and other biomolecules can be attached via the reactions of exposed amines to SAMs with terminal COOH groups. SAMs may also be modified 'post deposition' in a more site-selective manner by exploiting the high specificity of biomolecular reactions such as biotin–avidin, the reconstitution of avin or nicotinamide adenine dinucleotides (NADs), with their enzymes, or through complementary strand binding of oligonucleotide sequences.

Electrostatic binding also represents an important approach to introducing specific chemical functionalities within an SAM. For polyions, both naturally occurring and synthetic polymers with charged groups, the binding is sufficiently strong that the functionalized SAM can be used in aqueous media. In addition, as discussed below in Section 4.6, this electrostatic binding approach can be used to assemble multilayers with reasonable long-range order.

For some applications, conventional SAMs are insufficiently stable towards thermal desorption, solvent stripping, electrochemical desorption, and exchange with solution-phase thiols. Creating a polymerized SAM in which the individual adsorbates are covalently linked represents an important approach to creating highly stable interfacial supramolecular assemblies. While topochemical polymerization has not been proven in all cases, i.e. the polymerization reaction appears to trigger a structural change, the conversion of diacetylenes to a conjugated ene–yne structure, cross-linking of a terminal vinyl group, hydrolysis and cross-linking of a trimethoxysilane, dehydration of a boronic acid to a borate glass, and electrochemical polymerization of a terminal pyrrole, have all been explored.

4.4 Structural Characterization of Monolayers

The functioning of a true supramolecular assembly is not dictated simply by the electrochemical and photochemical properties of its component building blocks. Rather, its overall function is more than the sum of its parts and is intimately linked with the structure, organization and interactions present within the assembly. Therefore, it is essential to probe the physical structure of the assembly, to elucidate how this structure arises from the interaction of the building blocks, and to understand the structure–function relationship of the assembly.

4.4.1 Packing and Adsorbate Orientation

Spectroscopic studies reveal that for alkane thiol monolayers, crystalline lattices are formed on single-crystal electrodes, which are commensurate with the metal lattice. The thickness of this interfacial two-dimensional crystal depends linearly on the chain length of the alkane thiol used as the building block. Although formally limited to film thicknesses greater than 50 Å, ellipsometry represents perhaps the most direct optical technique for determining the thicknesses and refractive indices of monolayers, multilayers and thin films. The refractive index depends on the nature of the adsorbate as well as the water and ion content of the monolayer. For densely packed alkane thiol monolayers, e.g. docosanethiol ($n = 21$) [11], the refractive index, μ, is typically 1.50. This value is generally applicable to densely packed simple alkyl chain adsorbates for $n > 10$. The film thickness calculated from ellipsometric data, taking the film refractive index, η_F as 1.50, is 31.6 Å for a docosane thiol monolayer.

On the most common metal substrate, Au(111), alkane thiols form a ($\sqrt{3} \times \sqrt{3})R30°$ hexagonal lattice with an average spacing of 5.0 Å between the alkane chains. The latter are in the *trans*-conformation with very few *gauche* defects. The average tilt of the alkane chains with respect to the surface normals is typically less than or equal to 30°. The tilt angle is controlled by the headgroup spacing, combined with a minimization of the free volume in the alkane chain domain. A detectable

$C(4 \times 2)$ unit cell of four thiols within the lattice arises from variations in the twist of the alkane chains. In highly ordered SAMs, the orientation of the terminal group can depend on the number of methylenes separating the terminal group from the thiol, i.e. 'odd–even' effects can be observed. Domains of SAMs with nanometer dimensions are separated by grain boundaries corresponding to changes in chain tilt and/or registry with the substrate surface atoms. Even when the surface of the substrate is not atomically flat, or when a large group in the chain perturbs the packing, the exposed alkane chains often adopt a hexagonal close-packed structure. Increasing the temperature of the contacting medium can trigger structural changes within the monolayer. For example, chain 'melting' to a more liquid-like structure has been found, although other observations indicate only a slow increase in the degree of chain disorder at higher temperatures.

As discussed above in Chapter 2, scanning tunneling microscopy provides electronic and structural information with sub-angstrom resolution, thus making it an invaluable tool for studying the formation, structure and behavior of a wide variety of monolayers. In particular for monolayers which include transition metal complexes, theoretical models predict an enhanced tunneling through the redox-active centers because of resonant tunneling modes. Specifically, reorganizations of the inner-sphere vibrational modes will significantly contribute to the tunneling current. Hudson and Abruña [13] have used electrochemical scanning tunneling microscopy (ECSTM) to probe the formation and structure of spontaneously adsorbed $[Os(bpy)_2$ trimethylene-4,4'-bipyridine $Cl]^{2+}$ monolayers adsorbed on a Pt(III) substrate. As illustrated in Figure 4.10(b), during the initial stages of deposition, the images obtained appeared virtually unchanged from the image of a clean Pt(III) surface (Figure 4.10(a)). However, consistent with adsorption of $[Os(bpy)_2$ (trimethylene-4,4'-bipyridine) $Cl]^{2+}$ onto the electrode surface, the voltammetric response increases. As the coverage increased, small, high-contrast regions appeared in the image which are aligned along the microscope's fast-scan axis (Figure 4.10(c)). These features appear only in the presence of adsorbate and are considered to arise because of clusters of highly mobile adsorbate on the sample surface. It appears that these clusters become oriented because of interactions with the scanning tip. As illustrated in Figure 4.10(d), the adsorbate clusters continue to grow until, at coverages of about two thirds of a monolayer, they dominate the image. While defects were observed for intermediate surface coverages, as shown in Figure 4.10(e), the full monolayer was observed to be essentially free from large defects. However, long-range-ordered molecularly resolved images were not observed under these conditions.

Images obtained after emersion and drying of a Pt(III) crystal after modification with $[Os(bpy)_2$ (trimethylene-4,4'-bipyridine) $Cl]^{2+}$ suggest that approximately half of the surface is covered with 'islands' of crystalline complex. These regions are largely disordered and consist of crystallites with dimensions ranging from a few to tens of nanometers in width. Molecularly resolved images were obtained of the structure within the larger crystallites, showing a rectangular close-packed array with unit cell dimensions of 9.3 ± 0.3 by 12.4 ± 0.4 Å.

Spectroscopy plays an important role in characterizing the structure and integrity of proteins adsorbed on electrodes. For heme proteins, the Raman spectrum is a sensitive probe of the spin state, thus providing an insight into the integrity of

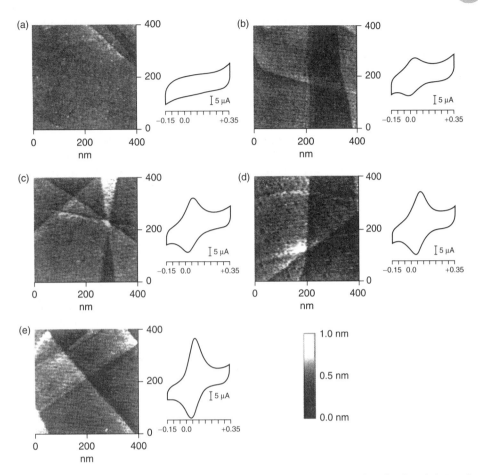

Figure 4.10 ECSTM images of a Pt(111) surface during the deposition of an [Os(bpy)$_2$(trimethyl-ene-4,4′-bipyridine)Cl]$^{2+}$ monolayer from a 0.1 M KClO$_4$ electrolyte solution, with the corresponding cyclic voltammograms taken immediately after withdrawing the tip from the sample surface. The integrated charge corresponds to (a) 0, (b) 25, (c) 49, (d) 66, and (e) 100 % of a full monolayer: tip bias, 50 mV; tunneling current, 2.5 nA; scan rate, 8.3 Hz. Reprinted with permission from J.E. Hudson and H.D. Abruña, *J. Phys. Chem.*, **100**, 1036 (1996). Copyright (1996) American Chemical Society

the active site. Surface Enhanced Resonance Raman Spectroscopy (SERRS) studies of cytochromes adsorbed at a bare silver electrode reveal that spin-state marker bands are shifted, hence indicating that the protein undergoes a conformational change upon adsorption which is likely to affect the functioning of the active site. Time-resolved SERRS has been used to measure the kinetics of electron transfer and coupled conformational changes in cytochrome C552 adsorbed at a silver surface. In contrast to the conformational changes observed at bare electrodes, both oxidized and reduced forms of cytochrome C remain in their native conformations when adsorbed at a silver electrode coated with a monolayer of bis(4-pyridyl) disulfide or 4-mercaptopyridine. The protein becomes immobilized by interacting with the pyridine lone pair of these organic monolayers which are sufficiently thin to allow resonant energy transfer between the electrode and protein.

As discussed above in Chapter 3, ellipsometry and quartz crystal microbalance (QCM) approaches provide a useful insight into the adsorption of both the supporting interfacial assembly and the proteins themselves. Beyond monitoring the adsorption dynamics and the structural integrity of the biomolecule, the orientation of the active site is of prime importance. For example, if the active site itself binds to the self-assembled monolayer, transport of the substrate or co-enzyme may be blocked.

Absorption linear dichroism and emission anisotropy have been used to probe the orientation of molecules adsorbed on various surfaces. The results obtained indicate that proteins typically adsorb onto SAMs in a disordered manner, thus resulting in a quite broad distribution of orientations. As discussed below in Section 4.5.1, this distribution of orientations results in non-ideal voltammetric responses, including peak broadening.

Two-component SAMs

Two-component SAMs, composed of molecules with similar chain lengths, are homogeneously mixed rather than phase-separated and only in high-resolution STM images is there any evidence of imperfect mixing. In contrast, when the lengths of the two components are very different, phase-separation into microscopic domains is detectable. The importance of stabilizing lateral interactions in creating ordered assemblies is demonstrated by these systems. For example, the longer-chain component exhibits a greater degree of disorder since it is not stabilized by packing over its entire surface.

4.4.2   Surface Properties

Contact angle measurements can provide a powerful insight into monolayer wetting and are among the most sensitive techniques for evaluating surface composition and structure. As detailed in Table 4.1, the advancing contact angle e.g. for water on for a smooth gold surface modified with a methylene-terminated alkane thiol

Table 4.1   Advancing contact angles on derivatized thiol monolayers on gold. HD = hexadecane.

Thiol system	$\theta_a(H_2O)$	$\theta_a(HD)$
$HS(CH_2)_2(CF_2)_5CF_3$	118	71
$HS(CH_2)_{21}CH_3$	112	47
$HS(CH_2)_{17}CH=CH$	107	39
$HS(CH_2)_{11}Br$	83	0
$HS(CH_2)_{11}Cl$	83	0
$HS(CH_2)_{11}OCH_3$	74	35
$HS(CH_2)_{10}CO_2CH_3$	67	28
$HS(CH_2)_8CN$	64	0
$HS(CH_2)_{11}OH$	0	0
$HS(CH_2)_{15}COOH$	0	0

SAM [14] is between 111 and 115°, while the hexadecane angle is only 45–46°. For disordered or liquid-like monolayers and for films terminated with hydrophilic groups, the contact angle tends to decrease. While experiments of this kind provide a useful insight into the surface composition and structure, it is important to note that surface contamination, roughness, heterogeneity and restructuring can complicate these measurements. However, contact angle measurements will continue to play pivotal roles in the technological exploitation of interfacial supramolecular assemblies, e.g. in developing biocompatible devices, including prosthetics and implantable sensors.

4.5 Electrochemical Characterization

4.5.1 General Voltammetric Properties of Redox-Active Monolayers

Inspired by its nearly ideally reversible electrochemistry in solution and chemical stability, ferrocene is the most common redox center that has been incorporated within SAMs. For single-component monolayers assembled from ferrocene labeled alkane thiols, the maximum coverage is consistent with a close-packed layer of ferrocenes. The voltammetry of ferrocene alkane thiol SAMs is characterized by non-ideal responses for high surface coverages because of the strong lateral interactions between the redox centers, double-layer effects and the formation of a solid phase film. For low coverages, or where the monolayer is diluted with electro-inactive thiols, nearly ideal voltammetry is observed. As indicated in Table 4.2, the formal potential can be shifted over a 0.5 V range, depending on the presence of electron-donating species and the depth to which the redox center is buried within the SAM.

The $[Ru(NH_3)_5 \text{ pyridine}]^{2+/3+}$ redox center exhibits well-defined, nearly ideal voltammetry across a wide range of electrolyte compositions, scan rates and temperatures. Despite the headgroups having comparable sizes, the saturation surface coverage is less than that found for ferrocene alkane thiol monolayers. This behavior is thought to reflect electrostatic repulsion between the polycationic redox centers and the need to incorporate counterions into the monolayer in both

Table 4.2 Properties of redox centers within alkane-thiol-based SAMs.

Redox center	Γ^a (mol cm^{-2})	$E^{0'}$ vs. SCE (V)	Comments on peak shapes
Ferrocenes	4–5	+0.2 to +0.7	Ranges from narrow to ideal to broad
$[Rupy(NH_3)_5]^{2+}$	1–2	0	Nearly ideal at all coverages
Cytochrome c	0.2	0	Slightly broader than ideal
Microperoxidase	2	−0.4	Broad cyclic voltammograms
Viologens	4	−0.3 to −0.5	Two waves, broad, distorted by dimerization reaction
Quinones	3–5	pH-dependent	Very broad
Azobenzenes	3–5	pH-dependent	Very broad, sensitive to packing

^{a}Maximum $\times 10^{10}$.

oxidation states. While the metal complex is generally quite inert toward chemically and photochemically triggered ligand substitution, decomposition is observed upon oxidation in alkaline electrolytes. In sharp contrast to the ferrocene systems, nearly ideal behavior is obtained for all coverages in aqueous electrolytes. Weak lateral interactions and a high degree of solvation contribute to this ideal response.

Viologens, i.e. derivatives of 4,4′-bipyridinium, can be reduced in two well-defined one-electron steps at negative potentials. In SAMs, the viologens yield high coverages, suggesting that a close-packed structure can be created. However, the reduced form undergoes a following chemical reaction involving dimerization of the radical anion, hence making the slow scan voltammetry non-ideal. Provided that heterogeneous electron transfer is sufficiently rapid, it ought to be possible to outrun this following reaction by using high-speed electrochemical techniques [15].

While spontaneously adsorbed monolayers containing quinones are considered in greater detail later, alkane thiol monolayers incorporating quinones have been successfully formed and characterized. Since both the oxidized, quinone, and reduced, hydroquinone, forms are neutral in acidic electrolyte, these investigations provide an important insight into charge effects within monolayers. SAMs incorporating quinones and azobenzenes are generally physically stable and can undergo repeated voltammetric cycling without significant loss of material. However, the rate of heterogeneous electron transfer can be very slow and, apart from highly acidic solutions, may be coupled to mass transfer of protons. As predicted by the Nernst equation, the formal potentials shift in a negative potential direction as the pH of the contacting electrolyte solution is increased. The magnitude of this shift is generally consistent with the 59 mV dec^{-1} expected for a $2e^- - 2H^+$ reaction.

4.5.2   Measuring the Defect Density

One of the most important applications of SAMs is to control the reactivity of electrode surfaces. SAMs play an important role in this regard since they can block access to the electrode surface either completely or selectively on the basis of analyte size or charge. The ability of SAMs to block redox reactions arises because the dense structure of the layer efficiently impedes mass transport and the large LUMO/HOMO separation means that only electron tunneling across the monolayer is possible. This blocking property can be exploited to inhibit corrosion of the metal itself, as the basis of nano-scale lithography and in electroanalysis by achieving a charge- or size-selective response.

However, SAMs are rarely structurally perfect and typically contain defects where crystalline domains meet, at step-edges, and where the electrode is not coated with the SAM. Defects of this kind all facilitate mass transport to the electrode surface where efficient electron transfer can take place. A key objective in characterizing SAMs is to map out the nature, size and distribution of the pinholes and other defects. Undoubtedly, scanning probe microscopy, such as the AFM and STM techniques discussed earlier in Chapter 3, play important roles in this area. However, voltammetry is an extremely powerful approach for detecting defects in SAMs when in contact with solution. This extraordinary sensitivity arises from the ability to routinely detect currents at the nanoamp and picoamp levels which correspond to redox reactions occurring at a tiny fraction of the electrode area. Moreover, the rate of mass transfer to small 'hot spots' on an otherwise insulated electrode is significantly larger than that found for an unmodified macroelectrode and is reminiscent of the behavior found for microelectrode arrays.

Figure 4.11 illustrates the cyclic voltammograms obtained for ferrocenylmethyltrimethylammonium dissolved in solution at a bare gold electrode and at electrodes coated with imperfect $HSC_{17}CH_3$ SAMs [16]. The solution-phase probe exhibits current peaks or small current plateaus near the formal potential, $E^{0'}$. Curve (a) illustrates the electrochemically reversible response obtained for a bare electrode, while curves (b) and (c) show that for the $HSC_{17}CH_3$-coated electrodes, the current peaks or plateaus greatly exceed the extremely small tunneling currents found on electrodes with perfect SAMs. Curve (b) shows only a slow rise in current with no visible plateau, thus indicating that while defects are present, the pinhole area fraction is extremely low. In curve (c), a sharp rise in the Faradaic current appears at low overpotentials, which is characteristic of pinholes. The ratio of the currents

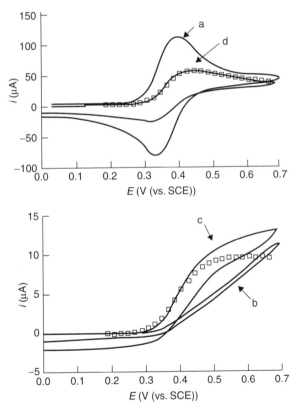

Figure 4.11 Effect of SAM formation on the cyclic voltammetry of ferrocenylmethyltrimethylammonium on a polycrystalline gold electrode; the supporting electrolyte is 0.5 M H_2SO_4, with a scan rate of 0.1 V s^{-1}. Curve (a) is the reversible cyclic voltammogram obtained on bare gold, while curves (b)–(d) are obtained on the same electrode with different monolayers (for details see text). The symbols represent theoretical fits to a microarray electrode model. Reprinted with permission from H.O. Finklea, D.A. Snider, J. Fedyk, E. Sabatani, Y. Gafni and I. Rubinstein, *Langmuir*, **9**, 3660 (1993). Copyright (1993) American Chemical Society

found at bare and modified electrodes can be used as a qualitative assessment of the defect density within the SAM. A more quantitative insight can be obtained by fitting calculated voltammograms to the experimental responses. Fitting such responses to theory allows the average radius of the defect, R_a and the radius of the insulating plane surrounding the microdisk, R_0, to be estimated. In the case of curve (c), R_a and R_0 were 1.5 and 27 µm, respectively. However, these models typically assume that the pinholes are of uniform size and separation. An alternative to voltammetry is to use scanning probe microscopy (SPM), to deposit metal within the defects, or to etch the defect and then image with SPM. These approaches suggest that the defects within high-quality SAMs range from atomic to nanometer length scales.

4.6 Multilayer Formation

4.6.1 Electrostatically Driven Assemblies

Spontaneously adsorbed and self-assembled monolayers represent attractive platforms upon which to create multilayer assemblies. These adsorbed monolayers can be made either anionic, e.g. by using a carboxylic-acid-terminated alkane thiol, or cationic, e.g. by using an osmium metal complex. A second layer can be adsorbed onto this surface through electrostatic attraction, e.g. a polyion such as quaternized poly-4-vinyl pyridine can be adsorbed onto a carboxylic-acid-terminated SAM at high pH levels. Because a bound polyion reverses the surface charge on the ionized SAM, it is possible to build up multiple layers of polyions by sequential immersion of the substrate in aqueous solutions of each component. Multilayers can also be constructed on SAMs via sequential deposition of a metal cation and then a difunctional molecule. Alternatively, if a highly ordered second monolayer is required, Langmuir–Blodgett techniques can be used to immobilize subsequent layers.

Strongly adsorbed monolayers of heteropolyanions (HPAs) are known to attract and bind multiply charged cations to surfaces. Faulkner and co-workers [17] have taken advantage of this property to construct multiple layers of HPA salts on electrode surfaces that were immersed alternatively in solutions of HPAs and multiply charged cations. The procedure required drying of the coatings between each immersion step, and coatings consisting of more than three or four layers were less stable.

Kuhn and Anson have recently exploited the strong electrostatic binding properties of HPAs to create heterostructures [18]. Specifically, these authors described a procedure in which a layer of a multiply charged heteropolyanion, $[P_2Mo_{18}O_{62}]^{6-}$, was adsorbed irreversibly on the surfaces of a variety of electrodes and then used as an electrostatic attractant to bind cations such as $[Os(bpy)_3]^{2+}$ to the same surface. Significantly, these investigations reveal that stable multiple layers can be grown by exposing the electrode surface repeatedly to a solution of the heteropolyanion and subsequently to a solution of the $[Os(bpy)_3]^{2+}$ cation. Figure 4.12 illustrates the cyclic voltammograms obtained from glassy carbon electrodes on which the $[P_2Mo_{18}O_{62}]^{6-}$ anion was adsorbed before the electrode was transferred to a pure supporting electrolyte to record the voltammogram. It is significant that

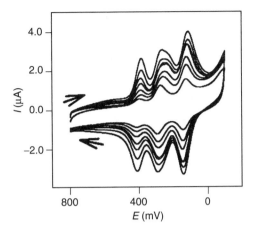

Figure 4.12 Deposition of multiple layers of heteropolyanions (HPAs) on glassy carbon electrodes, as shown by cyclic voltammograms of multilayer deposits containing $[P_2Mo_{18}O_{62}]^{6-}$ and $[Os(bpy)_3]^{2+}$ prepared by the alternating adsorption procedure. Each voltammogram was recorded just after an additional layer of the HPA had been added to the coating. The scan rate was 50 mV s^{-1}, with a supporting electrolyte of 0.5 M NaHSO$_4$. Reprinted with permission from A. Kuhn and F.C. Anson, *Langmuir*, **12**, 5481 (1996). Copyright (1996) American Chemical Society

the potentials of the three reversible responses are close to those observed for the same anion in solution. This observation suggests that the free energies of adsorption for the reduced and oxidized forms are similar and that the anions remain highly solvated when adsorbed. Solution-phase measurements suggest that each peak corresponds to a two-electron reduction/oxidation. As discussed previously in Chapter 3, the symmetrical shape of the peaks is consistent with that expected for a redox reaction involving a surface-confined reactant. An important objective is to create interfacial assemblies that are stable toward dissolution and can be reversibly switched between the oxidized and reduced states over many cycles. For this system, the peaks remain unchanged if the films are repeatedly cycled for tens of minutes, thus indicating that the redox responses are highly reversible and that desorption of the anion from the surface is slow. The total charge under the three reduction peaks corresponds to ca. 6×10^{-11} mol cm^{-2} of a reactant which undergoes an overall six-electron reduction. With this approach, it was possible to construct coatings containing over 20 layers relatively rapidly without any need to dry the coatings between each immersion step. This behavior contrasts with the multilayer HPA films described by Faulkner and co-workers [17] who found that the layers needed to be dried between depositions and that coatings consisting of more than three or four layers were less stable. The advantage of the $[P_2Mo_{18}O_{62}]^{6-}$ system appears to be its relatively stronger binding to the electrode surface.

Heterostructures involving alternate layers of HPA and a transition metal complex were prepared by first adsorbing a monolayer of $[P_2Mo_{18}O_{62}]^{6-}$ onto a glassy carbon electrode, followed by soaking in pure water for 5 min. Next, the coated electrode is transferred to a 0.5 M NaHSO$_4$ supporting electrolyte solution and its potential was cycled once from 0.8 to -0.3 to 0.8 V. Then, the electrode is transferred for 10 min to a solution containing $0.5 - 3$ mM $[Os(bpy)_3]^{2+}$ or $[Ru(bpy)_3]^{2+}$, but with no other electrolyte. Finally, the electrode is soaked for 5 min in pure water. Repeating

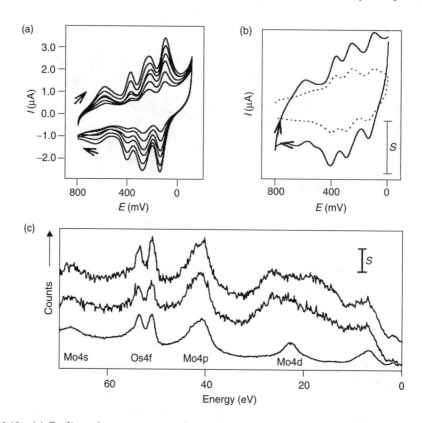

Figure 4.13 (a) Cyclic voltammograms of multilayer deposits containing $[P_2Mo_{18}O_{62}]^{6-}$ and $[Os(bpy)_3]^{2+}$ recorded just after an additional layer of $[Os(bpy)_3]^{2+}$ was added to the coating. (b) Cyclic voltammograms for a multilayer deposit consisting of seven layers of $[P_2Mo_{18}O_{62}]^{6-}$ and seven layers of $[Os(bpy)_3]^{2+}$. Continuous curve: scan rate, 20 mV s^{-1}; $S = 1$ A. Dotted curve: scan rate 2 V s^{-1}; $S = 100$ μA. (c) X-ray photoelectron spectra. Lowest curve: the insoluble salt formed by mixing $[P_2Mo_{18}O_{62}]^{6-}$ and $[Os(bpy)_3]^{2+}$ in solution; $S = 2000$ counts. Middle and upper curves: multilayer deposits containing (middle) four layers of $[P_2Mo_{18}O_{62}]^{6-}$ and three layers of $[Os(bpy)_3]^{2+}$, or (upper) four layers of each type of ion on GC electrodes; $S = 200$ counts. Reprinted with permission from A. Kuhn and F.C. Anson, *Langmuir*, **12**, 5481 (1996). Copyright (1996) American Chemical Society

the HPA deposition step allows multilayer structures to be assembled. Figure 4.13 illustrates voltammograms obtained during the formation of a multilayer of this kind. The increases in the areas of the three peaks for the $[P_2Mo_{18}O_{62}]^{6-}$ anions after each cycle demonstrate the growth of the multilayer. The average increase in the area after each cycle corresponded to approximately 2×10^{-11} mol cm^{-2} of the HPA, which corresponds to about one monolayer of close-packed anions on a smooth surface. The behavior remains essentially the same during the deposition of up to 20 monolayers. If the coating was not exposed to the solution of $[Os(bpy)_3]^{2+}$ or $[Ru(bpy)_3]^{2+}$ between exposures to the solution of $[P_2Mo_{18}O_{62}]^{6-}$, no increase in the peak magnitude occurred in the subsequent exposure to the solution of the HPA. Thus, while there is strong evidence that the electroactive cation is incorporated into the multilayer, the absence of an additional peak at approximately 0.600 V indicates that it is electro-inactive. However, the multilayers represent a true supramolecular assembly in which the components interact and influence one another's properties. A response from the $[Os(bpy)_3]^{3+/2+}$ couple does appear near 0.6 V in voltammograms recorded at the point in the deposition cycle after the electrode is exposed to the solution of $[Os(bpy)_3]^{2+}$ to incorporate this cation. For example, Figure 4.13 illustrates the continual increase in the responses from both the $[Os(bpy)_3]^{2+}$ couple and the couples centered on the HPA in the coating. The areas encompassed by the peaks for the $[Os(bpy)_3]^{2+}$ couple corresponded to ca. 2.5×10^{-11} mol cm^{-2} of $[Os(bpy)_3]^{2+}$ deposited in each successive layer which, by comparison with the corresponding areas of the peaks for the HPA, corresponds to less than three $[Os(bpy)_3]^{2+}$ cations for each $[P_2Mo_{18}O_{62}]^{6-}$ anion. In this sense, they consist of more than a simple precipitate or salt formed by reaction between the two oppositely charged ions. This approach also allows redox-active organic cations, such as the pyridinium dication, to be immobilized.

4.6.2 Ordered Protein Layers

The creation of ordered multilayer assemblies of proteins is important for applications ranging from sensors and reactors to models of natural systems. Layer-upon-layer approaches have the advantage that relatively ordered thick films can be created which can be characterized as the assembly is formed. This field is dominated by four distinct approaches, namely the building up of successive layers of an enzyme by reaction of the latter with bifunctional linkers [19], the use of a biotinylated enzyme linked in successive layers using avidin [20], the linking of successive layers using electrostatic immobilization [21], and the linking of successive layers using antibodies [22]. X-ray diffraction (XRD) studies of these layer-upon-layer assemblies reveal significant long-range order for those structures involving approximately four or more individual layers. In this sense, they are 'quasi-ordered'. A distinctive feature of these interfacial supramolecular assemblies is that the mobility of the species, e.g. a mediator for an oxidoreductase enzyme, can be controlled. For instance, the biocomponents can be electrostatically or covalently bound which will significantly change their abilities to diffuse.

The binding of a bi-functional bridge which is capable of immobilizing a protein or enzyme to an electrode can block the access to the electrode surface by solution-phase reactants, thus reducing the voltammetric current. Although somewhat outside of the scope of this work, it is useful to note that recent developments in producing nanostructure domains, e.g. photopatterned lines on the nanometer length scale, have been used to address this problem. For example, functionalized lines of biocomponents on electrodes, e.g. acid-terminated alkane thiols or biotin, allow selective chemistry to be performed at the interface while leaving substantial portions of the electrode unmodified at which species in the solution can react.

4.6.3 Surfactant-Based Multilayer Assemblies

The use of surfactant films which resemble lipid bilayers and multilayers represents an important approach to preparing electroactive films of protein molecules.

Rusling [23] has developed procedures for preparing surfactant films into which proteins are embedded. These films may consist of single or multilayer assemblies in which protein molecules are entrapped but exhibit significant diffusional motion. The surfactants are water-insoluble because of their long-chain alkyl components, e.g. didodecyldimethylammonium bromide (DDAB), dimyristoylphosphatidyl cholate (DMPC) and dihexadecyl phosphate (DHP). Another approach is to mix an aqueous vesicle dispersion of surfactant with protein solution, and then evaporate this onto the electrode. In common with solvent-cast films, this procedure gives multiple layers and the quantity immobilized can be conveniently monitored by using the QCM. The power of the approach is elegantly illustrated for myoglobin, a heme-containing oxygen binding protein that exhibits only slow electron transfer at bare electrodes. In contrast, when immobilized within a DDAB film at basal-plane graphite, Rusling [23] found that myoglobin gives reversible voltammetry. This approach has also been used with cytochrome P450, so allowing the redox transitions of the heme group to be directly detected. The electron density on the active site is influenced by the electrostatic interactions of the protein with the surfactant, e.g. the reduction potentials are different for DDAB- or DMPC-based films. The enzyme film catalyzes reduction of oxygen and reductive dechlorination of trichloroacetic acid. UV–visible, EPR and reflectance FTIR spectroscopic studies all confirm that the redox proteins retain their structure within films of this kind. For example, the C=O stretching band of the amide groups, which are sensitive to the protein conformation, do not alter significantly when myoglobin is transferred from aqueous solution to the DDAB film.

The physical structure of these DDAB surfactant films on highly ordered pyrolitic graphite (HOPG) has been probed by using atomic force microscopy. These studies reveal that cytochrome C rapidly adsorbs in a random manner to give a dense monolayer. In contrast, myoglobin adsorption onto clean HOPG is slow, but eventually results in chain-like structures about 60 nm long and 6 nm wide. The equilibrium voltammetric response exhibits a reversible redox couple with a potential close to that expected for myoglobin. Adsorption at a preformed DDAB film is significantly more random, suggesting that strong DDAB–protein interactions break up the protein–protein lateral interactions which drive the formation of a relatively ordered assembly at a bare electrode. The lack of order in the DDAB films is reflected in the voltammetry, which is less reversible than at the bare electrode.

Alternating layer-by-layer film formation is important since it may allow both the film composition and thickness to be controlled. In addition, depending on the film components it may be possible to block movement of the protein from one layer to the next. Stacked bilayers have been formed, starting from a layer of mercaptopropanesulfonic acid on gold or quartz, followed by successive layers of protein (e.g. myoglobin or cytochrome P450), polyanion–DNA, protein, etc. Significantly, the mobility of the redox proteins within these films is significantly diminished and the Faradaic current observed in voltammetry appears to be dominated by the proteins closest to the electrode surface.

For some enzymes, especially cytochrome C, redox-driven conformational changes are believed to be important in their roles as proton pumps. Therefore, it is important to develop immobilization methods that create stable structures and orient the active center of the enzyme away from the electrode surface, yet provide sufficient free volume to allow the enzyme to undergo a conformational change. The approach of Hawkridge and co-workers [24] to this challenging issue involves forming a partial monolayer of octadecylmercaptan (OM) on a metal surface. This layer templates the assembly of a bilayer using L-phosphatidylethanolamine or L-phosphatidylcholine. The formation of the bilayer is supported by QCM measurements. The membrane-mimetic electrode is prepared by entrapping an aliquot of solution containing detergent, deoxycholate-solubilized enzyme and amphiphile against the OM-modified electrode surface with a dialysis membrane. As the detergent dialyzes out, the enzyme becomes incorporated into the bilayer. Cyclic voltammetry reveals that the immobilized enzyme remains electrochemically active. An enhanced catalytic current is observed when cytochrome C is present in solution, thus suggesting that the binding site of the enzyme is oriented away from the electrode. Significantly, the charge that is passed during potential sweeps is far greater than that expected for a monolayer. This result is consistent with capacitive charging triggered by a redox-coupled conformational change which increases ion transport in the lipid bilayer.

4.7 Polymer Films

While from the deposition and synthetic point of view polymers are a very attractive option for the modification of surfaces, from the organizational point of view there are several drawbacks. Unlike, for example, for the case of self-assembled monolayers, the organization of polymer layers is much more limited. For instance, although multilayer assemblies can be prepared by alternating anionic and cationic polymer layers, substantial organization is only obtained perpendicular to the surface. A true three-dimensional molecular-scale organization cannot be obtained easily. However, there are many reasons to investigate the capacity of macromolecules as components in supramolecular assemblies. For example, studies of block copolymers of styrene and methylmethacrylate show that nanodomains with well-defined orientations can be obtained in these systems. Recent developments have also shown that the conductivity of polyalkylthiophene films depends strongly on the orientation of the polymer chains with respect to the solid substrate. Finally, one can compare the processes occurring in synthetic macromolecules with those observed for biological systems such as proteins. The function of proteins is to regulate reactions of reactive centers (co-enzymes). This regulation may be based on either kinetic or thermodynamic principles. Often, metal centers in enzymes are forced into particular coordination modes by the steric confirmation of the protein. In other cases, while the metal center in the co-enzyme might control the thermodynamics of a process (for example, by its redox potential or excited state energy), the protein controls the kinetics. The correct functioning of natural photosynthetic and metabolic systems relies on the communication of electrons and energy over distances much larger than the molecular scale. Redox-active polymer films are ideally suited to address these issues [25]. The properties of many of these electroactive polymers, e.g. their conductivity, charge distribution, shape, etc., can be changed in a controlled and reproducible way in response to environmental stimuli, e.g. a change in the nature of the contacting solution, an applied voltage, light intensity, or mechanical stress. A key issue in developing supramolecular redox systems of this kind is determining the exact mechanism of charge transport through the film. In common with proteins, the redox potential of the coatings is defined by the metal center or the nature of the conducting polymer, while the charge transport through the layer depends strongly on the nature of the polymer layer and its interaction with the contacting electrolyte. A better understanding of how the structural features of macromolecules affect processes such as electron transfer and the movement of ions and solvent is therefore of great importance for modeling biological processes.

Once the structure–function relationships are understood in these complexes (often heterogeneous, structures), the properties of polymers can be tailored for applications such as sensors, electrocatalysts, polymeric sonar materials, light-emitting polymers for flexible displays, energy conversion devices and the creation of biomimetic systems, e.g. the immobilization of enzymes onto electrode surfaces and their 'electrical' wiring, as well as the use of conducting polymers to emulate the analog processing of the retina.

Against this background, the discussion of thin polymer films in this and the following chapter will concentrate on the preparation and structural features of such films and on their structure–properties relationships. It will be shown that synthetic techniques can be used to modify structural features such as the molecular composition and that by such modifications the three-dimensional structures of films, the extend of organization, and their physical properties can be controlled quite satisfactorily. Some selected examples will be given of how information about structural features of thin, often amorphous, polymer films can be obtained by using techniques such as electrochemical measurements, neutron reflectivity and quartz crystal microbalance studies. Finally, the relationship between the structures of the layers and their properties will be illustrated in the case of conducting polymers and for the design of electrochemical sensors.

In the following section, the advantages and disadvantages of polymeric materials as modifying materials for solid substrates will be discussed. The discussion will range from deposition techniques to the discussion of synthetic procedures which can yield ordered materials.

4.7.1 Film Deposition Methods

One of the major attractions for the use of polymer coatings is the wide range of deposition techniques that can be employed to obtain relatively well-defined thin films with thicknesses of the order of microns and less. The methods that have been used include the following:

- Solvent evaporation
- Dip coating
- Spin coating
- Langmuir–Blodgett techniques
- Electrostatic self-assembly

- Electrochemical polymerization and deposition

- Chemical bonding and grafting

The first three techniques have been used extensively. However, although their application is straightforward, it is often difficult to achieve homogeneous coatings with a good reproducibility in layer thickness and three-dimensional structure. In the solvent evaporation technique, a known amount of solvent is placed on the surface and the solvent is then slowly evaporated under controlled conditions. In this manner, the amount of material deposited can be controlled. However, factors such as the composition of the coating solution (concentration of polymer and solvent), drying time, environment and temperature need to be controlled carefully so as to achieve a reproducible structure and strongly adhering films. Film adhesion can be improved by *in situ* cross-linking and by the use of high-molecular-weight materials. For the spin coating technique, the rotation speed used is important, while for dip coating the rate of withdrawal of the substrate affects the properties of the layers obtained.

In the Langmuir–Blodgett technique, amphiphilic monolayers, formed at a liquid–air interface, are transferred to a solid substrate by horizontal or vertical transfer. The thickness of such monolayers is of the order of a few nanometers, depending on the materials being used. With this method, multilayer structures can also be produced, either by repeated deposition of the same layer or by the deposition of alternate layers. In this manner, multilayers containing several hundred individual components can be obtained. In general, the thermal and mechanical stabilities of such layers are, however, limited.

Another way to produce polymeric multilayer systems is through the use of electrostatic self-assembly. With this technique, alternating positively and negatively charged layers can be deposited. This method leads to well-ordered multilayer systems, but when only polymeric components are used a substantial amount of interpenetration occurs. This problem can be reduced by the use of inorganic materials.

Electrochemical methods have also been widely used to produce surface-immobilized layers. A prerequisite for the application of these techniques is the presence of an electroactive group in the precursor material. Materials that have been widely studied include polyanilines, polypyrroles and polythiophenes and the study of conducting polymers of this type has attracted much attention since the original work by Shirakawa, Heeger and MacDiarmid in the 1970s [26,27]. With these compounds, thin conducting layers can be obtained for which the structure and layer thickness can, to a certain extent, be controlled by the electrochemical methods used to produce them. Although again the deposition methods used to produce these conducting polymer layers are very accessible, the variation on the electrochemical methods used has resulted in the production of very many different polyanilines, polythiophenes and polypyrroles. Electrodeposition has also been used to produce thin layers of compounds containing electroactive substituents such as vinyl and amino groupings.

Self-assembly and grafting techniques have been used to attach both polymeric and monomeric materials to surfaces. The result of this approach is well-ordered monolayers in the case of monomeric precursors (See Section 4.3 above). Polymers with active endgroups have also been immobilized by using this technique. Interactions that have been studied include thiols on gold, pyridines on gold, carboxy groups on metal oxide surfaces and silanes on silicon. Spontaneously absorbed polymeric monolayers obtained in such a manner can also be used as templates for the growth of multilayers systems.

An attractive route to creating stable multilayers is electrostatic self-assembly using polyelectrolytes. Since the interaction between the polymer layer and the substrate, or between alternating layers, is based on electrostatic interactions, the chemical nature of the electrostatic components can be varied systematically. As a result, there is an extensive literature based on, among others, polymer–polymer, polymer–organic, polymer–inorganic, and polymer–biomolecular assemblies.

As an illustration of this approach, the formation of two types of assemblies will be discussed here. The first example is the assembly of multilayers based on alternating polyelectrolyte films containing poly(styrene sulfonate) (PSS) and poly(allylamine hydrochloride) (PAH) (see Figure 4.14) on flat silicon surfaces [28]. In these studies, the composition of the repeating units was varied systematically, as well as the number of layers applied. In order to obtain a surface with a homogeneous surface charge, the silicon surface was precoated with 4-aminobutyldimethylmethoxysilane. On this modified surface, alternating anionic PSS and cationic PAH layers were deposited. In this manner, assemblies of up to 80 layers were constructed. In some typical assemblies, the repeating units are [PSSd$_7$–PAH] (polymer I), [PSSh$_7$–PAH–PSSd$_7$–PAH] (polymer 2), and [(PSSh$_7$–PAH)$_2$–PSSd$_7$–PAH] (polymer 3). In these assemblies, deuterated PSS, i.e. PSSd$_7$, was used to determine the layer thickness by neutron reflectivity (see Section 4.8 below). In such coatings, the number of repeat units was 10 for polymer 1 and 8 for polymers 2 and 3.

As outlined below in Section 4.8, a disadvantage of the use of organic polyelectrolytes as components for the self-assembly of multilayer films is a substantial interdigitation of subsequent layers. This direct interaction between different layers will not allow for the application of such multilayers in devices based on efficient charge separation. In order to address this issue, Mallouk and co-workers have adapted the electrostatic self-assembly methods by introducing inorganic components [29]. With these mixed inorganic–organic multilayers, effective charge-separation is obtained, as shown in the next two examples. The inorganic polyelectrolytes used were α-Zr(HPO$_4$)$_2$.H$_2$O (α-ZrP) or KTiNbO$_5$. These materials

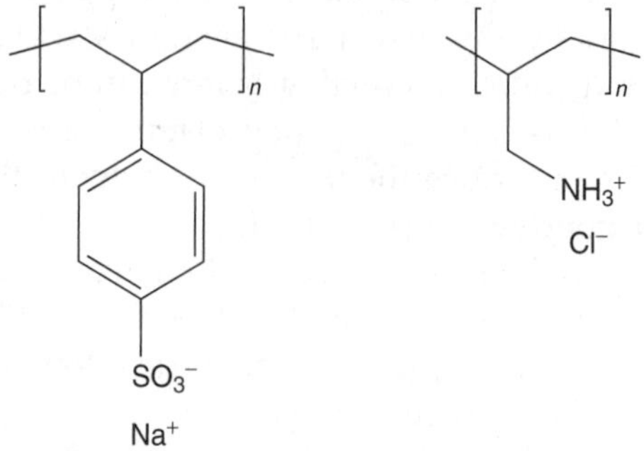

Figure 4.14 Structures of (a) the sodium salt of poly(styrene sulfonate), and (b) poly(allylamine hydrochloride)

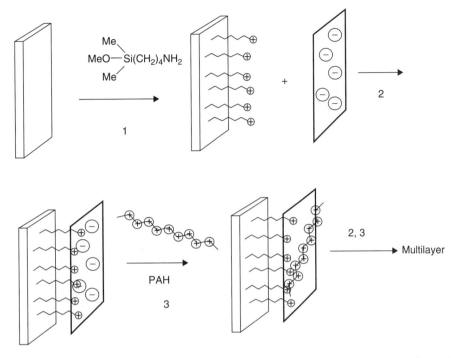

Figure 4.15 Schematic representation of the sequential steps taken for the formation of multilayers based on electrostatic self-assembly using cationic polymers and anionic α-ZrP sheets (see text for further details). Reprinted from *Coord. Chem. Rev.*, **185–186**, D.M. Kaschak, S.A. Johnson, C.C. Waraksa, J. Pogue and T.E. Mallouk, 'Artificial photosynthesis in lamellar assemblies of metal poly(pyridyl) complexes and metalloporphyrins', 403–416, Copyright (1999), with permission from Elsevier Science

are polyanionic sheets of well-defined layer thickness and it was anticipated that such materials will prevent interpenetration of alternating anion–cation layers. Organic–inorganic multilayers were grown by sequential adsorption reactions, as outlined in Figure 4.15. The substrate (glass or silica) is primed with an aminoalkyl-silane derivative, which at neutral pH contains protonated amino groups. Upon dipping this modified surface in a solution containing α-ZrP, anionic sheets of the polyanion are adsorbed as a monolayer. In turn, this anionic surface can absorb a monolayer of polycations, such as poly(allylaminehydrochloride) (PAH), as in step 2 in Figure 4.15. The first two steps can be repeated many times and can be used to produce a well-defined multilayer assembly. The photophysical properties of such assemblies will be discussed further in Chapter 5.

4.7.2 Synthetic Procedures for the Preparation of Redox-Active Polymers

Redox polymers are electroactive polymers for which the redox centers are localized on pendent, covalently attached redox centers. The electrochemical properties of such materials depend not only on both the loading and the nature of the redox-active center but also on the type of polymer backbone. The electroactive groups are typically metal complexes, which are covalently attached to a polymer backbone. Such materials can be obtained by two different methods, i.e. either by the electropolymerization of mononuclear precursors at an electrode surface, or by the polymerization of the appropriate polymer precursor, followed by the attachment of redox active groupings. The latter method has the advantage that the composition of the polymer backbone and its molecular weight can be controlled and that the material obtained is soluble if cross-linking is prevented. This solubility enables a full characterization of the redox materials to be obtained by using spectroscopic techniques. Furthermore, the concentration of the electroactive centers can be controlled effectively by manipulation of the metal-center-to-polymer reaction ratio. The polymers obtained by this method can be used to coat a variety of surfaces by using dip coating or solvent evaporation techniques. The disadvantage of electropolymerization is that it can only be carried out on electroactive surfaces and needs the presence of specific polymerizable monomeric precursors. The structures of thin layers obtained by these two methods are often quite different. Electropolymerization leads to heavily cross-linked films, while polymer films produced by spin coating of preformed soluble polymers have open flexible structures which interact strongly with contacting electrolytes and allow for the efficient penetration of ions and solvent. As a result, the structural features of thin films based on redox polymers produced by the chemical route based on pre-formed polymers can be manipulated very effectively by changes in the nature of a contacting electrolyte. Cross-linked films produced by electropolymerization are far less flexible.

Chemical synthesis

Polymer backbones of homo- and copolymers based on poly(4-vinylpyridine) and poly(N-vinylimidazole) containing redox-active centers such as ruthenium- and osmium-based polypyridyl complexes have shown some very interesting electrochemical and photochemical properties. The synthetic approach taken for the preparation of these redox-active materials is outlined in Figure 4.16 for poly(4-vinylpyridine) (PVP) [30]. In this approach, PVP is produced from 4-vinylpyridine by radical polymerization. The polymer obtained is purified and fractionated to obtain a narrow molecular weight distribution. In general, molecular weights of the order of 100 000 are most suitable since the thin layers produced, e.g. by spin coating techniques, from shorter-chain-length molecules tend to be less physically stable.

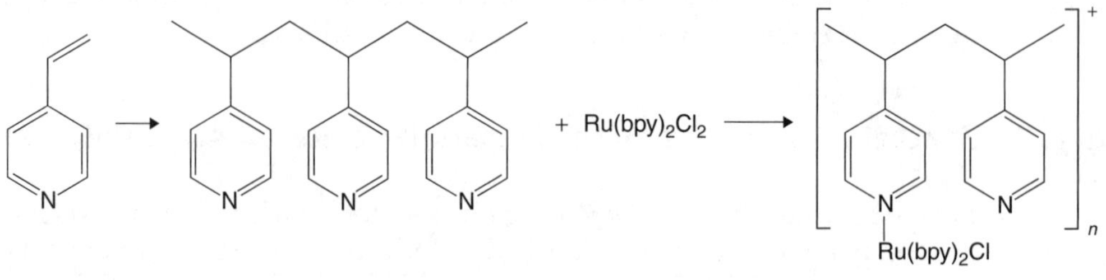

Figure 4.16 Synthetic approach to the preparation of redox polymers based on preformed polymers

The polymeric materials obtained are subsequently reacted with the redox-active metal complex (see Figure 4.16). This reaction takes place in two steps, as follows:

$$[Ru(bpy)_2Cl_2] + S \longrightarrow [Ru(bpy)_2Cl\,S]^+ \tag{4.15}$$

$$[Ru(bpy)_2Cl\,S]^+ + nPVP \longrightarrow [Ru(bpy)_2(PVP)_nCl]^+ + S \tag{4.16}$$

In the first step, the precursor, typically a ruthenium or osmium bis(2,2'-bipyridyl) (bpy) complex, reacts with solvent (S) to produce a solvated complex. When solvents such as dry methanol and ethanol are used, only one chloride is exchanged and the species $[Ru(bpy)_2(PVP)_nCl]^+$ is obtained as the sole product. The nature of the coordination sphere around the metal center can be determined by UV–visible (UV/Vis) spectroscopy (λ_{max}, 496 nm) and by its redox potential, (about 0.65 V (vs. SCE), depending on the electrolyte being used). By a systematic variation of the ratio of monomer units to redox-active centers, the loading of the polymer backbone (n) can be varied systematically. (Here, n stands for the number of monomer units in the polymer per redox-active center, e.g. in a PVP-based, $n = 10$ polymer, there are 10 pyridine units for every redox center.

With this synthetic approach, the coordination mode of the pendent group can be controlled by a variation of the reaction conditions. By the addition of water to the reaction mixture, the second chloride can be removed and a different coordination around the metal center is obtained. For example, the material obtained in Figure 4.16 can be modified to produce an aquo species, as follows:

$$[Ru(bpy)_2(PVP)_{10}Cl]^+ + H_2O \longrightarrow [Ru(bpy)_2(PVP)_{10}H_2O]^{2+} + Cl^- \tag{4.17}$$

This species can be identified by its UV/Vis absorption spectrum, which in ethanol shows a λ_{max} of 460 nm, and its very specific electrochemical behavior. Detailed electrochemical studies of this redox polymer and of mononuclear analogues have shown that two different electrochemical processes occur, both of which are pH-dependent [31]. In the first step, the Ru(II) center is oxidized to Ru(III). Below the pK_a of the Ru^{III}–H$_2$O species of 0.85, the Ru$^{II/III}$ couple may be described by the following equation:

$$[Ru^{II}(bpy)_2(PVP)_{10}H_2O]^{2+} \longrightarrow [Ru^{III}(bpy)_2(PVP)_{10}H_2O]^{3+} + e^- \tag{4.18}$$

while for a pK_a > 0.85, the Ru$^{II/III}$ couple is represented by the following:

$$[Ru^{II}(bpy)_2(PVP)_{10}H_2O]^{2+} \longrightarrow [Ru^{III}(bpy)_2(PVP)_{10}OH]^{2+} + H^+ + e^- \tag{4.19}$$

At higher potentials, a further redox process is observed that can be described as a coupled electrochemical/chemical process involving the disproportionation of the initial redox product; this Ru$^{III/IV}$ redox process is best described by the following reaction sequence:

$$[Ru^{II}(bpy)_2(PVP)_{10}H_2O]^{2+} + [Ru^{IV}{=}O(bpy)_2(PVP)_{10}]^{2+} + 2H^+$$
$$\longrightarrow 2[Ru^{III}(bpy)_2(PVP)_{10}H_2O]^{3+} \tag{4.20}$$

$$[Ru^{II}(bpy)_2(PVP)_{10}H_2O]^{2+} + [Ru^{IV}{=}O(bpy)_2(PVP)_{10}]^{2+}$$
$$\longrightarrow 2[Ru^{III}(bipy)_2(PVP)_{10}OH]^{2+} \tag{4.21}$$

where Equations (4.20) and (4.21) are applicable at pH < pK_a and pH > pK_a of the Ru(III) aquo complex, respectively. These reactions have been studied in solution for [Ru(bpy)$_2$(py)(H$_2$O)]$^{3+}$. In solution, the equilibrium constant, $K(= k_f/k_r)$, for the process shown in Equation (4.21) is 72 and $k_f = 2.1 \times 10^5$ mol^{-1} dm^3 s^{-1}. For the polymer, the equilibrium lies strongly to the right and Ru$^{II/IV}$ coproportionation is rapid. However, the disproportionation reaction, which is important in the Ru$^{III/IV}$ oxidation process, is relatively slow, as indicated by the stability of the aquo species. These observations show that the physical properties of mononuclear model compounds can be transferred to their polymeric analogues and that the redox reactions and chemical behaviors of the metallopolymers produced can be predicted and tuned. In agreement with the behavior observed for mononuclear model compounds, the aquo species can further react to produce the disubstituted metal center [Ru(bpy)$_2$(PVP)$_{10}$]$^{2+}$, in which two monomer units are coordinated.

These results indicate that the properties of the redox polymers, such as redox potentials and spectroscopic properties, can be varied systematically and, more importantly, can be predicted from those observed for mononuclear model compounds. As an example of the transfer of photochemical properties from monomeric analogues to the corresponding polymers, the photochemical behavior of the redox polymer [Ru(bpy)$_2$(PVP)$_5$Cl]Cl will be considered. This polymer contains one metal center for every five-monomer units. Photolysis of a thin layer of this material on a glassy carbon surface leads to a change in the redox potential of the material from about 650 to 850 mV (See Figure 4.17) [32]. The voltammetric process affected is associated with a metal-center-based Ru(II/III) redox process. By analogy to the behavior observed for the mononuclear species [Ru(bpy)$_2$(py)Cl]$^+$ (py = pyridine),

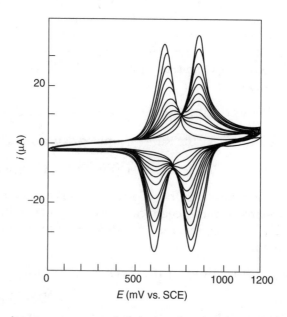

Figure 4.17 Cyclic voltammograms recorded during the photolysis of [Ru(bpy)$_2$(PVP)$_5$Cl]$^+$ in 1.0 M HClO$_4$, at a scan rate of 100 mV s^{-1}. Reprinted with permission from O. Haas, M. Kriens and J.G. Vos, *J. Am. Chem. Soc.*, **103**, 1318 (1981). Copyright (1981) American Chemical Society

the change in the cyclic voltammograms can be interpreted as a ligand exchange reaction, as shown in the following equation:

$$[Ru(bpy)_2(PVP)_5Cl]^+ + H_2O \xrightarrow{h\nu} [Ru(bpy)_2(PVP)_5H_2O]^{2+} + Cl^- \qquad (4.22)$$

As outlined above, the electrochemical properties of this redox species are strongly pH-dependent and this behavior can be used to illustrate the supramolecular nature of the interaction between the polymer backbone and the pendent redox center. The cyclic voltammetry data shown in Figure 4.17 are obtained at pH = 0, where the polymer has an open structure and the free pyridine units are protonated $(pK_a(PVP) = 3.3)$. The cyclic voltammograms obtained for the same experiment carried out at pH 5.7 are shown in Figure 4.18. At this pH, the polymer backbone is not protonated and upon aquation of the metal center the layer becomes redox-inactive, since protons are involved in this redox process. This interaction between the redox center and the polymer backbone is typical for these types of materials. Such an interaction is of fundamental importance for the electrochemical behavior of these layers and highlights the supramolecular principles which control the chemistry of thin films of these redox-active polymers. Finally, it is important to note that the photophysical properties of polymer films are very similar to those observed in solution. Since the layer thickness is much more than that of a monolayer, deactivation by the solid substrate is not observed.

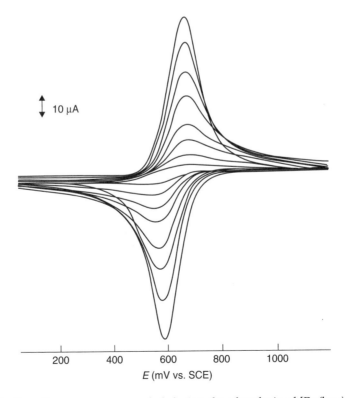

Figure 4.18 Cyclic voltammograms recorded during the photolysis of $[Ru(bpy)_2(PVP)_5Cl]^+$ in 1.0 M ClO_4^-: pH, 5.7; scan rate, 100 mV s^{-1}

The analogous osmium polymers have also been studied in great detail. The synthetic procedures required for these metallopolymers are the same as those described above for ruthenium; however, the reaction times are longer. The similarity between the analogous mononuclear and polymeric species is further illustrated by the fact that the corresponding osmium polymers have considerably lower redox potentials and are also photostable, as expected on the basis of the behavior observed for osmium polypyridyl complexes.

Although the above discussion is centered on the synthesis of polymeric osmium and ruthenium complexes, the methods employed are also very successful in the preparation of mononuclear complexes. In this context, the preparation of ruthenium or osmium complexes which are suitable for the formation of self-assembled monolayers (see Section 4.3 above) can be prepared by using the same approach. Starting from the precursor $[M(bpy)_2Cl_2]$, one chloride atom can be replaced to yield complexes of the type $[M(bpy)_2Cl\,L]^+$, where L is the surface active ligand. In the presence of water, species of the type $[M(bpy)_2(L)_2]^{2+}$ are obtained.

In conclusion, with this synthetic approach a wide range of redox-active polymers can be obtained where factors such as the nature of the polymer backbone (e.g use of copolymers), the loading of the metal center, and the nature and coordination sphere of the metal center can be controlled systematically. This has been used to great effect in the design of electrochemically driven chemical and biochemical sensors.

Electropolymerization

An alternative approach for the preparation of thin films of electroactive polymers is electropolymerization. This method requires the use of metal complexes containing substituents that can be polymerized by reductive or oxidative electropolymerization. Some typical structures are shown in Figure 4.19. Upon electropolymerization of solutions containing such monomeric precursors, thin layers of polymerized materials attached to electrode surfaces are obtained. The best adhering films are obtained with metal compounds containing more than one polymerizable unit. This is consistent with considerable cross-linking of the polymeric film. As a result, the coatings obtained tend to be rigid, with little flexibility of the polymer chains and little or no swelling. The films produced show well-defined redox potentials which are normally very similar to those observed for the monomeric precursors. The advantage of the electrochemical technique over the chemical synthesis method is that by controlling the current the layer thickness can also be controlled very effectively. Since the films obtained are strongly cross-linked, access of ions and solvent is often more restricted than for the redox polymers obtained by chemical synthesis. This limits their applications, for example, as sensors, since substrates are not able to penetrate the film and as a result mediation can only occur at the film–electrolyte interface. Neither approach yields materials with significant long-range order and therefore amorphous materials are usually obtained.

4.7.3 Synthetic Methods for the Preparation of Conducting Polymers

Since the pioneering work of MacDiarmid and co-workers [27], there has been an ever-growing interest in the chemistry of conducting polymers. The interest in these

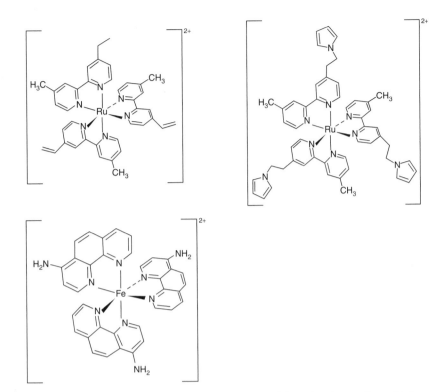

Figure 4.19   Examples of some typical electropolymerizable ruthenium polypyridyl complexes

materials stems from their potentially high conductivities, which are based on the presence of a conjugated polymer backbone. Because of the instability of polyacetylene, materials such as polypyrrole, polythiophene and polyaniline have attracted much attention for device applications, such as electrochromic devices, batteries, light emitting diodes, sensors and also as anti-corrosion and anti-static coatings. The similarity between the structures of polyacetylene and such polyheterocycles is shown in Figure 4.20. The investigation of conducting polymers was greatly facilitated by the development of electrochemical polymerization methods. The oxidative polymerization of polypyrrole was later extended to other heterocyclic compounds, such as thiophene, furan, carbazole pyrene and aniline. A wide variety of both chemical and electrochemical methods has been developed for their syntheses.

However, it is important to note that different synthetic techniques produce materials with widely varying electrochemical, conductive and morphological

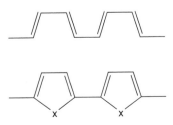

Figure 4.20   Structural similarities between polyacetylene and polyheterocycles

properties. For example, the structure and intrinsic properties of polypyrrole obtained by oxidative electropolymerization depend strongly on factors such as the solvent, electrolyte, the current applied, and the electrochemical technique used. Polypyrrole obtained by chemical methods shows different properties again. Therefore, one of the main problems with the study of these materials is the morphological complexity of the polymer films that are obtained. Although calculations carried out on such systems often produce long-range ordered structures, very few well-defined supramolecular structures have been reported and most of the materials studied can be classified as amorphous. This lack in control of the structural features of such materials has greatly hindered the study of their fundamental properties. Since in this section the main interest is in the supramolecular aspects of thin films on surfaces, our discussion here of the synthetic aspects of conducting polymers will focus on conducting polymers for which the structural features are well understood. The discussion will concentrate on polyalkylthiophenes for which elegant synthetic methods have been developed that produce soluble, regioselective polymeric materials which can be used to produce thin films with well-defined long-range ordering and show clear structure–function relationships.

Electrochemical synthesis of conducting polymers

In oxidative electropolymerization, monomers such as pyrrole, thiophene, alkylthiophenes or aniline are dissolved in an appropriate solvent containing an electrolyte that can act as a source for the anions needed to neutralize the cations formed during the oxidation process [33,34]. The nature of this electrolyte is of great importance to the structural features obtained for the electropolymerized layer since the dopants become an intrinsic part of the polymer layer structure. A general outline of a mechanism describing the electropolymerization process is shown in Figure 4.21.

The first step in this mechanism consists of the oxidation of the monomer to produce a radical cation. In the following chemical step, two of these cations react to form a dimer, via a dihydrodimer cation intermediate. Since the dimer formed

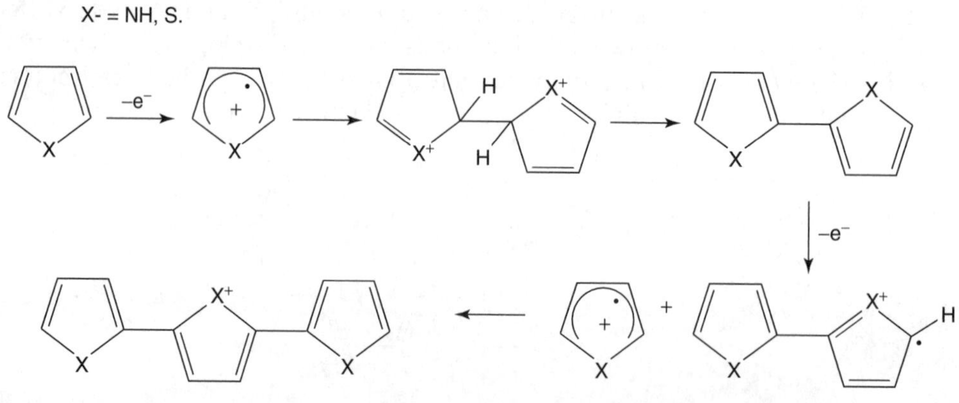

Figure 4.21 Electropolymerization pathways for the formation of polyheterocycles: X = NH or S

can be oxidized at a lower potential then the monomer, a chain-growth mechanism is established. The propagation of this electrochemical/chemical polymerization process eventually leads to the precipitation of polymer chains on the electrode surface, governed by the solubility of the material being formed. Because of its experimental ease, the electropolymerization route has become very popular. One of the advantages of the electropolymerization method is that the layer thickness of electrode coatings can be controlled by the amount of current passed.

As already pointed out above, electropolymerization has many variables which are difficult to control, and thus the structural parameters of the materials obtained tend to be variable. For example, depending on the electrochemical procedures used for the preparation of polythiophene, the conductivity can vary from 0.1 to 1000 S cm^{-1}. This may be due to chain defects, the orientation of chains and the molecular weight; the shorter the chain length, then the lower the conductivity is expected to be, since the conjugation is broken.

Chemical synthesis of polyalkylthiophenes

Chemical preparation methods for the synthesis of conducting polymers have been widely used [35]. It has become clear that to fully exploit the potential of the conducting polymer, better-defined soluble materials with a clear correlation between structure and properties need to be prepared. In this section, monosubstituted alkylthiophenes will be discussed as an example of how, via the development of well-documented systematic methods, ordered polymer layers can be obtained with improved conductive properties.

One of the reasons that polyalkylthiophenes have been investigated is the limited solubilities of unsubstituted thiophenes. The synthetic methods used for the preparation of both types of compounds are very similar. For example, both the parent thiophene and alkylthiophenes can be polymerized by reaction with FeCl$_3$ in chloroform. Other synthetic approaches involve the reaction of dihalogen precursors in the presence of magnesium, zinc or nickel. Some examples of these approaches are shown in Figure 4.22. In all of these methods, the monomers are linked through the C2 and C5 carbon atoms rather than via 2,4 linkages.

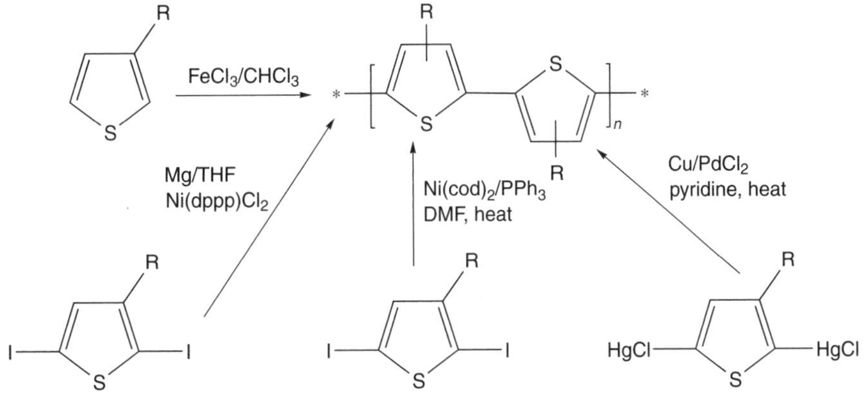

Figure 4.22 Examples of various chemical polymerization methods for preparing polyalkylthiophenes

It is important to point out that although monosubstitution of alkylthiophenes promises an increased solubility, the problems of obtaining a well-defined polymer structure are increased. Since in monosubstituted thiophenes, the C2 and C5 positions are no longer equivalent, three different types of coupling can be expected, i.e. the 2,5′ or head-to-tail (HT) coupling, the 2,2′ or head-to-head (HH) coupling, and finally the 5.5′ or tail-to-tail (TT) coupling. These three coupling possibilities will lead to the formation of four different types of trimeric regioisomers (see Figure 4.23).

The standard methods discussed above yield materials with mainly HT coupling, as expected from steric considerations. Materials with up to 70 % of HT coupling can be obtained by using the $FeCl_3$/THF method (see Figure 4.23). Electrochemical methods yield a material with about 62 % HT coupling. These variations in composition also affect the conjugation length, which itself controls the extent of delocalization of the polymers obtained. UV/Vis spectroscopy indicates that the longest conjugation lengths are obtained for those materials with the highest HT contents. These variations in the nature of the polymer backbone are also expected to affect the conductivities of the materials.

There is therefore a considerable interest in the design of synthetic methods which will produce regioregular polymers. One approach taken has been the polymerization of HH dimers. This leads to well defined HH–TT polymers (for the structure, see Figure 4.23). This improved regioselectivity does, however, not improve the conjugation length of the material obtained. Comparison of the UV/Vis spectrum of the polymer produced by chemical polymerization of 3,3′-dihexyl-2,2′-dithiophene with that of the polymer obtained by chemical polymerization, by the same route, from 3-hexylthiophene, shows that for the latter material the λ_{max} of the lowest energy $\pi-\pi^*$ transition at 508 nm is considerable lower than observed for the former (λ_{max} of 398 nm), thus indicating a longer conjugation length in the material directly obtained from 3-hexylthiophene. This latter material is irregular with about

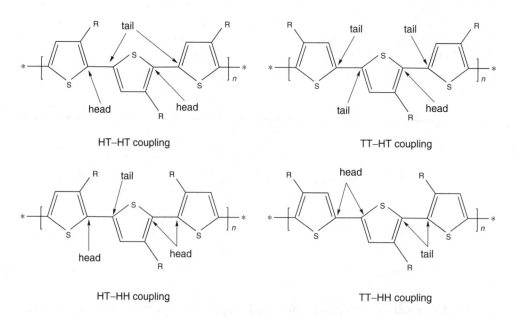

Figure 4.23 Illustrations of different coupling possibilities for alkylthiophenes

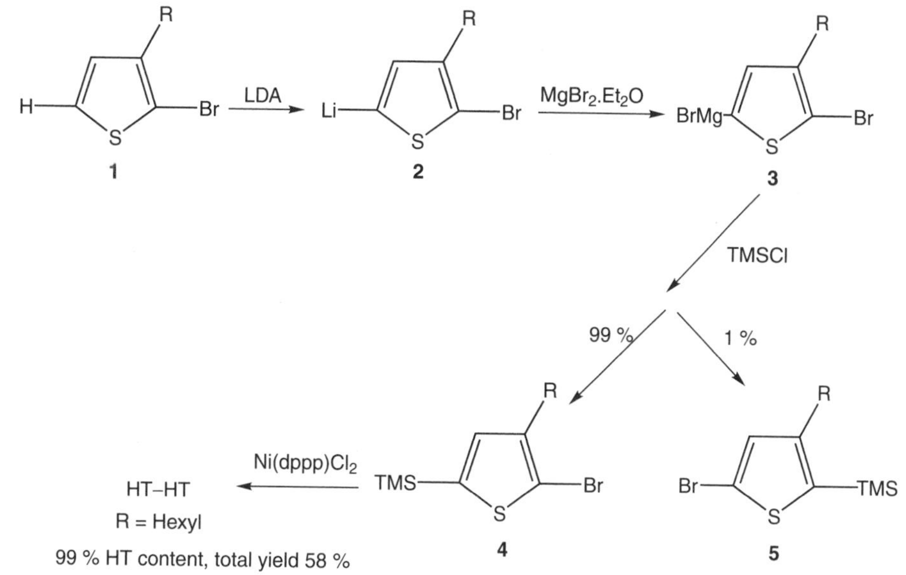

Figure 4.24 Illustration of the regioselective synthesis of polyalkylthiophenes

80 % HT couplings. The conductivity of the irregular material is about three times better at 15 S cm^{-1}, thus indicating that regioregularity does not necessarily improve the conjugation length and conductivity.

The result obtained from this and other studies suggests that the nature of the coupling may play an important factor. Calculations have show that HT coupling allows for *trans* coplanar orientation of the thiophene rings and this would be expected to increase both the conjugation length and conductivity substantially. The same studies indicate that HH coupling leads to twisted conformations of the thiophene rings; although the rings are still *trans*, they are no longer coplanar but twisted by as much as 40° from planarity. HH bonds are therefore expected to reduce the conjugation in the system. Several synthetic methods have been developed for obtaining regioregular HT polymers. One successful approach is shown in Figure 4.24. The important step in this reaction scheme is the selective metallation of compound **1** with lithium diisopropylamide (LDA) to yield **2**. The latter is stable at −78 °C and does not undergo halogen exchange. Further reactions occur without any 'scrambling', so that the HT–HT dimer is obtained of high purity. With this method, polymers are obtained which contain 98–100 % HT couplings for a range of substituents R such as *n*-dodecyl, *n*-octyl, *n*-hexyl and *n*-butyl. UV/Vis spectroscopy shows that the conjugation lengths of the regioregular materials are longer than those observed for irregular non-HT-specific materials. In a later section, the importance of the HT aspect of these polymers, and also of the effects of their orientations on the conductivities of thin films cast from them, will be discussed further (see Section 4.8).

Synthetic and structural aspects of block copolymers

Block copolymers consist of two or more chemically distinct units. A particular well-studied range of materials are rod–coil block copolymers, which consist of a rigid (rod) part and a more flexible coil unit. The rigid components can either be incorporated in the polymer backbone or as side chains (see Figure 4.25). These materials have been shown to be promising systems for the formation of nano-sized self-assembled structures with well-defined morphologies. The organizational properties of such materials are dependent on a careful choice of the coil-to-rod ratio.

The potential of block copolymers in forming supramolecular structures on surfaces has been studied in great detail by Stupp and co-workers [36]. Their studies have concentrated on the supramolecular properties of diblock and triblock rod–coil polymers which consist of one or two components that are conformationally flexible and one component with a more rod-like character at one of the ends of the polymer chain ('c' in Figure 4.25). This latter component can be crystallized and is expected to yield well-defined structural arrangements. The 'Stupp-type' copolymers are based on diblock styrene-*b*-isoprene coils combined with various rod-like components. The synthesis of such materials is carried out in two stages. In the first step, the diblock coil part is produced, as shown in Figure 4.26. The

(a)   (b)

(c)

Figure 4.25   Schematic representations of rod–coil block copolymers

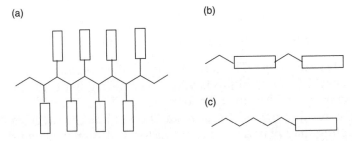

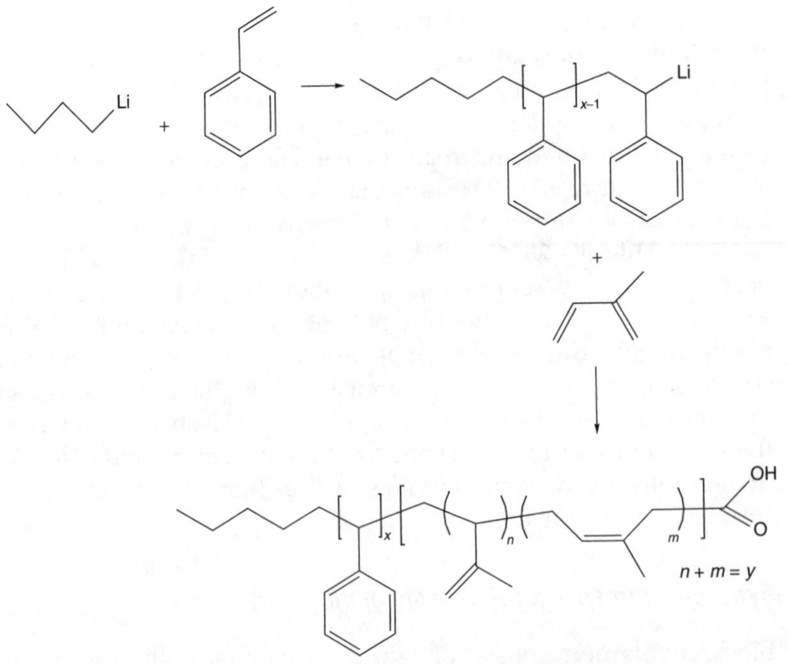

Figure 4.26   Illustration of the synthetic approach to producing diblock copolymers

approach taken is based on *living* anionic polymerization. With this polymerization method, the average chain lengths of the molecules obtained can be controlled, which itself allows for a control of the composition of the polymer backbone (ratio of x and y shown in Figure 4.26) through variation of monomer and initiator ratios. For the synthesis of such polymers, the polarity of the solvent employed can be used to manipulate both reaction rates and the nature of the isoprene reaction products. With this approach, the polydispersity index for the polymers obtained is between 1.04 and 1.12, while the molecular weight obtained for a polymerization reaction of nine equivalents of styrene and isoprene is typically 1478. After the purification of the diblock materials, the latter are further reacted to attach the rod component consisting of phenylene vinylene (PV) groupings. The approach by which rod–coil materials can be obtained which incorporate dimeric or pentameric PV rods is outlined in Figure 4.27. In this way, a range of materials can be prepared in which the composition of the polymer backbone can be systematically varied. Not only can the ratio of styrene to isoprene be controlled, and thus the composition and size of the coil part of the molecule, but also the length of the rod component. Characterization of the polymeric materials obtained shows that the styrene block is atactic, containing a random sequence of meso and racemic dyads, while the isoprene block contains mostly 1,4 and 3,4 additions (n and m) with possibly a small amount of 1,2 addition products. The rods of these triblock structures are dimers or pentamers (Figure 4.28) of phenylene vinylene. Using this approach, it is possible to design the length and composition of the coil and rod segments in a controlled manner. The effects of their composition on the supramolecular structure of thin films of these materials will be discussed further in the next section.

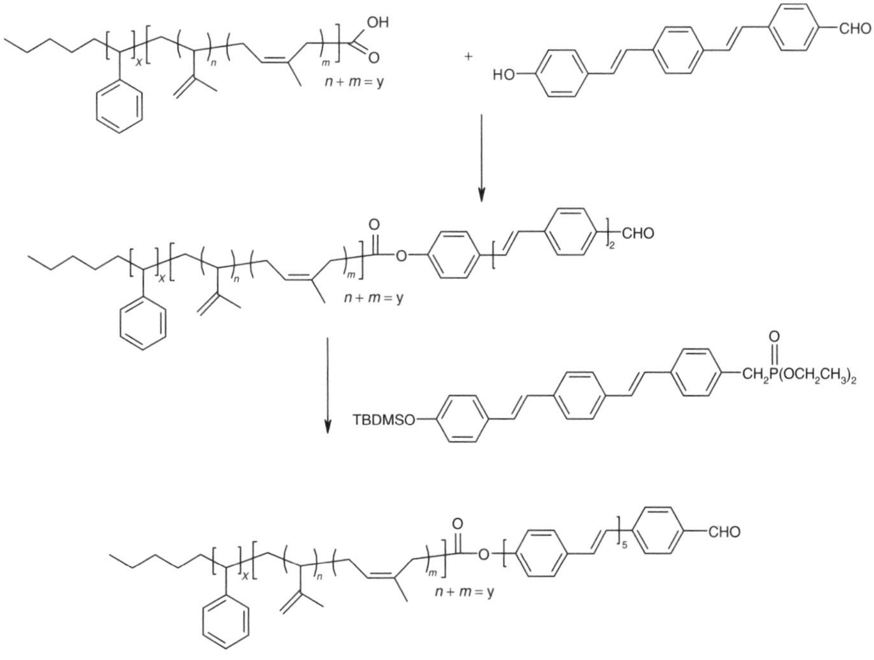

Figure 4.27 Illustration of the synthetic approach to producing rod–coil copolymers

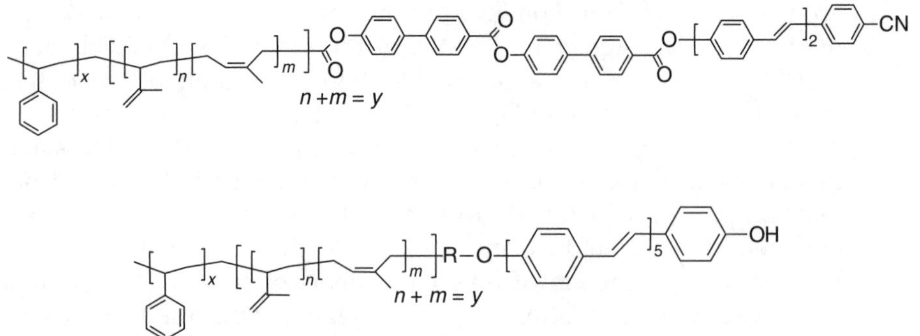

Figure 4.28 and associated chemical structures:

$n+m=y$

$n+m=y$

R = CO, x = 9, y = 9

Figure 4.28 Two examples of triblock copolymers based on oligostyrene, oligoisoprene and phenylene vinylene oligomers

4.8 Structural Features and Structure–Property Relationships of Thin Polymer Films

As with all supramolecular structures, one of the most important issues is whether a direct relationship between the structure of a material and its function or properties can be established. In the following, some examples of polymer systems which show such a correlation will be discussed. The materials addressed will include block copolymers, polyalkylthiophenes and a multilayer system based on the self-assembly of polyelectrolytes. Detailed studies on the electrochemical properties of redox-active polymers, based on poly(vinyl pyridine) modified with pendent osmium polypyridyl moieties, have shown that electrochemical, neutron reflectivity and electrochemical quartz crystal microbalance measurements can yield detailed information about the structural aspects of thin layers of these materials.

4.8.1 Structural Assessment of Redox Polymers using Neutron Reflectivity

An important emerging technique used to investigate polymer structure is neutron reflectivity. The fact that this technique can be applied *in situ* is important, since as already shown the structures of polymeric systems are greatly affected by their interactions with contacting electrolytes and solvents and that the value of the study of dry polymer films is limited. The layers used in these experiments were prepared by immobilization of the osmium polymer by using the spin coating technique on the gold working electrode. The basic principle of neutron reflectivity is to irradiate a sample with a collimated neutron beam at well-defined angles and to measure the reflected intensity, R, as a function of the momentum transfer. As discussed earlier in Chapter 3, in this manner, a scattering length density profile can be obtained which is dependent on the composition of the interface, since the reflectivity profile is representative of the composition of the interface. The scattering length profiles obtained for electrodes modified with the osmium polymer in perchlorate and in *p*-toluene sulfonic acid at different pH values are given in Figure 4.29 [37]. Clearly these profiles depend strongly on both the nature of the electrolyte and on the pH

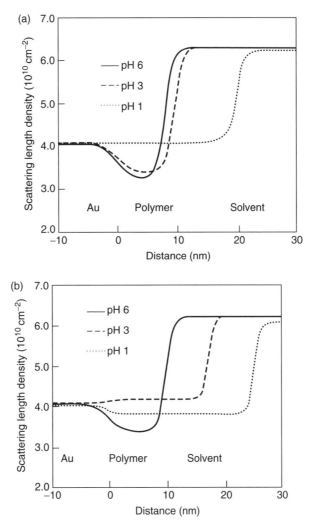

Figure 4.29 Scattering length density profiles for an [Os(bpy)$_2$(PVP)$_{10}$Cl]Cl film in (a) perchloric acid, and (b) *p*-toluene sulfonic acid at different pH levels. From R.W. Wilson, R. Cubitt, A. Glidle, A.R. Hillman, P.M. Saville and J.G. Vos, 'A neutron reflectivity study of [Os(bpy)$_2$(PVP)$_{10}$Cl]$^+$ polymer film modified electrodes: effect of pH and counterion', *J. Electrochem. Soc.*, **145**, 1454–1461 (1998). Reproduced by permission of The Electrochemical Society, Inc

of the solution. These data therefore provide direct evidence that the swelling of the layers is dependent on the nature of the contacting electrolyte. By a careful analysis of the data, additional information about the composition of this layer can be obtained. The scattering length density is a function of the volume fractions of the different components present and detailed modeling of the reflectivity profile will allow deconvolution of the various components responsible for the reflectivity profile being measured. In this manner, the composition of the interface can be determined as a function of the underlying electrode surface, the thickness of the polymer layer, and the penetration of solvent into the polymer structure. Figures 4.30 and Figure 4.31 show the volume-fraction profiles obtained in this manner for gold

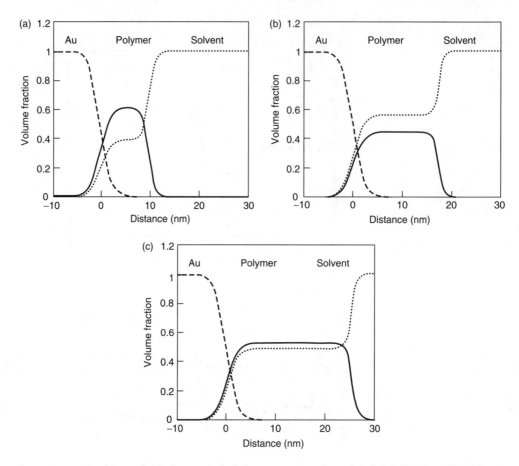

Figure 4.30 Neutron reflectivity study of the structure of an $[Os(bpy)_2(PVP)_{10}Cl]Cl$ film in D_2O/p-toluene sulfonic acid at (a) pH 6, (b) pH 3, and (c) pH 1. Volume fractions: (- - - -) gold; (———) polymer; ($\cdots\cdots$) electrolyte: surface coverage of 2×10^{-9} mol cm^{-2}. From R.W. Wilson, R. Cubitt, A. Glidle, A.R. Hillman, P.M. Saville and J.G. Vos, 'A neutron reflectivity study of $[Os(bpy)_2(PVP)_{10}Cl]^+$ polymer film modified electrodes: effect of pH and counterion', *J. Electrochem. Soc.*, **145**, 1454–1461 (1998). Reproduced by permission of The Electrochemical Society, Inc

electrodes coated with an $[Os(bpy)_2(PVP)_{10}Cl]ClO_4$ polymer layer as a function of pH in p-toluene sulfonic acid and perchloric acid, respectively. Important here is that these data were obtained *in situ*, under realistic electrochemical conditions. In this manner, the influence of the electrolyte on the structure of the polymer layer can be studied in detail and the information obtained can then be used to reinterpret the electrochemical data obtained in other independent studies. From electrochemical, as well as neutron reflectivity measurements, the surface coverage was estimated as 4×10^{-10} mol cm^{-2}. A layer of this coverage should have a thickness of about 6 nm. The data obtained show that in perchloric solutions a slight swelling of the layer is observed in going from the dry film to films in contact with pH 6 and pH 3 electrolytes. The swelling at pH 3 is about 50 %. A similar layer in pH 3 p-toluene sulfonic acid shows a much increased swelling of about 300 %. For both layers at pH 3, the roughness factor, which indicates the diffuse nature of the layer at the

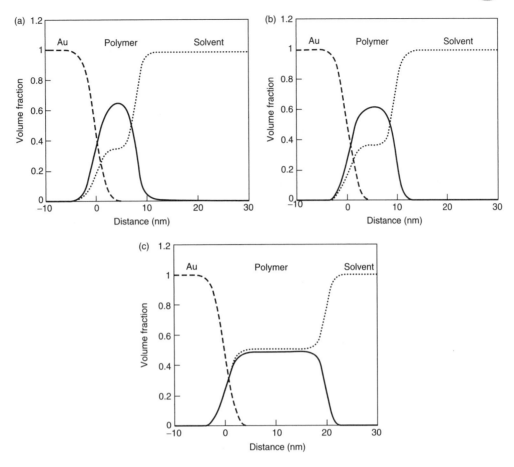

Figure 4.31 Neutron reflectivity study of the structure of an $[Os(bpy)_2(PVP)_{10}Cl]Cl$ film in D_2O/perchloric acid at (a) pH 6, (b) pH 3, and (c) pH 1. Volume fractions: (- - - -) gold; (———) polymer; (······) electrolyte: surface coverage of 2×10^{-9} mol cm^{-2}. From R.W. Wilson, R. Cubitt, A. Glidle, A.R. Hillman, P.M. Saville and J.G. Vos, 'A neutron reflectivity study of $[Os(bpy)_2(PVP)_{10}Cl]^+$ polymer film modified electrodes: effect of pH and counterion', *J. Electrochem. Soc.*, **145**, 1454–1461 (1998). Reproduced by permission of The Electrochemical Society, Inc

polymer–electrolyte interface, is about 1.5 nm, i.e. about 20 % of the layer thickness. This value is not significantly dependent on the pH of the contacting electrolyte. The films are quite homogeneous, and apart from the small diffuse outer layer, can be treated as a single component. What is immediately clear is that more substantial swelling is taking place when the pH is lowered further to pH 1. This is not unexpected as at this pH the free pyridine units in the PVP polymer backbone are expected to be protonated. When comparing with the dry layer, the layer thickness, and therefore the polymer volume, has increased by a factor of 300 to 400 % at this pH, depending on the electrolyte being employed.

Two processes contribute to this volume increase. The first of these is the protonation of the polymer backbone, which is accompanied by ingress of an additional counterion into the polymer film. The second factor is the accompanying ingress of water molecules into the layer. The latter process will reduce electrostatic interactions within the polymer structure. From the scattering length profiles obtained, the volume fractions of polymer and water can be obtained, as shown in Figures 4.30 and 4.31. This analysis shows that for $HClO_4$ at pH 6 the water content of the polymer is about 35 %, while at pH 1 a much thicker layer with a 50 % water content is obtained. For the *p*-toluene sulfonate case, the water content is estimated at about 50 % at pH 3.

These experiments address a number of issues which are of importance for immobilized polymer layers in general. First of all, the layer structure and thickness is strongly dependent on the nature of the contacting electrolyte. Therefore, in cases where liquid–polymer interfaces are envisaged to allow for particular applications, this interaction needs to be considered in detail. As discussed below in Chapter 5, in order to understand the charge transport properties of these layers the interplay between polymer layer and contacting liquids needs to be considered seriously.

4.8.2 Structural Features of Electrostatically Deposited Multilayer Assemblies

One of the attractions of polymer coatings is the ease with which they can be applied to solid surfaces either as monolayers or as multilayers. One particularly attractive route for creating stable multilayers is the use of electrostatic self-assembly involving polyelectrolytes. Since the interaction between the polymer layer and the substrate, or between the alternating layers, is based on electrostatic interactions, the chemical nature of the electrostatic components can be varied systematically. As a result, there is an extensive literature based on, among others, polymer–polymer, polymer–organic, polymer–inorganic and polymer–biomolecular assemblies. Decher and co-workers [28] have investigated the formation of thin multilayer films by alternate deposition of cationic or anionic compounds. This layer-by-layer molecular self-assembly process exploits electrostatic attraction to produce complex layered structures in which it may be possible to precisely control the layer composition and thickness. With multilayer polyelectrolyte films, the same problems as already identified with other polymer coatings have to be addressed. Of interest is how the layer thickness and structure of the layers are affected by the deposition conditions and also whether the thickness and structure of subsequently deposited layers in multilayer assemblies remains constant during the deposition process. Finally, an important issue is whether interpenetration of layers is taking place.

These questions have been addressed by Decher and co-workers, who have investigated the structural features of alternating polyelectrolyte films containing poly(styrene sulfonate) (PSS) and poly(allylamine hydrochloride) (PAH) (see Figure 4.14) by using neutron reflectivity. In these studies, the composition of the repeating units was varied systematically, as well as the number of layers applied. In order to obtain a surface with a homogeneous surface charge, the silicon surface was precoated with 4-aminobutyldimethylmethoxysilane. On this modified surface, alternating anionic PSS and cationic PAH layers were deposited. In this manner, systems of up to 80 layers were assembled. Some typical repeat units used in these assemblies include [$PSSd_7$–PAH] (polymer I), [$PSSh_7$–PAH–$PSSd_7$–PAH] (polymer 2) and [($PSSh_7$–PAH)$_2$–$PSSd_7$–PAH] (polymer 3). In such coatings, the number of repeat units was ten for polymer 1 and eight for polymers 2 and 3. Neutron reflectivity does not allow differentiation between PAH and PSS layers because of the lack of contrast between these two types. However, the technique does differentiate between deuterium and hydrogen. Therefore, deuterated PSS (PSSd$_7$) layers were introduced in the repeat units. The inclusion of a deuterated layer at controlled intervals allows one to determine the distance between these deuterated layers and as a result, indirectly, the overall layer thickness and composition.

By using this approach, the thicknesses of the repeat units in polymers 2 and 3 were determined as 104.3 and 159 Å, respectively. The overall layer thicknesses obtained for polymers 1–3 were 475, 822 and 1225 Å, respectively. This indicates an average thickness for a PSS–PAH layer pair of about 50 Å, with the layer thicknesses obtained for the three assemblies being in good agreement with the coating regimes used for these polymers. Detailed analysis shows, however, that the layer thickness is not uniform throughout, i.e. layers deposited immediately on the surface are thinner than those further away. Typically, the thickness of the first polyelectrolyte layer is only 50 % of those further out. This is explained by the presence of free-moving polymer chains at the solution–polymer interface in subsequent layers, where these provide an increasing number of adsorption units for polyelectrolytes present in the coating solution, which will result in the thicker layer. These free-moving and dangling polymer chains do, however, promote interdigitation of the layers. This interdigitation causes a complication in the analysis of the neutron reflectivity data. The presence of interdigitation does not allow the use of a model where the layers are seen as discreet isolated units, since the boundaries of particular layers can no longer be determined accurately. To allow for this mixing of layers, a composition–space model was applied to the data. In such a model, every layer consists of polymer chains as well as associated water and salt; these components are each given their own volume fraction. For example, the volume of the PSS polyanion chain, V_{pa}, is given as follows:

$$V_{pa} = V_{PSS} + n_w^{PSS} V_{water} + n_s^{PSS} V_{Na^+} \qquad (4.23)$$

where V_{PSS} is the molecular volume of the styrene sulfonate monomer, n_w^{PSS} is the number of water molecules associated with this monomer, and n_s^{PSS} is the number of sodium cations. The volumes of water and cation are given by V_{water} and V_{Na^+}, respectively. As a result, the scattering length density of the PSS layer, b_{pa}, becomes a composite of the scattering length densities of the film components, as follows:

$$b_{pa} = b_{PSS} + n_w^{PSS} b_{water} + n_s^{PSS} b_{Na^+} \qquad (4.24)$$

By using this approach, the diffuse nature of the layer interfaces can be taken into account and the structure of the individual layers can be determined. This is illustrated in Figure 4.32, where the structure of the multilayer film, as obtained by using this method of analysis, is given in terms of the volume fraction of the constituents and as a function of the distance to the substrate–polymer interface. In this figure, the volume fractions of PSS and PAH and water are given at the left, while at the right-hand side of the diagram the volume fraction of water associated with either PSS or PAH is given. The films are clearly layered, although there is a significant amount of interdigitation. This interlinking of the cationic and anionic

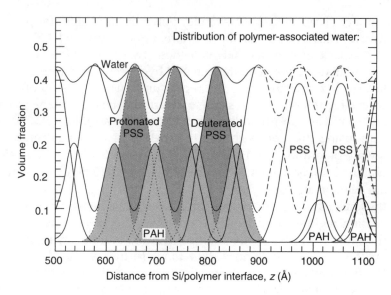

Figure 4.32   Structure of the stratified film of the polymer as obtained from neutron reflectivity experiments shown in terms of volume fractions: distances are normal to the substrate surface; [(PSS-h$_7$-PAH)$_3$-PSS-d$_7$-PAH]. Above 900 Å, the distribution of water associated with the different polyelectrolytes is indicated. Reprinted with permission from M. Losche, J. Smitt, G. Decher, W.G. Bouwman and K. Kjaer, *Macromolecules*, **31**, 8893 (1998). Copyright (1998) American Chemical Society

layers is of the same order as the actual thickness of the individual layers. As a result, there is direct interaction between polyelectrolytes that are four layers apart. These studies further show that for films of more than eight layers, the layer thickness, which can be determined with an accuracy of 3 %, remains constant. Also observed is a clear dependence, although not linear, of the thickness of a layer pair on the ionic strength of the dipping solution. The molecular weights of the polyelectrolytes have a less important influence. The amount of water within the layers has also been estimated. In fully hydrated films, water occupies about 40 % of the volume within the film. About twice as much water is associated with PSS as with PAH. The incorporated inorganic salt only plays a minor role; however, the ionic strength of the deposition solution is an important factor, which greatly influences the layer equilibrium thickness of the layer.

These results show that by using multilayers of alternating polyelectrolytes, predictable and well-defined surface structures can be obtained, provided that the coating conditions and environmental conditions such as humidity are controlled. A substantial amount of information about the three-dimensional structure of the films is obtained and, although the structures cannot be defined with atomic precision, their compositions are well-defined in general terms.

4.8.3   Self-Assembled Monolayer Films of Thiol-Derivatized Polymers

Investigations of self-assembled monolayers have mostly concentrated on the use of mononuclear components. However, the self-assembly of polymers has also

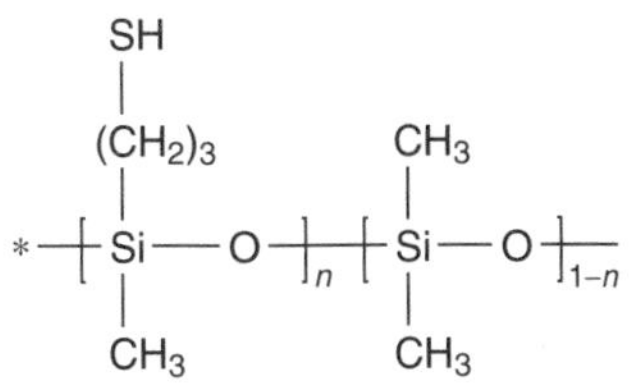

Figure 4.33 General structure of thiol-derivatized siloxane polymers

been studied. As with their mononuclear analogues, the formation of SAMs from polymers is limited to macromolecules which contain functional groups capable of interacting with a surface. These groups can be introduced as either endgroups or as side chains. Ringsdorf and co-workers have investigated the structures of self-assembled monolayers of polydimethylsiloxane (PDMS) [38] and poly(methyl methacrylate) (PMMA) [39], modified with the surface active groups, −SH and −SCH$_3$, respectively. These binding groups were incorporated as side chains (for the general structures, see Figure 4.33). The structural features of the layers obtained with such materials are strongly dependent on the amount (loading) of surface active groups in the polymer being used. For PMMA and PDMS polymers that are fully loaded, only about 50 % of the surface active groups are bound to the surface, with the rest being located within the polymer layer. Such high loadings result in a rigid structure with reduced chain mobility. As a result, highly loaded PDMS films are the thinnest systems. Ellipsometry and contact angle measurements indicate that films cast from polysiloxanes which are fully modified have a thickness of 19 Å, while films cast (under the same conditions) from polysiloxanes only containing 1–3 % SH groups have a thickness of 42 Å. For loadings of between 1–10 %, the behavior observed for modified PMMA and PDMS is significantly different. For the PMMA samples, the surface active groups are evenly distributed through the polymer layer, while in PDMS polymers with a similar loading the groups were predominantly surface-bound. These differences could be due to the different mobilities of the polymer backbones and the difference in surface active groups. The PDMS polymers are well above their glass transition temperatures and are liquid-like in nature, while PMMA is more rigid, being below its glass transition temperature. The significant chain mobility in PDMS films allows also for preferential orientation of the polymer in which the Si–O backbone lies parallel to the surface. With this immobilization method, the surface structures of the films can therefore be manipulated by variations in the nature of the surface active groups and their loadings. Clearly, the amount of ordering that can be obtained by this approach is limited; however, systematic variation of the synthetic and deposition techniques may lead to the formation of better-defined films.

4.8.4 Structural Properties of Block Copolymers

As already shown above in Section 4.7, the composition of block copolymers can be changed in a systematic manner. Stupp and co-workers [36] have used this flexibility to investigate the ability of thin films of such materials to form well-defined structural arrangements. The structures of the polymers that will be discussed in this context are shown above in Figure 4.28. Films of these rod–coil polymers were investigated by using transmission electron microscopy and wide-angle electron diffraction techniques. These studies demonstrated that films cast from the polymer shown in Figure 4.28(a), with $x = y = 9$, contain small nano-sized ordered structures with crystalline domains. Such domains are present because of the crystallization of the rod components of the polymers. Detailed analysis shows that polydisperse nanostructures are obtained with an average diameter of about 3.5 nm. In addition, small areas can be seen where the domains are 'strip-like' in form. Orientation among the nanocrystalline units is observed, with the rods being oriented perpendicular to the surface. After annealing the films at 125 °C for several hours, the average size of the domains is about 5 nm and the small strip-like domains have disappeared. However, the electron diffraction pattern of the annealed film is identical to that observed for the original film, thus indicating that orientation among the supramolecular units is preserved. Small-angle X-ray scattering measurements indicate that the layer spacing is 6.0 nm less than the fully extended length of the average size molecule. This layer thickness in increased to 7.8 nm after annealing, possibly due to a change in the solvent content of the film. The measurements also suggest that ordering in the $x–y$ direction is improving at higher temperatures. The temperature-induced changes are irreversible and suggest that the structures observed at higher temperatures are controlled by thermodynamics, while the films cast at lower temperatures are controlled by kinetic factors. The film structure obtained changes substantially when the coil section of the block copolymer is lengthened ($x = 11$ and $y = 15$). This change in coil volume yields nanostructures of more regular size and with sharply defined boundaries. More long-range ordering is also observed for this polymer and may be explained by a narrower polydispersity, which is related to a larger entropic contribution. The layer spacing for this material is 8.0 nm. The layer is, however, substantially less crystalline, which indicates that with these longer coils there is less order in the rods.

Polymers of type shown on in Figure 4.28(b) were also investigated. For materials with five PV units, well-defined nanostructures were observed which are spaced 8 nm from center to center and have a length of 80 nm. Electron diffraction measurements show that the rods are packed into the same structure as the poly(*p*-phenylene vinylene) homopolymer. The rods are again perpendicular to the surface. When only two PV units are present, no nano-scale organization is observed and glassy solids were obtained instead. This latter observation shows that in order to obtain ordered nanostructures, a rod containing two PV units is not sufficient.

These results show that the supramolecular structures of thin films of block copolymers can be manipulated by varying the rod-to-coil ratios. Variables such as the polydispersity, the nature of the structures, and their crystallinity can be controlled in this manner. The factors that govern the formation of ordered structures from these copolymers are, however, complex. Important factors include entropy effects associated with the flexible coil segments, crystallization of the rods, and steric considerations. Upon crystallization of the rods, the entropies of the coil blocks may be increasingly compromised as a result of increasing steric repulsion. This may effect the sizes of the aggregates that are formed. The organization of ordered structures can furthermore be controlled by non-specific interactions such as $\pi-\pi$ stacking and van der Waals interactions. The data obtained also show that the structures can be modified substantially by annealing of the layers. This temperature-driven rearrangement highlights the many variables which control the structure of a thin polymer film. The drying conditions and temperature, plus the manner in which the layer is stored, are factors that need to be regulated in order to produce reproducible and well-defined surface morphologies.

4.8.5 Domain Control with Styrene–Methyl Methacrylate Copolymers

The ability of block copolymers to arrange themselves on solid substrates and also the ability to manipulate such arrangements has been illustrated for diblock copolymers of styrene and methyl methacrylate. When thin films of such materials are cast, microdomains are formed, with the orientation of these domains being parallel to the surface. This is explained by the specific segregation of one of the components to the interface. However, it was demonstrated that when this specific interaction is removed the microdomains orientate perpendicular to the substrate [40]. It was also shown that preferential interfacial segregation of the polymer components can be eliminated by anchoring random copolymers containing the same monomers (P(S-*r*-MMA)) to the diblock copolymer at both the solid–polymer and polymer–air interfaces. P(S-*r*-MMA) was anchored to the silicon substrate by chemical grafting and located at the polymer–air interface by end-linking the polymer with perfluorinated groups. The low surface energy of the perfluorinated groups forces the random copolymer to the solid–air interface.

Atomic force spectroscopy shows that when only the polymer–substrate surface is modified, parallel orientation of the nanoscopic cylinders is observed in the 137 nm thick film (see Figure 4.34(a)) because of the lower surface energy of the polystyrene block. However, the presence of a random copolymer at both the polymer–substrate and the polymer–air interfaces eliminates the preferential

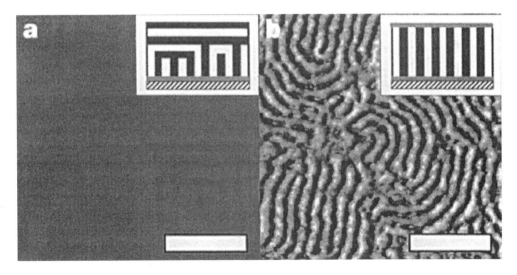

Figure 4.34

segregation of the polystyrene block and as a result a normal orientation of the cylinders can be deducted from atomic force microscopy (see Figure 4.34(b)). This approach shows that interfaces can play an important role in the formation of surface structures and that systematic manipulations of such interfaces may lead to the formation of novel structural features.

4.8.6   Structure–Conductivity Relationships for Alkylthiophenes

In this section, the relationships between the structures and conductivities of thin polyalkylthiophene films are addressed. As outlined earlier in Section 4.7, polyalkylthiophenes can be obtained by either electropolymerization or by chemical polymerization in solution. While the materials obtained have relatively high conductivities, the values are, however, dependent on the synthetic method being used, with differences of several orders of magnitude having been observed. Furthermore, the conductivities obtained are well below the expected theoretical values. Such high conductivities are predicted based on assuming an efficient intrachain electron hopping process. However, in practical terms the conductivity is limited by slow interchain processes and by the presence of microdomain boundaries. An improvement in the regularity of the structure would be expected to improve the charge-carrying properties of thin layers of these materials. Such organized structures have been produced from solutions of highly stereoregular polyhexylthiophene, containing about 98 % HT linkages (see Section 4.7 above) [41]. X-ray diffraction studies on films cast from this polymer point to the formation of lamellar structures in which parallel polymer chains are separated by alkyl chains. An important structural feature of these films is the occurrence of interdigitation of the alkyl chains. A typical schematic structure is given in Figure 4.35 for poly(3-hexylthiophene). Intermolecular π-stacking leads to a distance between the thiophene rings of 3.8 Å and the alkyl chains are organized in a regular manner with a distance of 16.0 Å between neighboring chains. This regular structure enhances the conductivities of the films. Films cast from these stereoregular polymers have conductivities between 1000 and 150 S cm^{-1}, while analogous films using polyalkylthiophenes produced by reaction with FeCl$_3$, yield conductivities in the range of 0.1–20 S cm^{-1}. This strongly suggests that the increased order in the stereoregular materials greatly enhances the conductivities of the latter.

Another parameter which may be expected to influence the conductive behavior of the films, and in particular their interactions with the solid substrate, is the orientation of the polymer chains with respect to this surface. This has indeed been observed in a study where thin films of poly(3-hexylthiophenes) with different stereoregularities were deposited on field-effect transistor (FET) substrates [42]. Depending on the amount of regioregularity in the polymer, the orientation of the polymer chains was either normal or perpendicular to the substrate. Wide-angle X-ray scattering experiments show that for spin-coated films of samples of high regioregularity (>91 %) and low molecular weights, the polymer chains are perpendicular to the FET plane, while for materials having low regioregularity (<81 %) and high molecular weights the polymer chains are normal to the substrate surface (see Figure 4.36). The charge–carrier mobilities of the two orientations are

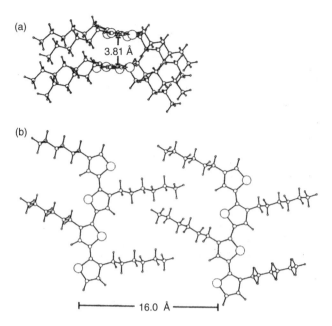

Figure 4.35 Calculated structure for an HT 3-hexylthiophene tetramer obtained by using molecular mechanics modeling, where the globally minimized tetramers have been docked in an idealized manner to X-ray structural parameters. (a) Intermolecular p-stacking between the thiophene rings as inferred from a (90 °C) X-ray pattern of the film. (b) Lamellar stacking as inferred from X-ray scans of intensity versus 2θ data. Reprinted from R.D. McCullough, S. Tristram-Nagle, S.P. Williams, R.D. Lowe and M. Jayaraman, *J. Am. Chem. Soc.*, **115**, 4910 (1993). Copyright (1993) American Chemical Society

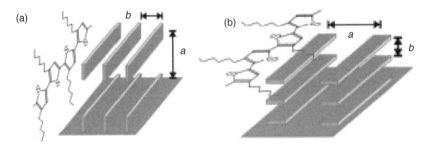

Figure 4.36 Two different orientations of ordered poly(3-hexylthiophene) domains with respect to the substrate, relating to films with regioregularities of (a) 96, and (b) 81%. Reprinted with permission from H. Sirringhaus, P.J. Brown, R.H. Friend, M.M. Nielsen, K. Bechgaard, B.M.W. Langeveld-Voss, A.J.H. Spiering, R.A.J. Janssen, E.W. Meijer, P. Herwig and D.M. de Leeuw, 'Two-dimensional charge transport in self-organized high-mobility conjugated polymers', *Nature (London)*, **401**, 685–688 (1999). Copyright (1999) Macmillan Magazines Limited

quite different. At room temperature, the highest mobilities of 0.05–0.1 cm^2 V^{-1} are observed for samples with the highest regioregularity (96%). For spin-coated samples of 81% regioregularity, in which the 010 axis is normal to the surface, the mobility is about three orders of magnitude less, at 2×10^{-4} cm^2 V^{-1}, even though there is pronounced in-plane crystallinity in the (100) direction with particle sizes of 130 Å. It was also observed that the coating method used to modify the FETs affects the organization obtained. For solution-cast films of the low-regioregular polymer, the mobility is about one order of magnitude higher than that observed for a spin-coated layer of the same material, and is similar to the values obtained for spin-coated films of high-regioregular polymers.

These results suggest that the high mobility observed for the parallel orientation can be explained by interchain charge transport, resulting from strong interchain interactions. These observations clearly suggest that a high mobility can be promoted by an improved organization of the polymer films. Future research in further optimizing the self-assembly properties of these materials may lead to the development of ordered materials with dramatically improved charge-transport properties.

4.9 Biomimetic Assemblies

The ultimate relationship between structure and function is observed in nature where the interaction between proteins and active centers results in the formation of structural features which facilitate very specific reactions. For example, the cell membrane plays a vital role in sensing changes in the chemical and physical environments of cells. Here, environmental stimuli are translated into intracellular signals and then the signals are transmitted to a response system. Mimicking the structure of biological membranes is essential for developing membrane science and technology and for understanding the elementary processes involved in living systems. Two approaches for mimicking the intelligent functions of biological membranes are illustrated in Figure 4.37. This section focuses on the creation of interfacial supramolecular assemblies whose structures and functions mimic living biosystems.

Functional modeling of the membranes found in living biological systems can provide critical insights into developing high-performance artificial membranes. For

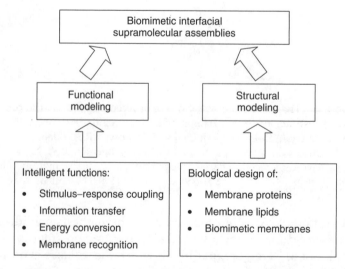

Figure 4.37 Two different approaches for developing biomimetic interfacial supramolecular assemblies

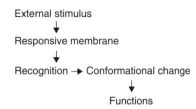

External stimulus
↓
Responsive membrane
↓
Recognition → Conformational change
↓
Functions

Figure 4.38 Schematic representation of the key functions of a biomimetic interfacial supramolecular assembly

example, chemoreception is a typical example of a stimulus–response system. In this case, chemical stimuli, such as ions, hormones or neurotransmitters, bind to specific receptor proteins on the surfaces of their target membranes. These interactions are then transformed via conformational changes in the protein molecules into electrical signals across the membranes. It is essential to understand the fundamental mechanisms involved, e.g. host–guest interactions, electronic communication, the generation of a significant transmembrane potential and controlled switching of molecular conformation. As illustrated in Figure 4.38, in seeking to create artificial membranes which reproduce the functions of natural systems, highly ordered assemblies are required in which there are well-defined molecular mechanisms for physical and chemical stimulation, transduction and response.

Beyond the complete assembly of biomimetic membranes, interfacial supramolecular assemblies which incorporate biocomponents represent an important approach to replicating the biological functions outside of living systems. For example, the ability to link or 'wire' otherwise electro-inactive enzymes to electrodes so that they can efficiently transport electrons allows sensitive and selective sensors to be developed for important bioactive molecules, e.g. glucose, lactate, urea, etc.

4.9.1 Protein Layers

The design and structural characterization of interfacial supramolecular assemblies in which biomolecules retain their natural function, e.g. proteins or enzymes that remain redox-active, is an area of intense research effort. Protein adsorption is no longer viewed as an undesirable experimental difficulty. In contrast, adsorbed species now represent important model systems for understanding these complex systems, many of which seek to replicate the kinds of interfaces found in natural systems. Therefore, a major drive in this area is to create assemblies in which proteins are strongly adsorbed but retain their native properties so that the assembly provides a well-defined environment in which electron transfer or spectroscopic features can be probed. Figure 4.39 illustrates the most important kinds of protein–electrode interface. Building blocks for these assemblies can now be synthesized so that the strength of the interactions between protein and electrode can be tailored over a wide range. For example [43], weak interactions may give rise to reversible electrochemical responses for solution-phase reactants where irreversible behavior would be observed at an unmodified electrode. In contrast, strong interactions allow highly stable protein films, suitable for sensing applications, to be created.

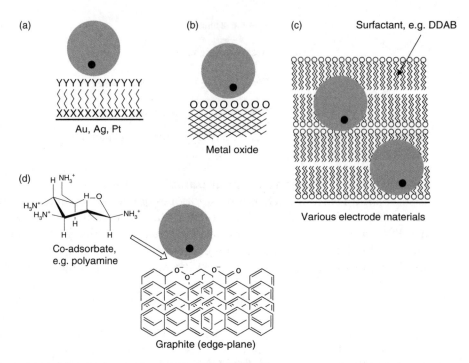

Figure 4.39   Schematic representations of the various types of electrode surfaces for which protein voltammetry is commonly observed: (a) a metal electrode modified with an 'XY' SAM; (b) a metal oxide electrode; (c) an electrode modified with a surfactant layer in which protein molecules are embedded; (d) a pyrolitic graphite 'edge' electrode, often used in conjunction with mobile co-adsorbates such as aminocyclitols. Reprinted from *Electrochim. Acta*, **45**, F.A. Armstrong and G.S. Wilson, 'Recent developments in faradaic bioelectrochemistry', 2623–2645, Copyright (2000), with permission from Elsevier Science

4.9.2   Biomolecule Binding to Self-Assembled Monolayers

Self-assembled monolayers are extremely important in this area, with a thiol often being the 'X-functionality' of Figure 4.39 and the 'Y-functionality' interacting with the solution and protein molecule. Selective binding of particular proteins can be achieved by changing the functionality Y. Thus, SAMs with hydrophilic surface groups, particularly an ethylene oxide oligomer, are very resistant to adsorption of proteins from solution. In sharp contrast, cytochrome C will bind strongly through lysine groups around the exposed heme edge to an acid-terminated thiol monolayer [44,45]. Acidic proteins, e.g. plastocyanin or ferredoxins, respond best if Y is a basic group such as a protonated amine. A significant focus is to elucidate the mechanism by which proteins interact with self-assembled monolayers and how the SAM mediates electron transfer to and from the protein. For example, how mobile are the proteins on the modified surface? Is the ability of the SAM to block the adsorption of contaminants important?

Although the structure of the interfacial supramolecular assembly is often less well defined, carbon and metal oxides offer certain advantages over SAM-modified metal electrodes. For example, pyrolitic graphite provides a very wide potential window, while metal oxides are optically transparent, thus allowing UV–visible spectral changes to be recorded as the redox state of the surface-bound protein is switched. Protein molecules may also be attached to electrodes by covalent bonding, e.g. through classical carbodiimide coupling, cytochrome C oxidase has been immobilized on a gold electrode by using a 3-mercaptopropionic acid bridge [46]. The 'chemisorbed' enzyme shows reversible voltammetry, and can transfer electrons to cytochrome C in solution. Not only does this approach allow an enzyme to be surface immobilized, it is also possible to orient the biomolecules. For example, by engineering a cysteine residue into the structure of azurin it is possible to immobilize it on gold in an orientation that allows electron transfer but restricts lateral movement of the biocomponent [47].

4.9.3 Redox Properties of Biomonolayers

Until the early 1980s, the idea that the oxidation states of biomolecules as large as cytochrome C could be switched at electrodes was regarded with skepticism. The pioneers of this area, including Niki, Hill, and Kuwana, demonstrated that redox-active proteins undergo electron transfer reactions efficiently at surfaces ranging from pristine metal electrodes to metal oxide electrodes, and those functionalized with organic 'promoter' monolayers. Voltammetry is now accepted as a powerful technique for elucidating the properties of biomolecular interfacial supramolecular assemblies. For instance, chemical titrants and mediators are not needed and, unlike some spectroscopic methods, e.g. EPR spectroscopy, which sometimes requires cryoscopic (cryogenic) temperatures, the investigations can be performed across a wide range of temperatures and pressures. For example, thermodynamic information such as reduction potentials can be directly measured *in situ* as a function of temperature, solvent composition, ionic strength and microenvironment. In addition, in voltammetric experiments the driving force can be tuned to more extreme values than are accessible when using chemical reductants. For example, as illustrated in Figure 4.40, the formal potentials of reduction processes that occur at quite negative potentials can be conveniently measured by using voltammetry, which could not be achieved with solution-phase reductants even as powerful as dithionite. Voltammetry is particularly useful in studying systems of this kind where redox switching is not accompanied by a significant change in the spectrum of the compound. Moreover, voltammetry can provide kinetic as well as thermodynamic information. For example, in biological systems, electron transfer often drives coupled chemical reactions, e.g. proton transfer or structural rearrangement of the reactive site, i.e. conformationally 'gated' electron-transfer reactions.

Important advantages are gained if proteins are immobilized at the electrode surface, hence giving an electroactive film that is ideally of monolayer coverage. This approach has been termed *protein film voltammetry* and one of its major advantages is that microscopic quantities of the protein are required and thermodynamic and kinetic information can be obtained with higher accuracy and resolution.

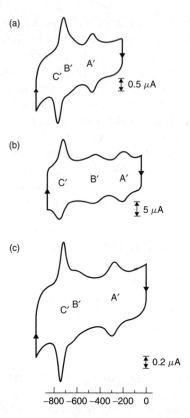

Figure 4.40 Voltammograms of films of different 7Fe ferredoxins on a pyrolitic graphite (edge) electrode, obtained at 0 °C: (a) *Azotobacter vinelandii* (Fd I), at pH 7.0, with a scan rate of 20 mV s^{-1}; (b) *Desulfovibrio africanus* (Fd III), at pH 7.0, with a scan rate of 191 mV s^{-1}; (c) *Sulfolobus acidocaldarius* (Fd), at pH 7.4, with a scan rate of 10 mV s^{-1}. In each case, an electroactive coverage of approximately one monolayer is obtained in the presence of polymyxin as the co-adsorbate. The signals, 'A', 'B' and 'C' refer to the redox couples $[3FE-4S]^{+/0}$, $[4FE-4S]^{2+/+}$ and $[3FE-4S]^{0/2+}$, respectively. Reprinted from *Electrochim. Acta*, **45**, F.A. Armstrong and G.S. Wilson, 'Recent developments in faradaic bioelectrochemistry', 2623–2645, (Copyright) 2000, with permission from Elsevier Science

4.10 Conclusions

The field of molecular self-assembly is currently dominated by the adsorption of a small range of materials, notably thiols and related structures, to give interfacial supramolecular assemblies that are stable, and have predictable electrochemical and photochemical properties. We can expect this range of materials to increase significantly. Moreover, the emphasis is likely to shift towards exploiting weaker, more subtle driving forces for organization, e.g. hydrogen bonding, that can be more easily addressed and controlled by using chemical, electrochemical and photochemical stimuli. The future looks bright in terms of new advances for probing the structures of interfacial supramolecular assemblies, e.g. femtosecond spectroscopy and x-ray methods for monitoring dynamic structural changes.

The discussion also highlights that polymer layers, although not crystalline in general, can nevertheless produce well-defined structural features that can be manipulated to obtained a particular behavior or function. The application of polymer films is therefore expected to be extended in areas such as sensor design, conducting polymers and solar energy devices.

Overall, the approaches outlined in this chapter show that molecular components, whether they are mononuclear or polymeric, can be treated as building blocks and organized into a variety of functional supramolecular assemblies. It is tempting to consider the impact of this on nanotechnology. The ability to manipulate ensembles of molecules represents a 'bottom-up' approach to constructing nanodevices from their constituent parts. The other possible route to nanodevices is to mechanically maneuver the atoms or molecules by means of laser tweezers or the tip of an atomic force microscope. However, these processes are laborious and can cause the building blocks to decompose. Not surprisingly, this approach has produced little more than words or 'logos' drawn with atoms on various surfaces. In sharp contrast, the chemical manipulation approach described here is more elegant and vastly more subtle because it relies on instructions programmed into the system to determine the ultimate location of each molecule.

References

[1]   Finklea, H.O. (1996). Electrochemistry of Organized Monolayers of Thiols and Related Molecules on Electrodes, in *Electroanalytical Chemistry*, Vol. 19, A.J. Bard and I. Rubinstein (Eds), Marcel Dekker, New York, pp. 109–120.

[2]   Finklea, H.O. (2000). Self-assembled Monolayers on Electrodes, in *Encyclopedia of Analytical Chemistry*, Vol. 11, R.A. Meyers (Ed.), Wiley, Chichester, UK, pp. 10090–10116.

[3]   Forster, R.J. and O'Kelly, J.P. (2001). *J. Electrochem. Soc.*, **148**, E31.

[4]   Adamson, A.W. and Gast, A.P. (1997). *Physical Chemistry of Surfaces*, Wiley, New York.

[5]   Goloud, T.P. and Koopal, L.K.(1997). *Langmuir*, **13**, 673.

[6]   Bain, C.D., Troughton, E.B., Yao, Y.T., Evall, J., Whitesides, G. and Nuzzo, R. (1989). *J. Am. Chem. Soc.*, **111**, 321.

[7]   Fieldne, M.L., Claesson, P.M. and Verrall, R.E. (1999). *Langmuir*, **15**, 3924.

[8]   Walcak, M.M., Popenoe, D.D., Deinhammer, R.S., Lamp, B.D., Chung, C. and Porter, M.D. (1991). *Langmuir*, **7**, 2687.

[9]   Forster, R.J. (1996). *Anal. Chem.*, **68**, 3143.

[10]   Forster, R.J., Keyes, T.E., Farrell, M. and O'Hanlon, D. (2000). *Langmuir*, **16**, 9871.

[11]   Porter, M.D, Bright, T.B., Allara, D. and Chidsey, C.E.D. (1987). *J. Am. Chem. Soc.*, **109**, 3559.

[12]   Smith, C.P. and White, H.S. (1993). *Langmuir*, **9**, 1.

[13]   Hudson, J.E. and Abruña, H.D. (1996). *J. Phys. Chem.*, **100**, 1036.

[14]   Bain, C.D., Evall, J. and Whitesides, G.M. (1989). *J. Am. Chem. Soc.*, **111**, 7155.

[15]   Forster, R.J. (2000). Ultrafast electrochemical techniques, in *Encyclopedia of Analytical Chemistry*, Vol. 11, R.A. Meyers (Ed.), Wiley, Chichester, UK, pp. 10142–10171.

[16]   Finklea, H.O., Snider, D.A., Fedyk, J., Sabatani, E., Gafni, Y. and Rubinstein, I. (1993). *Langmuir*, **9**, 3660.

[17]   Ingersoll, D., Kulesza, P.J. and Faulkner, L.R. (1994). *J. Electrochem. Soc.*, **141**, 140.

[18]   Kuhn, A. and Anson, F.C. (1996). *Langmuir*, **12**, 5481.

[19] Willner, I., Katz, E. and Willner, B. (1997). *Electroanalysis*, **9**, 965.

[20] Pantano, P., Morton, T.H. and Kuhr, W.G. (1991). *J. Am. Chem. Soc.*, **113**, 1832.

[21] Hodak, J., Etchenique, R., Calvo, E.J., Singhal, K. and Bartlett, P.N. (1997). *Langmuir*, **13**, 2708.

[22] Bourdillon, C., Demaille, C., Moiroux, J. and Savéant, J.-M. (1996). *Acc. Chem. Res.*, **29**, 529.

[23] Rusling, J.F. (1998). *Acc. Chem. Res.*, **31**, 363.

[24] Burgess, J.D., Rhoten, M.C. and Hawkridge, F.M. (1998). *Langmuir*, **14**, 2467.

[25] Forster, R.J. and Vos, J.G. (1992). Theory and analytical applications of modified electrodes, in *Comprehensive Analytical Chemistry*, Vol. XXVII, G. Svehla (Ed.), Elsevier, Amsterdam, p. 465–485.

[26] Ito, T., Shirakawa, H. and Ikeda, S. (1974). *J. Polym. Sci., Polym. Chem. Edn*, **12**, 11.

[27] Chiang, C.K., Park, Y.W., Heeger, A.J., Shirakawa, H., Louis, E.J. and MacDiarmid, A.G. (1977). *Chem. Phys. Lett.*, **39**, 1098.

[28] Losche, M., Smitt, J., Decher, G., Bouwman, W.G. and Kjaer, K. (1998). *Macromolecules*, **31**, 8893.

[29] Kaschak, D.M., Johnson, S.A., Waraksa, C.C., Pogue, J. and Mallouk, T.E. (1999). *Coord. Chem. Rev.*, **185–186**, 403.

[30] Forster, R.J. and Vos, J.G. (1990). *Macromolecules*, **23**, 4372.

[31] Clarke, A.P., Vos, J.G., Bandey, H.L. and Hillman, A.R. (1995). *J. Phys. Chem.*, **99**, 15973.

[32] Haas, O., Kriens, M. and Vos, H.G. (1981). *J. Am. Chem. Soc.*, **103**, 1318.

[33] Sadki, S., Schottland, P., Brodie, N. and Sabouraud, G. (2000). *Chem. Soc. Rev.*, **29**, 283.

[34] Roncali, J. (1992). *Chem. Rev.*, **92**, 711.

[35] McCullough, R.D. (1998). *Adv. Mater.*, **10**, 93.

[36] Tew, G.N., Pralle, M.U. and Stupp, S.I. (1999). *J. Am. Chem. Soc.*, **121**, 9852.

[37] Wilson, R.W., Cubitt, R., Glidle, A., Hillman, A.R., Saville, P.M. and Vos, J.G. (1998). *J. Electrochem. Soc.*, **145**, 1454.

[38] Tsao, M.-W., Pfeifer, K.-H., Rabolt, J.F., Castner, D.G., Häussling, L. and Ringsdorf, H. (1997). *Macromolecules*, **30**, 5913.

[39] Lenk, T.J., Hallmark, V.M., Rabolt, J.F., Häussling, L. and Ringsdorf, H. (1993). *Macromolecules*, **30**, 1230.

[40] Huang, E., Rockford, L., Russell, T.P. and Hawker, C.J. (1998). *Nature*, **395**, 757.

[41] McCullough, R.D., Tristram-Nagle, S., Williams, S.P., Lowe, R.D. and Jayaraman, M. (1993). *J. Am. Chem. Soc.*, **115**, 4910.

[42] Sirringhaus, H., Brown, P.J., Friend, R.H., Nielsen, M.M., Bechgaard, K., Langeveld-Voss, B.M.W., Spiering, A.J.H., Janssen, R.A.J., Meijer, E.W., Herwig, P. and de Leeuw, D.M. (1999). *Nature*, **401**, 685.

[43] Armstrong, F.A. and Wilson, G.S. (2000). *Electrochim. Acta*, **45**, 2623.

[44] Allen, P.M., Hill, H.A.O. and Walton, N.J. (1984). *J. Electroanal. Chem.*, **78**, 69.

[45] Taniguchi, Y., Toyosawa, K., Yamaguchi, H. and Yasukouchi, K. (1982). *J. Chem. Soc., Chem. Commun.*, 1032.

[46] Li, J., Cheng, G. and Dong, S. (1996). *J. Electroanal. Chem.*, **416**, 97.

[47] Davis, J.J., Halliwell, C.M., Hill, H.A.O., Canters, G.W., van Amsterdam, M.C. and Verbeet, M.P. (1998). *New J. Chem.*, **22**, 1119.

5 Electron and Energy Transfer Dynamics

'The realization of Integrated Communications Systems', the 'ICS' of electronics and photonics, is one of the principal goals in seeking to create functional interfacial supramolecular assemblies. Understanding those factors that control electron and energy transfer is not only of fundamental interest, but is vital for creating molecular electronic devices. This chapter describes selected case studies which illustrate the key factors that control electron and energy transfer within interfacial supramolecular assemblies and in particular across solid–film interfaces. In doing so, it seeks to identify those approaches which provide key fundamental insights and show the greatest promise for creating electrochemically and photochemically triggered molecular switches, sensors and biomimetic systems.

5.1 Introduction

Interfacial supramolecular assemblies play an important role in the field of molecular electronics since they allow molecules or nanoparticles to be integrated into scaleable, functional electronic devices which are connected to each other and to the outside world in a realistic and practical manner. Interfacial supramolecular assemblies can be designed to have individual molecules, or small assemblies of molecules, perform functions in electronic circuits currently carried out by semiconductor devices. The impact of these developments on our economy and society will be comparable to that of information technology, genomics or proteomics. Activity in this area is dominated by surfaces, often metals, that are modified with ordered arrays of adsorbed molecules with specific electrochemical or photochemical properties. Surfaces have the tremendous advantage over solution-phase reactants in that the molecular components can be directly addressed, e.g. their redox states can be switched (WRITE function) or probed (READ function) without the complication of slow diffusional mass transport. For example, the basic paradigm for electronic information storage is retention of charge in a capacitor, and, therefore, the most straightforward way to achieve to molecular-scale memory is to store charge in discrete molecules.

The driving force for molecular electronics is to use molecules to achieve further miniaturization, greater functionality and faster clock rates for advanced electronic systems which operate over a wide range of temperatures and preferably take advantage of three-dimensional (3-D) architectures. The techniques discussed above in Chapter 4, such as molecular self-assembly, are especially important in this regard since they offer the possibility of directing molecules to assemble into functioning devices and interconnects. For example, at the heart of the semiconductor industry is the semiconductor switch. The latter has become the fundamental device in all branches of modern electronics because combinations of switches can be made to perform all desired computational functions. However, in trying to have molecular assemblies perform these functions, not only is it necessary to have the synthetic control described in Chapter 4, but one must understand electron and energy transfer within the assemblies and across solid–film interfaces. This chapter presents selected case studies which illustrate the key features of electron and energy transfer within interfacial supramolecular assemblies.

5.2 Electron and Energy Transfer Dynamics of Adsorbed Monolayers

The field of molecular electronics ranges from well-defined and well-understood phenomena such as non-linear optical responses, to the tantalizing, and conceptually more difficult, areas of computing and information storage at the molecular level. Supramolecular assemblies, constructed by using single electroactive molecules as the building blocks, offer a striking way to create electrically conducting materials whose organized architectures makes them suitable for developing molecular electronic devices.

Progress toward this strategic goal demands not only the development of new synthetic approaches which will yield highly ordered materials but also careful attention to those elementary processes that dictate the rate of heterogeneous electron transfer across metal–monolayer interfaces. However, it is only recently that it has been possible to *directly* probe those factors which govern the rate of electron transfer processes that are complete within a few billionths of a second [1]. In fact, coupling recent advances in the design and fabrication of microelectrodes and electrochemical instrumentation which operate on a nanosecond timescale with chemical systems that are organized on the molecular level, promises to revolutionize investigations into electron transfer processes.

While experiments involving solution-phase reactants have provided deep insights into the dynamics of heterogeneous electron transfer, the magnitude of the diffusion-controlled currents over short timescales ultimately limits the maximum rate constant that can be measured. For diffusive species, the thickness of the diffusion layer, δ, is defined as $\delta = (\pi D t)^{1/2}$, where D is the solution-phase diffusion coefficient and t is the polarization time. Therefore, the depletion layer thickness is proportional to the square root of the polarization time. One can estimate that the diffusion layer thickness is approximately 50 Å if the diffusion coefficient is $1 \times 10^{-5} \mathrm{~cm^2~s^{-1}}$ and the polarization time is 10 ns. Given a typical bulk concentration of the electroactive species of 1 mM, this analysis reveals that only 10 000 molecules or so would be oxidized or reduced at a 1 µm radius microdisk under these conditions! The average current for this experiment is only 170 nA, which is too small to be detected with high temporal resolution.

One successful approach to eliminating this diffusion limitation is to immobilize the redox-active species within a supramolecular assembly. When immobilized on an electrode surface, the electroactive species no longer needs to diffuse to the electrode to undergo electron transfer. Moreover, the electroactive species is preconcentrated on the electrode surface. For example, in the situation considered above, there will be approximately 1.7×10^{-20} mol of electroactive material within the diffusion layer. Given that the area of a 1 μm disk is approximately 3.1×10^{-8} cm^2, this translates into an 'equivalent surface coverage' of about 5.4×10^{-13} mol cm^{-2}. In contrast, as discussed earlier in Chapter 4, the surface coverage, or number of moles per unit area, Γ, found for dense monolayers of adsorbates is typically more than two orders of magnitude larger, with coverages of the order of 10^{-10} mol cm^{-2} being observed. This higher surface coverage gives rise to much larger currents, which are easier to detect at short time-scales.

5.2.1 Distance Dependence of Electron Transfer

As discussed in Chapter 4, a wide variety of functionalized alkane thiols, $HS(CH_2)_n$-2, where $5 \leq n \leq 16$, form highly ordered self-assembled monolayers. As illustrated in Figure 5.1, redox-active species can be covalently bound to these bridges. The seminal work of Chidsey [2], Acevedo and Abruña [3] and Finklea and Hanshew [4] has demonstrated that electroactive adsorbed monolayers can exhibit close to ideal reversible electrochemical behavior under a wide variety of experimental conditions of time-scale, temperature, solvent and electrolyte. These studies have elucidated the effects of electron transfer distance, tunneling

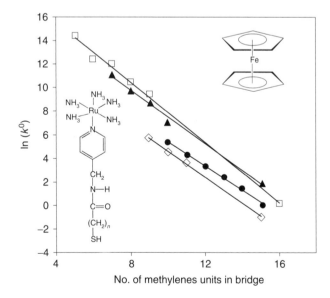

Figure 5.1 Semi-log plots of the standard heterogeneous electron transfer rate constant, k^0, versus the number of methylene units in the alkane thiol bridge for various materials electrostatically adsorbed on $HS(CH_2)_nCOOH$: ●, $[HS(CH_2)_nCONHCH_2py-Ru(NH_3)_5]^{2+}$; ▲, $HS(CH_2)_nNHCO$–ferrocene; □, $HS(CH_2)_nOOC$–ferrocene; ◊, cytochrome C

medium, molecular structure, electric fields and ion-pairing on heterogeneous electron transfer dynamics. Moreover, they have provided the detailed experimental data necessary to comprehensively test contemporary models such as the Marcus theory. These experimental investigations have revealed three major insights. First, as shown in Figure 5.1, a plot of the logarithm of the standard heterogeneous electron transfer rate constant, $k°$, versus the number of repeating units in the bridging ligand, typically the number of methylene units in an alkane chain, is linear [5–9]. This result agrees with the theory described previously in Chapter 2 which predicts that tunneling rates decay exponentially with distance. The slope of this plot yields the tunneling parameter, β, which for alkane thiol systems is 1.0–1.1 Å^{-1}. Secondly, as illustrated in Figure 5.2, unlike the predictions of the Butler–Volmer theory, a plot of the potential-dependent heterogeneous electron transfer rate constant, for the forward and backward processes, k_f and k_b, respectively, versus the overpotential, η, is not linear for all driving forces. As predicted by the Marcus theory, when the overpotential becomes comparable to the reorganization energy of the redox couple, curvature is observed and k eventually becomes independent of the driving force. Thirdly, by using temperature-dependent measurements of k and the formal potential, $E^{0'}$, the activation enthalpy and reaction entropy can be obtained. These values can then be used to calculate the free energy of activation and, provided that $k°$ is known, the pre-exponential factor can be determined. The experimentally

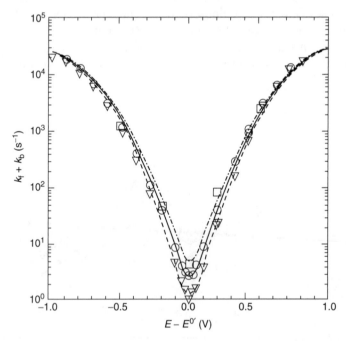

Figure 5.2 Tafel plots of ln k versus overpotential for a mixed self-assembled monolayer containing HS(CH$_2$)$_{16}$OOC–ferrocene and HS(CH$_2$)$_{15}$CH$_3$ in 1.0 M HClO$_4$ at three different temperatures: ∇, 1 °C; $\bigcirc$, 25 °C; $\square$, 47 °C. The solid lines are the predictions of the Marcus theory for a standard heterogeneous electron transfer rate constant of 1.25 s^{-1} at 25 °C, and a reorganization energy of 0.85 eV ($\equiv$ 54.8 kJ mol^{-1}). Reprinted with permission from C. E. D Chidsey, 'Free energy and temperature dependence of electron transfer at the metal–electrolyte interface', *Science*, **251**, 919–922 (1991). Copyright (1991) American Association for the Advancement of Science

determined reorganization energies have been compared with theoretical predictions to investigate the ability of contemporary theory to predict the thermodynamic parameters for electron transfer reactions.

Beyond synthetic systems, distance-dependent electron transfer is also important in biological systems and can occur across significant distances. Self-assembled and spontaneously adsorbed monolayers are important model systems for understanding these processes since they allow the physical structure and chemical composition of the assembly to be controlled. Alkane thiol monolayers terminated with suitable groups for binding *biomolecules* have been used to probe the dependence of the electron transfer rate constants on the length of the spacer. For example, Tarlov and Bowden [10] have used cyclic voltammetry to determine the electron exchange rate constants, k°, between cytochrome C and gold across medium-length $HS(CH_2)_nCOOH$ SAMs. Where $9 \leq n \leq 11$, the rate constants depend exponentially on chain length which is consistent with a through-bond tunneling mechanism and a decay factor of 1.1 per methylene group. This result is in excellent agreement with the data obtained for simple redox species attached to alkane thiol monolayers and confirms that for redox-active proteins the rate constant is dictated by the bridge rather than by the reorganization energy associated with switching the oxidation state of the protein. In sharp contrast, as the chain length is shortened, the rate constants do not increase as expected [8]. Rather, they approach a limiting value, suggesting that when the bridge is capable of supporting rapid electron transfer, the overall rate is controlled by the dynamics of redox-induced conformational changes. The implication is that the most stable orientation does not correspond with the most favorable orientation for electron transfer. However, it is perhaps important to note that investigations involving short alkane thiol monolayers labeled with redox-active groups, e.g. ferrocene, suggest that the situation for short bridges may be complicated by double-layer effects and quantum mechanical spillover of the electron density of the electrode.

5.2.2 Resonance Effects on Electron Transfer

Exploring the degree and mechanism of electronic interaction between remote redox centers and an underlying electrode surface on which they are adsorbed has been the subject of many experimental and theoretical investigations. The highly ordered structure of alkane-thiol-based self-assembled monolayers makes them useful model systems for probing fundamental issues such as the dependence of the heterogeneous electron transfer rate on the reaction free energy, the electron transfer distance, the molecular structure of the interface, and for investigating the role of ionic interactions in dictating electrochemical responses. However, redox-active components can also be used as bridging ligands, so offering the possibility of significant virtual coupling, resonant superexchange, if the redox potentials of the bridge and remote redox centers are similar. Electron transfer via superexchange may be the dominant mechanism if the lowest unoccupied molecular orbital (LUMO) of the bridge is close in energy to the donor and acceptor. This pathway predominates because the closeness of the bridge and redox center energies acts to reduce the activation barrier to electron tunneling. However, for long-range electron transfer reactions where the redox potentials of the bridge and reacting species are very different, through-space electron transfer pathways may become important.

Electroactive monolayers built by using adsorbates with multiple accessible redox states are perhaps the most useful systems for investigating the effect of the bridge–redox center energy separation on the electron transfer dynamics. For example, $[Os(bpy)_2py(p3p)]^{2+}$ monolayers undergo both metal- and bypyridyl-based electron transfer reactions where the supporting electrolyte is tetrabutylammonium tetrafluoroborate ($TBABF_4$) dissolved in acetonitrile [11]. As illustrated in Figure 5.3, well-defined voltammetric responses are obtained for the metal (positive potentials) and ligand-based redox process (negative potentials). The redox potentials for these metal and ligand-based processes are separated by at least 2.0 V, and while the redox potential of the metal-based oxidation is far from that of the bridging ligand, the bipyridyl-based reductions are within 0.3 V of the half-wave potential for the reduction of the uncomplexed bridging ligand.

The kinetic aspects of heterogeneous electron transfer to three of these redox couples, linking charge states 3+/2+, 2+/1+, and 1+/0, have been studied via chronoamperometry conducted on a microsecond time-scale. As illustrated in Figure 5.4, these responses are remarkably well-behaved over a wide range of time-scales, temperatures and potentials, thus allowing molecular bridge effects, in particular, mediating electronic states of the bridging ligand (superexchange), on distant charge tunneling to be investigated.

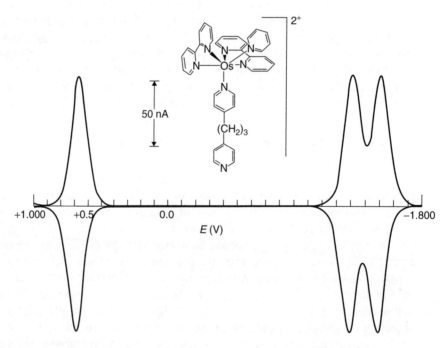

Figure 5.3 Cyclic voltammogram of a spontaneously adsorbed $[Os(bpy)_2py(p3p)]^{2+}$ monolayer, obtained by using a scan rate of $50\ V\ s^{-1}$, with a surface coverage of $9.5 \times 10^{-11}\ mol\ cm^{-2}$. The supporting electrolyte is 0.1 M $TBABF_4$ in acetonitrile, and the radius of the platinum microelectrode is 25 μm. The cathodic currents are shown as 'up', while the anodic currents are shown as 'down'. The complex is in the $[Os(bpy)_2py(p3p)]^{2+}$ form between +0.400 and −1.200 V; the initial potential was 1.000 V. Reprinted with permission from R. J. Forster, *Inorg. Chem.*, **35**, 3394 (1996). Copyright (1996) American Chemical Society

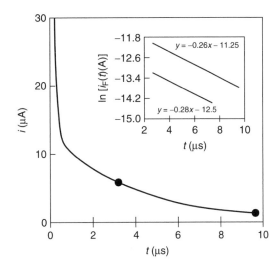

Figure 5.4 Current response for a 12.5 μm platinum microelectrode modified with a $[Os(bpy)_2$ $py(p3p)]^{2+}$ monolayer following a potential step where the overpotential η was -100 mV; the supporting electrolyte is 0.1 M TBABF$_4$ in acetonitrile. The inset shows $\ln[i_F(t)]$ versus t plots for the Faradaic reaction when using a 12.5 μm (top) and 5 μm (bottom) radius platinum microelectrode. Reprinted with permission from R. J. Forster, *Inorg. Chem.*, **35**, 3394 (1996). Copyright (1996) American Chemical Society

As illustrated in Figure 5.5, standard heterogeneous electron transfer rate constants, k^0, can be evaluated by extrapolating Tafel plots of $\ln k$ versus overpotential, η, to zero driving force to yield values of $4.8 \pm 0.3 \times 10^4$ s^{-1}, $2.5 \pm 0.2 \times 10^5$ s^{-1} and $3.3 \pm 0.3 \times 10^4$ s^{-1} for $k^0_{3+/2+}$, $k^0_{2+/1+}$ and $k^0_{1+/0}$, respectively. For large values of η, these Tafel plots are curved for all three redox reactions, and while those corresponding to metal-based electron transfer are asymmetric with respect to η, those corresponding to ligand-based reactions are symmetric. Significantly, temperature-resolved measurements of k reveal that the electrochemical activation enthalpy, $\Delta H^{\ddagger}$, decreases from 43.1 ± 2.8 kJ mol^{-1} for the 3+/2+ reaction to 25.8 ± 1.9 kJ mol^{-1} for the 1+/0 process. Probing the temperature dependence of the formal potential gives the reaction entropy, ΔS^0_{rc}. The latter depends on the state of charge of the monolayer, with values of 212 ± 18, 119 ± 9 and 41 ± 5 J mol^{-1} K^{-1} being observed for the 3+/2+, 2+/1+ and 1+/0 redox transformations, respectively. The corresponding free energies of activation are 11.5 ± 0.8, 14.2 ± 1.2 and 19.8 ± 1.3 kJ mol^{-1}, respectively, for the 3+/2+, 2+/1+ and 1+/0 charge states. That $\Delta G^{\ddagger}$ depends on the charge state emphasizes the importance of considering differences in reaction energetics when attempting to evaluate the degree of donor–acceptor electronic coupling from rate data. The electronic transmission coefficients, κ_{el}, describing the probability of electron transfer once the transition state has been reached, are all considerably less than unity, thus suggesting a non-adiabatic reaction involving weak electronic interaction between the electronic manifolds on the two sides of the interface. However, the κ_{el} values of 22.9 ± 8.3 and $24.4 \pm 9.9 \times 10^{-6}$ s^{-1}, observed for the first and second bpy reduction reactions, respectively, are more than an order of magnitude *larger* than those observed for the metal-based reaction, i.e. $1.5 \pm 0.7 \times 10^{-6}$ s^{-1}. These investigations suggest that two-state models, which consider just

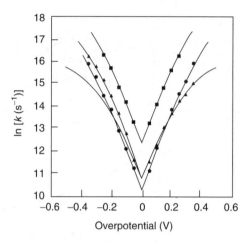

Figure 5.5 Tafel plots, [Os(bpy)$_2$py(P3P)]$^{2+}$ where ■, ▲ and ● denote experimental data for the 2+/1+, 3+/2+ and 1+/0 redox couples, respectively; the continuous lines represent theoretical fits. The 2+/1+ and 1+/0 data are modeled by using a through-bond tunneling approach in which the solvent reorganization energies are equal to 62 and 80 kJ mol^{-1}, respectively, while the 3+/2+ data are modeled by using a through-space approach in which the solvent reorganization energy is 48 kJ mol^{-1}. The heavy continuous line indicates the behavior expected for metal-based heterogeneous electron transfer occurring via a through-bond tunneling mechanism. The supporting electrolyte is 0.2 M TBABF$_4$ in acetonitrile; the errors on the rate constants are approximately equal to the size of the symbols. Reprinted with permission from R. J. Forster, *Inorg. Chem.*, **35**, 3394 (1996). Copyright (1996) American Chemical Society

the donor and acceptor, may not account for important factors that influence the rate of heterogeneous electron transfer across metal–monolayer interfaces.

Chemical control of bridge energies

Traditionally, the only approach to modulating the rate of electron transfer across a bridge linking molecular and bulk components was to synthetically change the structure of the bridge, the solvent, or to couple electron transfer to mass transport, e.g. proton-coupled electron transfer reactions. A particular challenge is to develop systems in which the electronic structure of the bridge can be reversibly changed in response to the local microenvironment, e.g. through a protonation reaction. The objective in systems of this kind is to tune the energy of the bridge states so as to control the electron transfer dynamics.

Monolayers in which electrochemically reversible redox centers are bound to an electrode surface using electrochemically 'non-innocent' bridges have the potential to provide significant insight into the issue of resonant tunneling. For example, Figure 5.6 illustrates a building block for spontaneously adsorbed monolayers incorporating [Os(bpy)$_2$ Cl]$^+$ moieties which are linked to an electrode surface through a redox-active 3,6-bis(4-pyridyl)-1,2,4,5-tetrazine (4-tet) bridge [12]. Bridges of this kind are attractive model systems, not only because they are redox-active, but

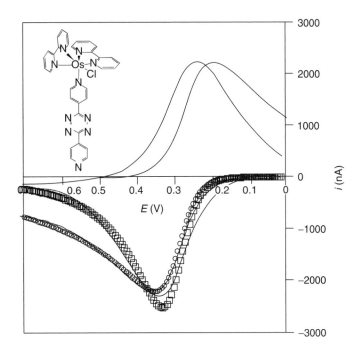

Figure 5.6 Cyclic voltammograms obtained for a spontaneously adsorbed $[Os(bpy)_2 \text{ 4-tet Cl}]^+$ monolayer on a 5 μm radius gold microdisk electrode where the scan rate is 1333 V s^{-1}. The theoretical fits to the data, using a non-adiabatic electron tunneling model at electrolyte pH values of 0.9 and 6.0, are denoted by $\bigcirc$ and $\square$, respectively. In both cases, λ is 27 kJ mol^{-1}, while k^0 is 1.1×10^3 and $1.1 \times 10^4 \text{ s}^{-1}$ at pH values of 0.9 and 6.0, respectively. Reprinted with permission from D. A. Walsh, T. E. Keyes, C. F. Hogan and R. J. Forster, *J. Phys. Chem.*, **105**, 2792 (2000). Copyright (2000) American Chemical Society

because they are also capable of undergoing protonation–deprotonation reactions depending on the pH of the contacting electrolyte solution. Voltammetry of these monolayers is well-behaved over a wide range of scan rates, electrolyte concentrations and pH values, thus allowing molecular bridge effects, in particular the role of superexchange on distant charge tunneling, to be investigated.

As illustrated in Figure 5.6, the voltammetry of the metal-based redox process, $Os^{II/III}$, is well-defined even at high scan rates. Modeling reveals that the redox switching mechanism is best described as a non-adiabatic, through-bond tunneling process. Significantly, while protonating the bridging ligand does not influence the free energy of activation ($10.3 \pm 1.1 \text{ kJ mol}^{-1}$), k^0 decreases by an order of magnitude from 1.1×10^4 to $1.2 \times 10^3 \text{ s}^{-1}$ on going from a deprotonated to a protonated bridge. Figure 5.7 illustrates the through-bond hole tunneling mechanism that can be used to explain these results. Protonation decreases the electron density on the bridge, so shifting the formal potential in a positive potential direction. This shift increases the energy difference between the osmium and the bridge states, thus reducing the strength of the electronic coupling between the metal center and the electrode. This weaker coupling causes a lower heterogeneous electron transfer rate constant to be observed.

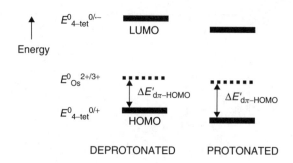

Figure 5.7 Schematic representation of the energy levels of a tetrazine-bridged osmium poly-pyridyl monolayer in the protonated and deprotonated states

5.2.3 Electrode Material Effects on Electron Transfer

Monolayers have helped to revolutionize our understanding of the role that distance, structure and the redox composition of the bridge play in dictating the rate and mechanism of electron transfer. However, the effect of changing the nature of the electrode material has not been probed with the same intensity. This situation is surprising given that traditional theory predicts that for electron transfer involving weakly coupled reactants the heterogeneous electron transfer rates should be directly proportional to the density of states, ρ_F, in the electrode. However, a recent theoretical model developed by Gosavi and Marcus [13] predicts that the electron transfer rate may be less sensitive to ρ_F. A particularly interesting issue is how the different orbitals of metals, such as gold and platinum, which contribute to the overall density of states, influence the electron transfer dynamics. For example, the s electrons dominate ρ_F for metals such as gold and silver, while the higher density of states for platinum arises predominantly because its d orbitals lie near the Fermi level. The Marcus–Gosavi model [13] predicts that the efficiency with which these different orbitals couple with the localized molecular states of the adsorbate may vary significantly. From a technological or device development perspective, this observation offers an important new approach to controlling the rate of non-adiabatic heterogeneous electron transfer.

Spontaneously adsorbed monolayers of $[Os(OMe(bpy))_2p3p\ Cl]^+$ (Figure 5.8) are useful for probing this issue because the complex forms stable, electrochemically reversible monolayers on a wide variety of electrode materials, including carbon fiber, mercury, platinum, gold, copper and silver (OMebpy is 4,4′-dimethoxy-2,2′-bipyridyl) [14]. As illustrated in Figure 5.8, monolayers on each electrode material exhibit well-defined voltammetry for the $Os^{II/III}$ redox reaction where the supporting electrolyte is aqueous 1.0 M $NaClO_4$. The high-scan-rate, >2000 V s^{-1}, voltammetric response has been modeled using a non-adiabatic electron transfer model. The standard heterogeneous electron transfer rate constant, k^0, depends on the identity of the electrode material, e.g. k^0 is 6×10^4 and 4×10^3 s^{-1} for platinum and carbon electrodes, respectively. Chronoamperometry, conducted on a microsecond time-scale, has been used to probe the potential dependence of the heterogeneous electron transfer rate constant, k. Figure 5.9 illustrates typical

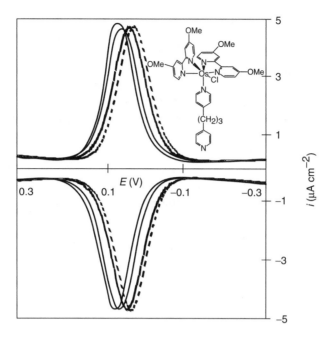

Figure 5.8 Cyclic voltammograms for the $Os^{2+/3+}$ redox reaction within spontaneously adsorbed $[Os(OMebpy)_2(p3p)Cl]^+$ monolayers. From right to left, the electrode materials are platinum, gold, carbon and mercury. The scan rate is $50\ V\ s^{-1}$, with a surface coverage of $1.0 \pm 0.1 \times 10^{-10}\ mol\ cm^{-2}$; the supporting electrolyte is aqueous 1.0 M NaClO$_4$. Reprinted with permission from R. J. Forster, P. J. Loughman and T. E. Keyes, *J. Am. Chem. Soc.*, **122**, 11948 (2000). Copyright (2000) American Chemical Society

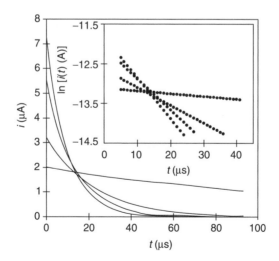

Figure 5.9 Current response for $[Os(OMebpy)_2(p3p)Cl]^+$ monolayers following a potential step where the overpotential, η, was 50 mV. From top to bottom on the left-hand side, the decays are shown for platinum, mercury, gold (5 µm) and carbon (12.5 µm) electrodes; the supporting electrolyte is aqueous 1.0 M NaClO$_4$. The inset shows plots of $\ln[i(t)]$ versus t for the Faradaic reaction. Reprinted with permission from R. J. Forster, P. J. Loughman and T. E. Keyes, *J. Am. Chem. Soc.*, **122**, 11948 (2000). Copyright (2000) American Chemical Society

examples of the current–time transients observed for the $Os^{II} - e^- \rightarrow Os^{III}$ redox reaction of $[Os(OMebpy)_2(p3p)Cl]^+$ monolayers on different electrode materials. In these experiments, the overpotential η ($\equiv E - E^{0'}$) was 0.05 V. In order to use these data to accurately measure k, the response time of the electrode must be shorter than the time-constant for heterogeneous electron transfer. The response times of the electrodes, RC, given by the product of the resistance, R, and capacitance, C, range from approximately 160 ns for the 5 μm radius gold, silver, platinum and mercury electrodes to 440 ns for a 12.5 μm radius carbon fiber electrode. These response times are sufficiently short such that double-layer charging will contribute to the currents observed in Figure 5.9 only at time-scales shorter than approximately 1.5 μs. Therefore, the heterogeneous electron transfer rate constant can be measured by analyzing the current–time transients at relatively longer time-scales. The standard heterogeneous electron transfer rate constants obtained from Tafel plots of $\ln k$ versus the overpotential, η, range from $4.0 \pm 0.2 \times 10^4$ to $3.0 \pm 0.3 \times 10^3$ s^{-1} on going from platinum to carbon electrodes. Temperature-resolved chronoamperometry and cyclic voltammetry reveal that the electrochemical activation enthalpy, $\Delta H^{\ddagger}$, and the reaction entropy, $\Delta S^{\ddagger}_{RC}$, are both independent of the electrode material, having values of 11.1 ± 0.5 kJ mol^{-1} and 29.6 ± 2.4 J mol^{-1} K^{-1}, respectively.

The electron hopping frequency, ν_{el}, is sensitive to the identity of the electrode material, with values of 7.6 ± 1.7, 5.0 ± 3.1, 2.6 ± 0.3, 2.1 ± 0.3 and $0.5 \pm 0.08 \times 10^5$ s^{-1} having been observed for platinum, mercury, gold, silver and carbon, respectively. The large difference in electron hopping frequencies between platinum and carbon might be expected given the known sensitivity of heterogeneous electron transfer to the density of edge-planes in carbon electrodes. More significant, perhaps, are the differences observed between Pt and Au where the $\Delta G^{\ddagger}$ values are indistinguishable. The density of states, ρ_F, of platinum is approximately 7.5 times that of gold which, according to traditional theory, is expected to cause a proportional increase in ν_{el}. However, the Marcus–Gosavi model [13] highlights the importance of considering the identities of the orbitals that are responsible for increasing ρ_F. For example, the d orbitals found in platinum that lead to an increased density of states are expected to couple less efficiently than the dominant sp states of gold. Thus, the Marcus–Gosavi model predicts a ratio of k^0_{Pt}/k^0_{Au} of 1.8. Superficially, the experimental value of 2.2 ± 0.4 found for this system compares favorably with the theoretical prediction.

These monolayers provide a significant opportunity to compare the extent of electronic communication across the p3p bridge when bound to a metal electrode as opposed to being coupled to a molecular species, e.g. within a dimeric metal complex. Electronic interaction of the redox orbitals and the metallic states causes splitting between the product and reactant hypersurfaces, which is quantified by $|H_{AB}|$, the matrix coupling element. The Landau–Zener treatment [15] of a non-adiabatic reaction yields the following equation:

$$\nu_{el} = (2|H_{AB}|^2/h)(\pi^3/\lambda RT)^{1/2} \qquad (5.1)$$

where h is the Planck constant. The matrix coupling element has been determined for each of the electrode materials and is less than 1 kJ mol^{-1} for monolayers on all of the electrode materials. This small $|H_{AB}|$ value confirms that the system is charge-localized and non-adiabatic. Significantly, the $|H_{AB}|$ values for the monolayers are approximately four orders of magnitude smaller than those found for dimers of similar structure. This behavior indicates that the metal center interacts much more weakly with the electrode surface than with another metal center.

5.2.4 Effect of Bridge Conjugation on Electron Transfer Dynamics

The production of 'molecular wires' which promote fast heterogeneous electron transfer between a remote redox center and a metal surface is important for high-speed molecular electronics applications, e.g. molecular-based computing. One notable strategy in this area is to use conjugated rather than aliphatic bridges to achieve faster electron transfer rates. For example, conjugated molecules of (1,1':4',1''-terphenyl)-4-methane thiol (TP) have been embedded in self-assembled monolayers of insulating *n*-alkane thiols, and the conductivity of single molecules measured [16]. In addition, Weber and Creager [17] have recently probed the electron transfer rates for ferrocene groups attached to gold electrodes via conjugated oligophenylethynyl bridges of variable length. Significantly, they found that the extrapolated rate constants at short distances were nearly the same for conjugated and aliphatic bridges. Moreover, for short electron transfer distances, the observed rates were consistent with those expected for an adiabatic electron transfer, thus suggesting strong electronic coupling between the redox centers and the electrode. However, directly comparing heterogeneous electron transfer rate constants will provide an insight into the strength of electronic coupling only if the free energies of activation are identical for conjugated and non-conjugated bridges. For example, switching from a non-conjugated to a conjugated bridging ligand may alter the structure of the monolayer, thus making the local dielectric constant and hence the outer-sphere reorganization energies different in the two cases.

The electron transfer dynamics of monolayers based on osmium polypyridyl complexes linked to an electrode surface through conjugated and non-conjugated bridges, e.g. *trans*-1,2-bis(4-pyridyl)ethylene (bpe) and 1,2-bis(4-pyridyl)ethane (p2p), respectively, have been explored [18]. The standard heterogeneous electron transfer rate constant, k^0, depends on both a frequency factor and a Franck-Condon barrier, as follows [19–21]:

$$k^0 = A_{et} \exp(-\Delta G^{\ddagger}/RT) \tag{5.2}$$

where A_{et} is the pre-exponential factor (equal to the product of Γ_n, the nuclear tunneling factor, κ_{el}, the electronic transmission coefficient and ν_n, the nuclear frequency factor) and $\Delta G^{\ddagger}$ is the electrochemical free energy of activation [22].

One approach to decoupling these two contributions is to use classical temperature-resolved measurements of k^0 to measure the free energy of activation, $\Delta G^{\ddagger}$, so allowing A_{et} to be determined. In this way, information about the strength of electronic coupling can be obtained [23,24]. A second method involves measuring electron transfer rate constants at a single temperature over a broad range of reaction driving forces. For example, Finklea and Hanshew [25] have assembled a model describing through-space electron tunneling which provides a good description of such tunneling in monolayers of this kind. In this model, the anodic rate constant, $k_{Ox}(\eta)$, is given by the integral over energy (E) of three functions, namely (a) the

Fermi function for the metal, $n(E)$, (b) a Gaussian distribution of energy levels for acceptor states in the monolayer, $D_{Ox}(E)$, and (c) a probability factor describing electron tunneling at a given energy, $P(E)$:

$$k_{Ox}(\eta) = A \int_{-\infty}^{\infty} D_{Ox}(E)n(E)P(E)\,d\varepsilon \qquad (5.3)$$

The zero point of energy is defined as the Fermi level of the metal at the particular overpotential of interest. The Fermi function describes the distribution of occupied states within the metal and is defined by the following:

$$n(E) = \left\{ \frac{1}{1 + \exp[(\varepsilon - \varepsilon_F)/k_B T]} \right\} \qquad (5.4)$$

where k_B is the Boltzmann constant. The density of acceptor states is derived from the Marcus theory [26,27] and is represented by the following equation:

$$D_{Ox}(E) = \exp\left[-\frac{(E + \eta - \lambda)^2}{4k_B \lambda T} \right] \qquad (5.5)$$

where λ is the reorganization energy. The probability of direct elastic tunneling [28,29] through a trapezoidal energy barrier of height E_B can be approximated by the following:

$$P(E) = (E_B - E + e\eta/2)\exp(-\beta d) \qquad (5.6)$$

where E_B is the average barrier height at zero overpotential, and d is the electron transfer distance.

Chidsey [30], Weber and Creager [31] and Murray and co-workers [32] have modeled non-adiabatic heterogeneous electron transfer for long-chain alkane thiol monolayers by using an expression similar to Equation (5.3), except that the energy-dependent prefactor in the tunneling probability expression is excluded.

The current for the reaction of an immobilized redox center following first-order kinetics is given by the following [32]:

$$i_F = nFA[k_{Ox}(\eta)\Gamma_{Red,\eta} - k_{Red}(\eta)\Gamma_{Ox,\eta}] \qquad (5.7)$$

where $\Gamma_{Red,\eta}$ and $\Gamma_{Ox,\eta}$ are, respectively, the instantaneous surface coverages of the oxidized and reduced species, and $k_{Ox}(\eta)$ and $k_{Red}(\eta)$ are the corresponding reaction rate constants, given by Equation (5.3), with or without a tunneling probability function.

This model has been applied to $[Os(bpy)_2(p2p)\,Cl]^+$ and $[Os(bpy)_2(bpe)\,Cl]^+$ monolayers in order to elucidate the effect of bridge conjugation on the rates of electron transfer. Figure 5.10 shows the experimental (background-corrected) cyclic voltammogram for a dense $[Os(bpy)_2(bpe)\,Cl]^+$ monolayer deposited on a 5 μm radius platinum microelectrode where the scan rate is 6000 V s^{-1}. This figure also illustrates the best fits obtained from Equations (5.3)–(5.7) for models which include and exclude through-space tunneling through a trapezoidal barrier. For both models, the optimum k^0 was $9.4 \pm 0.9 \times 10^3$ s^{-1} and $\Delta G^{\ddagger}$ was 11.4 ± 0.8 kJ mol^{-1}. Significantly, k^0 for the conjugated bpe ligand is approximately a factor of 30 *smaller* than that found for a comparable aliphatic bridge (p2p), i.e. $30.5 \pm 2.6 \times 10^4$ s^{-1}.

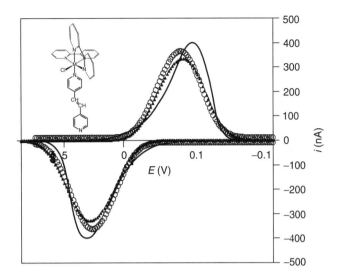

Figure 5.10 Cyclic voltammograms for a spontaneously adsorbed [Os(bpy)$_2$(bpe)Cl]$^+$ monolayer on a 25 μm radius platinum microdisk electrode at a scan rate of 6000 V s^{-1}. The continuous line represents experimental data, while ○ and ▲ denote the best-fit theoretical responses generated by using the Finklea and Chidsey models, respectively. Both theoretical responses correspond to a k^0 of 9.4×10^3 s^{-1} and a $\Delta G^{\ddagger}$ of 11.4 kJ mol^{-1}; the supporting electrolyte is aqueous 0.1 M LiClO$_4$. Reprinted with permission from R. J. Forster, P. J. Loughman, E. Figgemeier, A. C. Lees, J. Hjelm and J. G. Vos, *Langmuir*, **16**, 7871 (2000). Copyright (2000) American Chemical Society

The decreased rate constant observed for the conjugated bridge arises because the free energy of activation, $\Delta G^{\ddagger}$, is larger (11.4 ± 0.8 kJ mol^{-1}) for the bpe than for the p2p (8.7 ± 1.2 kJ mol^{-1}) bridge. Taken in conjunction with Equation (5.2), these values yield a pre-exponential factor for the non-conjugated p2p bridge ($11.1 \pm 0.5 \times 10^6$ s^{-1}) that is approximately an order of magnitude higher than that found for the conjugated bpe bridge ($0.9 \pm 0.3 \times 10^6$ s^{-1}). This observation indicates that the redox center and electrode are more strongly coupled for the non-conjugated system. Such a result is consistent with the through-space tunneling mechanism indicated by the cyclic voltammetry data in that the less rigid p2p bridge probably has sufficient flexibility to reduce the electron transfer distance below the value of 10 Å expected on the basis of a rigid-rod model. It is important to note that unlike a through-space pathway, the electron transfer distances for through-bond tunneling mechanisms are independent of the monolayer's conformation. Therefore, the *decrease* in the heterogeneous electron transfer rate constant observed in this study on going from a non-conjugated to a conjugated bridge, is not expected for a through-bond tunneling mechanism.

5.2.5 Redox Properties of Dimeric Monolayers

The development of spontaneously adsorbed and self-assembled monolayers has greatly facilitated investigations into those factors which influence the rate of heterogeneous electron transfer. However, important issues such as the effect of switching

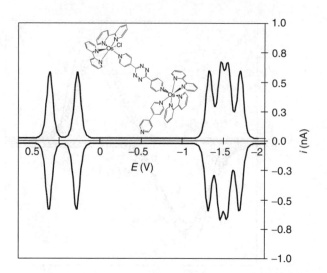

Figure 5.11   Voltammetric response for a 5 μm platinum electrode modified with a monolayer of $[(p0p)Os(bpy)_2(4\text{-tet})Os(bpy)_2Cl]^{3+}$, measured at a scan rate of 10 V s^{-1}. The supporting electrolyte is 1.0 M LiClO$_4$ in acetonitrile, and the monolayer surface coverage is 9.8×10^{-11} mol cm^{-2}. The cathodic currents are shown as 'up', while the cathodic currents are shown as 'down'. Reprinted with permission from R. J. Forster and T. E. Keyes, *J. Phys. Chem., B*, **105**, 8829 (2001). Copyright (2001) American Chemical Society

the oxidation state of a bridge unit on the dynamics of electron transfer, or comparisons of thermal versus photochemically driven reactions, cannot be explored by using traditional monolayers that contain a single redox center. In contrast, monolayers composed of surface active dimers allow these issues to be investigated.

Spontaneously adsorbed monolayers of the dimeric complex (Figure 5.11) $[(p0p) Os(bpy)_2 (4\text{-tet}) Os(bpy)_2 Cl]^{3+}$, where p0p is 4,4'-bipyridyl, bpy is 2,2'-bipyridyl and 4-tet is 3,6-bis(4-pyridyl)-1,2,4,5-tetrazine, have been assembled on platinum microelectrodes in an attempt to address these issues [33]. Significantly, as illustrated in Figure 5.11, the voltammetric response associated with the Os$^{II/III}$ reaction is unusually ideal for both metal centers. Studies using mononuclear model compounds reveal that the redox responses centered at approximately 0.620 and 0.300 V correspond to the 'inner' $[(p0p) Os(bpy)_2 (4\text{-tet})]^{2+}$ and 'outer' $[(4\text{-tet}) Os(bpy)_2 Cl]^+$ moieties, respectively. The observation of two well-defined voltammetric waves indicates that electron transfer can occur across the $[(p0p) Os(bpy)_2 (4\text{-tet})]^{2+}$ bridge to the outer $[Os(bpy)_2 Cl]^+$ moiety, i.e. charge trapping does not occur.

High-speed chronoamperometry reveals that the standard heterogeneous electron transfer rate constants, k^0, for the 'inner' $[(p0p) Os(bpy)_2 (4\text{-tet})]^{2+}$ and 'outer' $[(4\text{-tet}) Os(bpy)_2Cl]^+$ moieties are $1.3 \pm 0.2 \times 10^6$ and $1.1 \pm 0.1 \times 10^4$ s^{-1}, respectively. The reorganization energy is at least 0.6 ± 0.1 eV. The difference in rate constants for the inner versus the outer redox centers is remarkably small given that the difference in electron transfer distances for the two couples is approximately 14 Å, assuming that electron transfer occurs via a through-bond tunneling mechanism for both centers. For example, taking a value of 1.1 Å^{-1} for the distance-dependent tunneling factor, β, one would expect the rate constant for the outer site to be approximately four orders of magnitude smaller than what is experimentally observed! Significantly, where electron transfer proceeds via a coherent superexchange mechanism [34], the rate depends algebraically on the difference between the energy levels of the bridge and donor–acceptor, $\Delta E_{\text{Bridge–D/A}}$. For the p0p-Os-4-tet bridge considered here, reducing the inner metal center creates a new bridge state centered at the E^0 for the $\text{Os}^{\text{II/III}}$ couple (0.620 V). The larger k^0 observed for the inner redox couple arises because of a larger prefactor, i.e. coupling effects, in Equation (5.2), thus suggesting that converting the metal center from the oxidized to the reduced form changes the strength of electronic coupling between the outer metal center and the electrode. This ability to electrochemically modulate electronic coupling across a bridge is highly desirable for developing molecular switches based on resonant tunneling approaches.

Traditionally, it is assumed that the dynamics of electron transfer across electrode–monolayer interfaces is influenced by the structure of the bridge rather than the tunneling barrier between the electrode and the bridge, e.g. the gold–thiol link in alkane thiol monolayers. However, there have been remarkably few experimental investigations of this issue, i.e. comparisons of electronic coupling across a bridge linking two metal centers compared with the same bridge linking remote redox centers to a metal surface. In this $[(\text{p0p}) \, \text{Os(bpy)}_2 \, (\text{4-tet}) \, \text{Os(bpy)}_2 \, \text{Cl}]^{3+}$ dimer system, solution-phase transient emission measurements reveal that the rate of photoinduced electron transfer (PET) between the two metal centers is $1.6 \pm 0.1 \times 10^7 \, \text{s}^{-1}$. This rate constant is a factor of approximately 400 smaller than the ground-state electron transfer rate constant for monomeric $[(\text{4-tet}) \, \text{Os(bpy)}_2 \, \text{Cl}]^+$ monolayers when the driving forces are identical. This significant difference is interpreted in terms of the energy separation between the ground or excited states and the bridge. These data also reveal that the strength of electronic coupling across the tetrazine bridge is significantly greater for two metal centers than for a spontaneously adsorbed monolayer in which a metal center is replaced with a metal electrode.

5.2.6 Coupled Proton and Electron Transfers in Monolayers

There has been a resurgence of interest in proton-coupled redox reactions because of their importance in catalysis, molecular electronics and biological systems. For example, thin films of materials that undergo coupled electron and proton transfer reactions are attractive model systems for developing catalysts that function by hydrogen atom and hydride transfer mechanisms [4]. In the field of molecular electronics, protonation provides the possibility that electrons may be trapped in a particular redox site, thus giving rise to molecular switches [5]. In biological systems, the kinetics and thermodynamics of redox reactions are often controlled by enzyme-mediated acid–base reactions.

Quinones represent an important class of compound that undergo proton-coupled electron transfer reactions [35]. The order and kinetics of the two-electron/two-proton redox reactions of the quinone (Q)/hydroquinone (H_2Q) couple continue to be active subjects of investigation. The interconversion of Q to H_2Q can involve up to seven different intermediates depending on the pH of the solution and the solvent. However, in low-pH electrolytes electrochemically reversible behavior can be observed despite the significant changes that accompany redox switching. Beyond

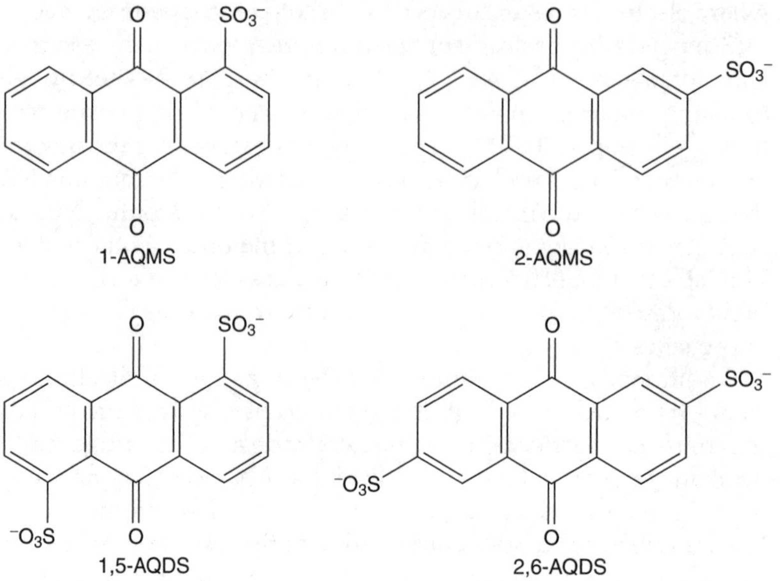

Figure 5.12 Structures of some anthroquinones used to form spontaneously adsorbed monolayers on mercury electrodes

extensive investigations into the thermodynamics and kinetics of solution-phase reactants, quinones are building blocks for assembling interfacial supramolecular assemblies. In a careful study, Faulkner and co-workers [36] developed a detailed description of the electrochemical behavior of 2,6-anthraquinone disulfonate (2,6-AQDS) at mercury surfaces in contact with aqueous electrolytes. In less detail, they also considered the behavior of 1,5-anthraquinone disulfonate (1,5-AQDS) and 2-anthraquinone monosulfonate (2-AQMS). The structures of these quinones are illustrated in Figure 5.12. These investigations reveal that the anthraquinone disulfonates, particularly 2,6-AQDS, exhibit a wide range of phenomena of interest to contemporary electrochemistry. For example, they form extremely tightly bound adsorbate layers, which are stable in both redox forms. As illustrated in Figure 5.13, under accessible, convenient conditions, these layers can be interconverted between redox forms with virtually ideal electrochemical responses. Thus, they are excellent model systems for studying coupled electron and proton transfer dynamics. In seeking to functionalize electrode surfaces, it is frequently implicitly assumed that the properties of the interface can be 'tuned' by simply immobilizing a molecule with the desired chemical, optical or electrochemical properties onto the electrode surface. However, there are several issues that may cause an adsorbed reactant to behave differently from its solution-phase counterpart. These effects include the effect of adsorption on the electron density of the reactant, lateral interactions between the adsorbates, or the differences in solvation that may exist between solution phase and adsorbed species. In order to address these issues, as illustrated in Figure 5.14, Faulkner and co-workers [36] probed the effect of changing the pH of the supporting electrolyte on the formal potential of the Q/H_2Q reaction for both solution-phase and adsorbed reactants. Both the adsorption peak and the diffusion peak shift in a negative potential direction as the pH is increased. The $E^{0'}$

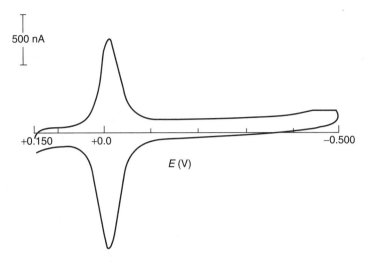

Figure 5.13   Cyclic voltammogram of a 10 μM solution of 2,6-AQDS in 0.1 M HNO₃. Reprinted with permission from P. He, R. M. Crooks and L. R. Faulkner, *J. Phys. Chem.*, **94**, 1135 (1990). Copyright (1990) American Chemical Society

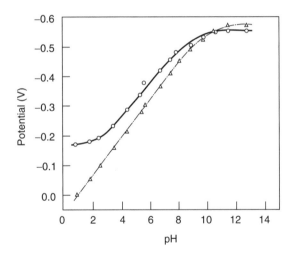

Figure 5.14   Relationship between peak potentials and solution pH for adsorbed (△) and freely diffusing (◯) 2,6-AQDS. Reprinted with permission from P. He, R. M. Crooks and L. R. Faulkner, *J. Phys. Chem.*, **94**, 1135 (1990). Copyright (1990) American Chemical Society

for the solution-phase reactant shifts linearly for $2 \leq pH \leq 7$ with a slope of 57 mV pH^{-1}, thus indicating that each electron transferred is coupled to the transfer of a single proton. For the adsorbed monolayer, the slope is 63 mV pH^{-1}, reflecting not only the proton dependence of the redox reaction but also pH-induced changes in adsorption strength of the reduced form relative to that of the oxidized form. That a 'super-Nernstian' slope is observed suggests that the relative adsorption strength of the reduced form decreases as the pH increases. However, this behavior appears to be sensitive to the position of the sulfonic acid substituents [37]. For example, as illustrated in Figure 5.15 the dependence of $E^{0\prime}$ on electrolyte pH is significantly

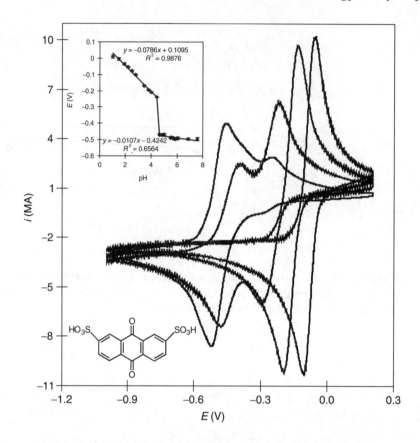

Figure 5.15 Cyclic voltammograms obtained for a mercury electrode immersed in a 5 mM solution of 2,7-AQDS as the pH of the unbuffered contacting electrolyte solution (0.1 M LiClO$_4$) is varied by using either HClO$_4$ or NaOH. The pH values, from left to right, are 2.8, 3.5, 4.8 and 6.1. The scan rate is 0.1 V s^{-1}, with an initial potential of -1.000 V. The inset shows the dependence of the formal potential on the solution pH in unbuffered solution. Reprinted from *J. Electroanal. Chem.*, **498**, R. J. Forster and J. P. O'Kelly, 'Protonation reactions of anthroquinone-2,7-disulfonic acid in solution and within monolayers', 127–135, Copyright (2001), with permission of Elsevier Science

different for 2,7-AQDS. In unbuffered solution, $E^{0'}$ shifts in a negative potential direction as the pH is increased, with a slope of 78 ± 5 mV pH^{-1} for $1.0 < \text{pH} < 4.7$. The slope observed is somewhat larger than the ideal value of 59 mV pH^{-1} expected for the two-proton, two-electron conversion of the quinone, Q, to hydroquinone, H$_2$Q, probably reflecting a decreasing relative adsorption strength for the reduced form as the pH increases. Two redox couples are observed in the pH range 4.7 to 5.9. The total charge under these two peaks is indistinguishable from that observed under the single peak found at low or high pH. Moreover, the peak shape and peak-to-peak separations remain consistent with redox reactions involving the transfer of two electrons. It appears that the two peaks arise because the concentration of both H$_3$O$^+$ and OH$^-$ ions is significantly less than that of the quinone. Specifically, for $4.7 < \text{pH} < 5.9$, the proton concentration is such that a significant fraction of the quinone can be reduced to hydroquinone, so giving rise to the first set of peaks at an E^0 dictated by the bulk pH of the solution. However, because electron and proton transfer are coupled in this system, electrolysis is accompanied by a significant change in the interfacial pH. Thus, a second set of peaks is observed as the remainder of the quinone is reduced to the deprotonated form of the hydroquinone. This changeover from proton-coupled, to proton-independent, redox switching causes the discontinuity in the plot of $E^{0\prime}$ versus pH illustrated in Figure 5.15.

These interfacial pH effects have been investigated by probing the voltammetry in buffered solutions. Figure 5.16 shows that for $1.0 \leq \text{pH} \leq 10.6$, $E^{0\prime}$ depends linearly on pH, with a slope of 63 ± 3 mV. This value is indistinguishable from the slope of 59 mV pH^{-1} expected for a coupled proton/electron transfer and indicates that the H_2Q species is produced when the monolayer is reduced. Between pH 10.6 and 12.0, the slope decreases to 25 ± 4 mV pH^{-1}, which compares favorably with the slope expected (29.5 mV pH^{-1}) for a two-electron, one-proton transfer reaction. Therefore, over this pH range Q is reduced to HQ$^-$ and the pK_a for the Q/HQ$^-$ couple is 10.6 ± 0.2. For pH ≥ 12.0, $E^{0\prime}$ is independent of pH, thus indicating that the pK_a of the HQ$^-$/Q^{2-} couple is 12.0 ± 0.2.

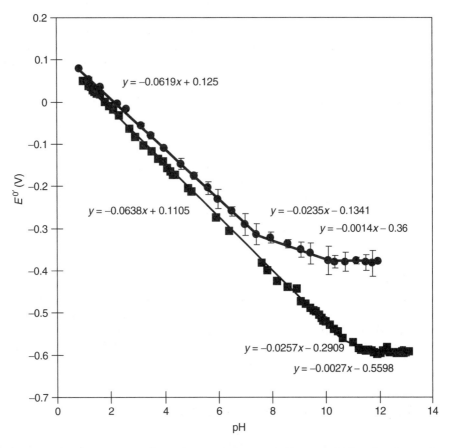

Figure 5.16 Formal potential of 2,7-AQDS as a function of pH in buffered 0.1 M LiClO$_4$: ■, solution-phase data; ●, monolayer data. The formal potentials are typically reproducible to within +5 mV, which is comparable to the size of the symbols. Reprinted from *J. Electroanal. Chem.*, **498**, R. J. Forster and J. P. O'Kelly, 'Protonation reactions of anthroquinone-2,7-disulfonic acid in solution and within monolayers', 127–135, Copyright (2001), with permission of Elsevier Science

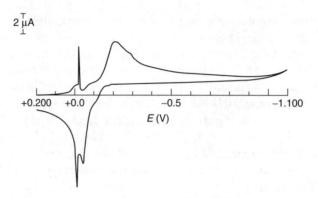

Figure 5.17 Cyclic voltammogram of 1 mM 2,6-AQDS in 0.1 M HNO$_3$. Reprinted with permission from P. He, R. M. Crooks and L. R. Faulkner, *J. Phys. Chem.*, **94**, 1135 (1990). Copyright (1990) American Chemical Society

An important objective is to probe the effect of surface confinement on the pK_as of these couples. The inflection points of Figure 5.16 suggest that the solution-phase pK_as are 7.6 ± 0.2 and 10.6 ± 0.2 for the H$_2$Q/HQ$^-$ and HQ$^-$/Q^{2-} couples, respectively. Both pK_as shift towards higher values when the anthraquinones are immobilized within a dense monolayer, thus indicating that both couples are weaker acids in the adsorbed state. Beyond the issue of double-layer effects, there are two dominant processes that could cause the quinones to behave as weaker acids upon immobilization. First, adsorption could change the electron density on the hydroquinone moieties. Secondly, the local microenvironment, especially the effective dielectric constant within the monolayer, may be very different from that found in the bulk solution. Typically, the formal potential shifts by less than 15 mV upon surface confinement and $E^{0\prime}$ does not depend significantly on the surface coverage of the adsorbates. That $E^{0\prime}$ shifts in a negative potential direction indicates that the adsorbed form is somewhat more easily oxidized and has a higher electron density. Therefore, one would expect the adsorbed anthraquinones to act as weaker acids.

As illustrated in Figure 5.17, at high surface coverages, corresponding to higher bulk concentrations, adsorbed 2,6-AQDS shows extremely sharp spikes in its voltammetry, apparently reflecting extensive reorganization. As shown in Figure 5.18, this film restructuring probably involves a phase transition in the adlayer which is driven by strong intermolecular interactions. Under these circumstances, the layers can prove useful as models for studies of redox-induced structural alterations in two-dimensional assemblies.

5.2.7 Redox-Switchable Lateral Interactions

Assemblies formed by the coadsorption of surfactants at the solid–liquid interface represent attractive model systems for probing the nature and strength of lateral interactions among surfactants. These studies reveal strong synergistic effects in

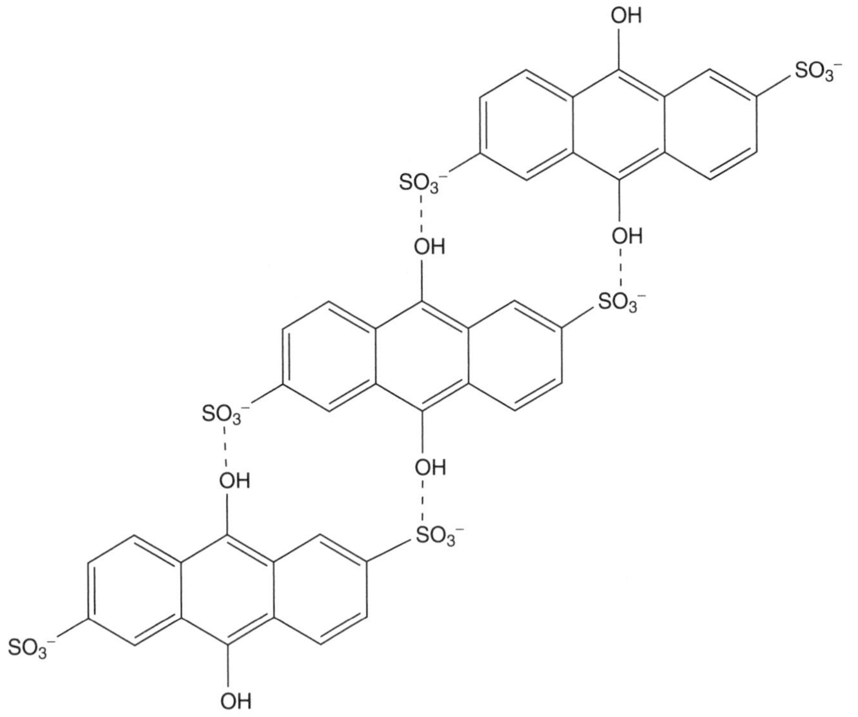

Figure 5.18 Proposed extended hydrogen-bonded structure for reduced adsorbed 2,6-AQDS at high surface coverage. Reprinted with permission from P. He, R. M. Crooks and L. R. Faulkner, *J. Phys. Chem.*, **94**, 1135 (1990). Copyright (1990) American Chemical Society

the adsorption of ionic–nonionic surfactant mixtures. Non-covalent interactions, such as hydrogen bonding, dipole–dipole, hydrophobic interactions, etc., have proven to be useful driving forces for the assembly of ordered structures in a way similar to that found in biological systems. Hydrogen bonding is a particularly attractive lateral interaction since it may be possible to control the bonding strength by changing the redox composition of the assembly. Moreover, for redox-active monolayers, changing the inter- and intramolecular hydrogen bonding offers the possibility of controlling the rate of heterogeneous electron transfer.

Quinone-based monolayers are especially attractive model systems for probing the impact of hydrogen bonding on the redox properties of monolayers because they can be reversibly switched between oxidized, quinone and reduced, hydro-quinone forms. Given that the strength of hydrogen bonding is sensitive to the nature of the functional groups present, the lateral interactions in these assemblies are expected to depend on the redox composition of the monolayer. Another attractive feature of quinonoid monolayers is that because electron and proton transfer are coupled, switching between Q and H_2Q does not change the charge state of the film. Thus, electrostatic effects are unlikely to significantly influence the nature or degree of lateral interactions. For example, as illustrated in Figure 5.19, single-component monolayers of anthraquinone 2,6-disulfonic acid (2,7-AQDS) and

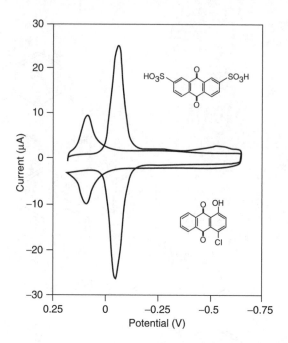

Figure 5.19 Cyclic voltammograms for a pure monolayer of 2,7-AQDS (left-hand side) and 1,4-AQClOH (right-hand side); the concentrations of each are 20 μM. The solution is 20 vol% DMF in H$_2$O containing 1.0 M HClO$_4$ as the supporting electrolyte, with a scan rate of 5 V s^{-1}. Reprinted with permission from D. O'Hanlon and R. J. Forster, *Langmuir*, **16**, 702 (2000). Copyright (2000) American Chemical Society

1-chloro-4-hydroxyanthraquinone (1,4-AQClOH), formed by spontaneous adsorption onto mercury, exhibit nearly ideal voltammetry [38]. The concentration dependence of the surface coverage is well described by the Frumkin adsorption isotherm over the concentration range from 1 to 30 μM for both 2,7-AQDS and 1,4-AQClOH. The saturation coverages, Γ_s, are $6.8 \pm 0.32 \times 10^{-11}$ and $1.9 \pm 0.15 \times 10^{-10}$ mol cm^{-2}, respectively, while the adsorption coefficients, β, are $4.9 \pm 0.25 \times 10^4$ and $7.5 \pm 0.5 \times 10^4$ M^{-1}. Both monolayers show stabilizing lateral interactions. However, the 2,7-AQDS adsorbates interact more strongly with a Frumkin interaction parameter, g, of -1.62 ± 0.21 being observed, compared to -0.21 ± 0.02 for the 1,4-AQClOH monolayers.

Mixed monolayers, formed from simultaneous adsorption of both anthraquinones, also show lateral interactions, with full widths at half maximum of 56 ± 2 mV being observed for both components. Moreover, as shown in Table 5.1, the saturation coverages, free energy adsorption coefficients and lateral interaction (Frumkin) parameters of both adsorbates differ significantly when the adsorbates are immobilized within two-component rather than single-component monolayers. The relationship between the composition of the deposition solutions and the two-component monolayers suggests that mixed monolayers exist as a single phase in which the adsorbates do not have a preference for being surrounded by their own kind. These stabilizing lateral interactions are consistent with intermolecular hydrogen bonding in which the 2,7-AQDS and 1,4-AQClOH adsorbates act as electron acceptors and donors, respectively.

Table 5.1 Saturation coverages (Γ_S), adsorption coefficients (β) and Frumkin interaction parameters (g) for single- and two-component monolayers of 2,7-AQDS and 1,4-AQClOH.

Parameter	Single-component	Two-component
$\Gamma_{S,\ 2,7-AQDS}$ (mol cm^{-2})	$6.8 \pm 0.32 \times 10^{-11}$	$1.1 \pm 0.05 \times 10^{-10}$
$\Gamma_{S,\ 1,4-AQClOH}$ (mol cm^{-2})	$1.9 \pm 0.15 \times 10^{-10}$	$1.2 \pm 0.04 \times 10^{-10}$
$\beta_{2,7-AQDS}$ (M^{-1})	$4.9 \pm 0.25 \times 10^{4}$	$8.7 \pm 0.4 \times 10^{3}$
$\beta_{1,4-AQClOH}$ (M^{-1})	$7.5 \pm 0.5 \times 10^{4}$	$1.3 \pm 0.1 \times 10^{5}$
$g_{2,7-AQDS}$	-1.62 ± 0.21	-2.10 ± 0.21
$g_{1,4-AQClOH}$	-0.21 ± 0.02	-1.71 ± 0.15
$g_{2,7-AQDS/1,4-AQClOH}$	$-$	-0.95 ± 0.1

5.2.8 Electron Transfer Dynamics of Electronically Excited States

Electronically excited states play pivotal roles in areas as diverse as dye sensitization of semiconductors for solar energy conversion to photosynthesis and display devices [39]. Photoexcited reactants have been used extensively to study electron transfer because excited states, created upon absorptions of photons, are simultaneously better electron donors and acceptors than their ground-state precursors [40]. However, while our understanding of the thermodynamics of ground-state electron transfer has evolved to a high degree on both theoretical and experimental fronts, there have been few reports on *direct* measurements of excited-state redox potentials [41]. Moreover, the direct measurement of oxidation and reduction kinetics of electronically excited states remains almost completely unexplored. This situation has arisen largely because electronically excited states are transient species and typically have sub-microsecond lifetimes. Electrochemical methods cannot provide a meaningful insight into the redox properties of these fleeting species for two reasons [1]. The first of these is that traditional voltammetry involving macroelectrodes is restricted to millisecond, or longer, time-scales. The second is that the time constant for diffusional mass transport greatly exceeds that of the electronically excited state. However, with the advent of interfacial supramolecular assemblies formed on microelectrodes, it is now possible to directly probe the properties of species having sub-microsecond lifetimes [1].

Electrochemical measurements on excited states within monolayers are not often performed [42,43] because, beyond the desired reductive or oxidative quenching of the excited state, other deactivation pathways exist. First, if the monolayers are formed on mirror-smooth surfaces, then as discussed above in Chapter 2, energy transfer from the excited state to the electrode is expected to be efficient for short tethers [44,45]. However, short bridges and strong electronic coupling are required to promote the fast heterogeneous electron transfer dynamics required if the excited state is to undergo heterogeneous electron transfer before decaying back to the ground state. These apparently contradictory demands, i.e. a short electron transfer distance for fast heterogeneous electron transfer versus a large separation of the excited state from the surface to prevent energy transfer, coupled with the extreme demands on the experimental time-scale, represent significant barriers to progress in this area. In addition, because the surface concentration of the adsorbates can be high, lateral energy or electron transfer may lead to quenching of the excited states. Therefore, the ability to control the surface coverage is important.

Ruthenium polypyridyl complexes are attractive building blocks for creating interfacial supramolecular assemblies suitable for probing excited-state redox processes. For example, they have reasonably long lived electronically excited states (of the order of 1 µs), they have multiple accessible redox states that are electrochemically reversible, and they are synthetically flexible, e.g. a surface active pyridine or thiol group can be incorporated within the complex. For instance, dense monolayers of [Ru(bpy)$_2$Qbpy]$^{2+}$), where bpy is 2,2′-bipyridyl and Qbpy is 2,2′:4,4″:4′4″-quarterpyridyl, have been formed by spontaneous adsorption onto clean platinum microelectrodes [46]. As illustrated in Figure 5.20, cyclic voltammetry of these monolayers is nearly ideal, and five redox states are accessible over the potential range from +1.3 to −2.0 V. Chronoamperometry conducted on a microsecond time-scale has been used to measure the heterogeneous electron transfer rate constant, k, for both metal- and ligand-based redox reactions. Heterogeneous electron transfer is characterized by a single unimolecular rate constant, k (s^{-1}). As illustrated in Figure 5.21, the standard heterogeneous electron transfer rate constants, k^0, have been evaluated by extrapolating Tafel plots of $\ln k$ versus overpotential (η) to zero driving force to yield values of $5.1 \pm 0.3 \times 10^5$ s^{-1}, $3.0 \pm 0.1 \times 10^6$ s^{-1} and $3.4 \pm 0.2 \times 10^6$ s^{-1} for $k^0_{3+/2+}$, $k^0_{2+/1+}$ and $k^0_{1+/0}$, respectively.

Beyond their ideal ground-state electrochemistry, following photoexcitation using a laser pulse at 355 nm, emission is observed from the monolayers with an excited state lifetime, 6.2 µs, which exceeds that of the complex in solution, 1.4 µs. It appears

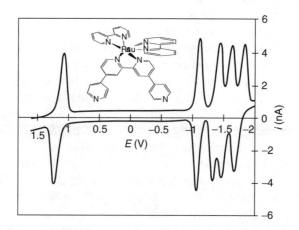

Figure 5.20   Cyclic voltammogram of a spontaneously adsorbed [Ru(bpy)$_2$Qbpy]$^{2+}$ monolayer, obtained at a scan rate of 1 V s^{-1}; the surface coverage is 1.04×10^{-10} mol cm^{-2}. The supporting electrolyte is 0.1 M TBABF$_4$ in acetonitrile, with the radius of the platinum microelectrode being 25 µm. The cathodic currents are shown as 'up', while the anodic currents are shown as 'down'. The complex is in the 2$^+$ form between approximately +1 to −1 V. The inset shows the structure of the surface active complex. Reprinted with permission from R. J. Forster and T. E. Keyes, *J. Phys. Chem., B*, **102**, 10004 (1998). Copyright (1998) American Chemical Society

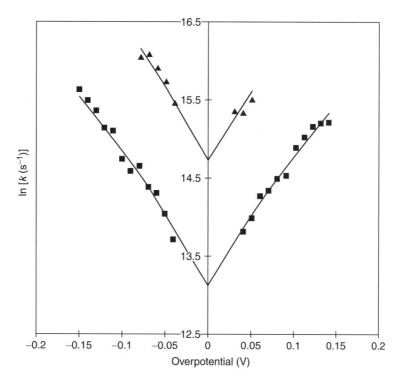

Figure 5.21 Tafel plots for the system illustrated in Figure 5.20, where ■ and ▲ denote data for the 3+/2+ and 2+/1+ redox couples, respectively. The supporting electrolyte is 0.2 M TBABF$_4$ in acetonitrile; the errors on the rate constants are approximately equal to the sizes of the symbols. Reprinted with permission from R. J. Forster and T. E. Keyes, *J. Phys. Chem., B*, **102**, 10004 (1998). Copyright (1998) American Chemical Society

that weak electronic coupling between the adsorbates and the electrode means that the excited states are not completely deactivated by radiationless energy transfer to the metal. Where the objective is to directly probe the dynamics of heterogeneous electron transfer from the electronically excited species, the excited state lifetime dictates the maximum time available for the electrochemical experiment. The advantage of microelectrodes is that the response time, is of the order of 50 ns for a 5 μm radius electrode, i.e. an ultrafast electrochemical experiment can be performed before the excited state decays. For example, Figure 5.22 shows the response observed at a scan rate of 3×10^5 V s^{-1} immediately after photoexcitation. In this experiment, the voltammetric time-scale is shorter than the excited-state lifetime and voltammetry characteristic of the electronically excited state can be obtained. This figure shows that on the first scan an oxidative current response is observed at approximately −0.42 V. The peak potential of this process is in approximate agreement with the excited-state redox potential of −0.67 V predicted by the Rehm–Weller equation. That this current response is not observed for the second or subsequent scans confirms the transient nature of the phenomenon. Beyond the useful insight into the energetics and dynamics of excited-state quenching, luminescent monolayers of this kind are likely to find application in fast-responding active displays and in trace analysis.

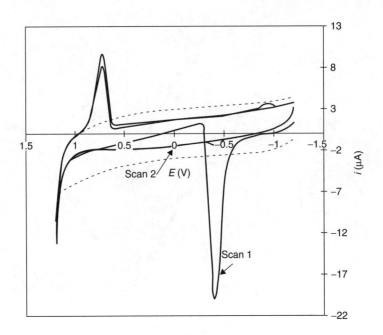

Figure 5.22 Cyclic voltammetry data obtained for a 5 μm radius platinum microelectrode modified with a $[Ru(bpy)_2Qbpy]^{2+}$ monolayer following laser excitation at 355 nm, with a scan rate of 3×10^5 V s^{-1}. The surface coverage is 1.1×10^{-10} mol cm^{-2}, the supporting electrolyte is 0.1 M TBABF$_4$ in acetonitrile, and the initial potential is -1.2 V. Reprinted with permission from R. J. Forster and T. E. Keyes, *J. Phys. Chem., B*, **102**, 10004 (1998). Copyright (1998) American Chemical Society

5.2.9 Conformational Gating in Monolayers

Miniaturization of devices to the molecular level or nanoscale dimensions represents one of the most challenging research subjects in modern science. Interfacial supramolecular assemblies have a key role to play in these developments in that the chemical components are integrated with a solid transducer so that the molecular assembly can communicate with its macroscopic environment. The objective is to create a 'molecular machine' that can be activated or blocked by external signals, e.g. electrical or photonic triggers, and where the state of the machine, e.g. sensory level, structural position and chemical state, is electronically transduced to the outside world. Molecular switches and shuttles have been created in solution in which external photonic, electrical or pH signals were used to achieve switching. However, as already noted, the disadvantage of solution-phase systems is that diffusional mass transport is slow which can make the response times of the devices impracticable. Creating a switchable interfacial supramolecular assembly means that rapidly responding devices can be created that are addressable by both electronic and photonic means. Specifically, electrical transduction of photonic information recorded by photosensitive monolayers on electrode supports has been used to create molecular optoelectronic systems [47,48].

As illustrated in Figure 5.23, Willner *et al.* [49] have developed a light-driven molecular shuttle. This device consists of a ferrocene-functionalized β-cyclodextrin

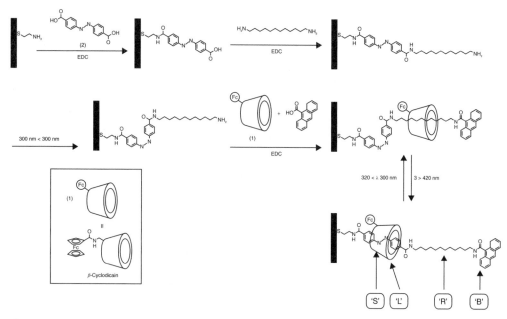

Figure 5.23 Schematic illustration of the organization of the 'molecular train 'monolayer' assembly on a gold electrode and its photoinduced translocation. Reprinted from *J. Electroanal. Chem.*, **497**, I. Willner, V. Pardo-Yissar, E. Katz and K. T. Ranjit, 'A photoactive "molecular train" for optoelectronic applications: light-simulated translocation of a β-cyclodextrin receptor within a stopped azobenzene–alkyl chain supramolecular monolayer assembly on an Au-electrode', 172–177, Copyright (2001), with permission of Elsevier Science

(Fc-β-CD) molecule threaded onto a surface-confined alkyl component containing a photoisomerizable azobenzene unit, and terminated with a bulky anthracene group. In the *cis*-azobenzene configuration, the stoppered 'L'-component, Fc-β-CD, rests on the alkyl chain, the 'R'-unit of the system. Photoisomerization of the monolayer ($\lambda = 420$ nm) to the *trans*-azobenzene configuration yields a high-affinity binding site, 'S'-component, for the 'L'-unit. By further isomerization of the monolayer to the *cis*-configuration, the 'S'-unit is distorted, leading to the translocation of Fc-β-CD to the alkyl chain 'R'-site. The key objective in systems of this kind is to physically transduce the switchable feature of the assembly, i.e. the physical location of the ferrocene, to the external macroscopic world. This objective can be achieved by using chronoamperometry to measure the rate of heterogeneous electron transfer for the oxidation of the ferrocene units of Fc-β-CD when docked in the two distinct locations. Figure 5.24(a), curve (1), shows the chronoamperometric decay of the ferrocene electroactive probe in the *trans*-azobenzene monolayer configuration. At short times, a fast current decay is observed. Curve (2) in this figure illustrates the chronoamperometric response after photoisomerization of the monolayer to the *cis*-azobenzene monolayer. The current decay is substantially slower, with a lower pre-exponential term, when compared to the *trans*-azobenzene monolayer-electrode system. These current responses are fully reversible, and irradiation of the *cis*-azobenzene-monolayer electrode ($\lambda = 420$ nm) regenerates the *trans*-azobenzene monolayer, with the characteristic transient being shown by curve (1). Similarly, further isomerization of the *trans*-azobenzene monolayer

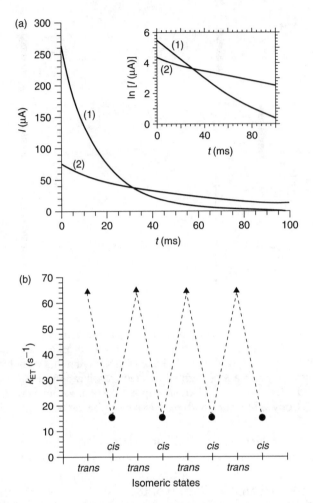

Figure 5.24 (a) Chronoamperometric responses of the Fc–β–CD 'molecular locomotive-shuttle' upon photoisomerization of the azobenzene monolayer, where curve (1) represents the monolayer in the *trans*-azobenzene configuration, and curve (2) the monolayer in the *cis*-azobenzene configuration. The inset shows the semi-logarithmic analysis of the chronoamperometric transients. The applied potential step was from 0.0 to 0.5 V (vs. SCE). (b) Cyclic variation of the electron transfer rate constant of Fc–β–CD upon photoisomerization of the azobenzene–alkyl chain monolayer assembly: ▲, monolayer in the *trans*-azobenzene configuration; ●, monolayer in the *cis*-azobenzene configuration. All measurements were carried out in 0.1 M phosphate buffer solution, at pH 7.3. Reprinted from *J. Electroanal. Chem.*, **497**, I. Willner, V. Pardo-Yissar, E. Katz and K. T. Ranjit, 'A photoactive "molecular train" for optoelectronic applications: light-simulated translocation of a β-cyclodextrin receptor within a stoppered azobenzene–alkyl chain supramolecular monolayer assembly on an Au-electrode', 172–177, Copyright (2001), with permission of Elsevier Science

($320 < \lambda < 380$ nm) restores the slow current response, as shown by, curve (2). Thus, the chronoamperometric responses of the functionalized monolayer upon irradiation agree well with the shuttling of the Fc-β-CD within the molecular network. In the presence of the *trans*-azobenzene unit, the Fc-β-CD is located close to the electrode surface, and a rapid interfacial electron transfer is observed.

Photoisomerization of the monolayer to the *cis*-azobenzene configuration removes the threaded Fc-β-CD to a spatially separated position which retards the electron transfer rate. Therefore, the assembly functions as a molecular optoelectronic system which records optical information and transduces it into an electronic signal.

5.2.10 Electron Transfer within Biosystems

Electron transfer rates within biosystems are exquisitely sensitive to both the local microenvironment and the nature of the protein–membrane interaction. The ability to control the chemical composition, e.g. the identity of the terminating groups, of self-assembled monolayers has opened up significant new opportunities for understanding these effects. For example, Bowden and co-workers have probed the electron transfer rates for horse heart and yeast cytochrome C at mixed SAMs [50]. With horse heart cytochrome C, the electron transfer rate constant is relatively insensitive to the monolayer composition, increasing by approximately a factor of five on going from a pure $HS(CH_2)_{10}COOH$ monolayer to a mixed SAM comprising equimolar amounts of $HS(CH_2)_{10}COOH$ and $HS(CH_2)_8OH$. In contrast, with yeast cytochrome C, electron transfer at a pure $HS(CH_2)_{10}COOH$ SAM is very slow but increases by a factor of 2500 when the two-component monolayer is used. Given that the reorganization energies are similar for the two proteins on both surfaces, this result suggests that the rate differences are caused by differences in the strength of electronic coupling between the protein and the bridge. It appears likely that the large differences in amino acid composition play an important role in this process.

The work of Armstrong and co-workers [51] has revealed the significant impact of subtle changes in structure on electron transfer rate constants. These investigations revealed that replacing the distal histidine, H64, by amino acids such as glycine or leucine dramatically increases the electron transfer (ET) rate for myoglobin at a pyrolitic graphite electrode. Often, these changes in ET rates are not proven to be correlated with changes in structure. However, in this case, the structures of many of these H64 mutants have been characterized by using X-ray crystallography. Histidine-64 stabilizes the aquo ligand which is bound to Fe(III) but not Fe(II), and also links this coordinated water molecule to bulk solvent water through hydrogen-bonding interactions. These H-bonding interactions are sensitive to the dielectric constant of the medium, i.e. the reorganization energy depends on the hydrophobicity of the immobilizing film.

5.2.11 Protein-Mediated Electron Transfer

Many proteins are exclusively involved in intra-protein electron transfer and typically function in ordered structures such as mitochondria. Under these circumstances, the redox-active centers are generally accessible on the outer surface of the protein. In contrast, the redox reactions catalyzed by oxidoreductases involve small molecules with the reaction involving two redox couples, i.e. the substrate and the co-factor or co-substrate. Because the catalytic center of the enzyme is often located deep within the protein, it often undergoes only a very weak electronic interaction with the electrode. This weak coupling results in very slow heterogeneous electron transfer. Driven in part by their ability to selectively catalyze reactions of analytical importance, e.g. lactate and glucose, a major objective is to create redox-active interfacial supramolecular assemblies capable of efficiently mediating electron transfer between the active site and the electrode.

For many applications, the sensor needs to be 'reagentless' i.e. to operate without the need for additional reagents to be added to the sample. This objective can be quite easily achieved by using oxidase enzymes because many test samples have sufficient oxygen present for the correct functioning of the sensor. The reaction sequence is as follows:

$$S_{Red} + E_{Ox} \rightleftharpoons S_{Ox} + E_{Red} \tag{5.8}$$

$$S'_{Ox} + E_{Red} \rightleftharpoons S'_{Red} + E_{Ox} \tag{5.9}$$

For example, in the case of glucose oxidase (GOx), S_{Red}, S_{Ox}, S'_{Ox} and S'_{Red} correspond to glucose, gluconic acid, oxygen and hydrogen peroxide, respectively. For optimum performance, Equation (5.8) must be rate-limiting, which requires a large excess of oxygen. The overall rate of the reaction can be monitored by following either the rate of consumption of oxygen or the formation of hydrogen peroxide. Monitoring oxygen consumption requires two measurements, i.e. one in the presence of enzyme and one in its absence. The advantage of measuring the oxygen concentration is that a selective response can be achieved by using a gas-permeable membrane. However, this additional barrier film has the disadvantage of adding structural 'complexity' to the device. Hydrogen peroxide can be monitored by applying a potential of around 600 mV versus Ag/AgCl as a reference electrode. While oxygen is not electroactive at this potential, numerous endogenous species, such as ascorbic acid, uric acid and a variety of biogenic amines, can be oxidized or reduced. Therefore, to be successful, the device must incorporate a perm-selective membranes to exclude interferences. In contrast, if an interfacial supramolecular assembly can be formed so that the enzyme is electronically coupled or 'wired' to the electrode, then, in principle, the sensor response would no longer be sensitive to fluctuations in S'_{Ox}. This condition can only be met if the rate of electron transfer via the wired route is very much larger than the rate of reaction of the enzyme with endogenous oxygen. While this approach has met with significant success, many mediator-based sensors involving oxidases fail to achieve sufficiently strong coupling between the active site and the electrode and also exhibit parasitic effects of oxygen on their responses.

Willner *et al.* [52] have created some elegant interfacial supramolecular assemblies to address this issue by removing the non-covalently bound flavin adenine dinucleotide (FAD) redox center from glucose oxidase and immobilizing the enzyme on a tether consisting of cystamine chemisorbed on a gold surface, a pyrroloquinoline quinone (PQQ) link and FAD. The mediator potential and electron transfer distances of this assembly were carefully chosen so that transfer of electrons from the FAD to the PQQ and to the electrode is very fast. A maximum rate of $900 \pm 150 \text{ s}^{-1}$ for the enzymatic reaction within this monolayer assembly was obtained, which is indistinguishable from the value of about 1000 s^{-1} obtained for the enzyme in solution. While monolayers can offer molecular-level control of the interfacial structure, the total amount of enzyme immobilized in this approach is very limited, which may affect the long-term stability of the device.

In contrast, 'wiring' enzymes by using redox-active polymers allows significantly greater loading of the enzyme, and three-dimensional catalysis, to be achieved. Heller and co-workers [53] have investigated a variety of approaches to redox wiring of enzymes so that the mediator reacts rapidly with the enzyme, but has a low potential so as to lessen the impact of interferences. For example, in the case of oxidases the formal potential of the flavin redox center is low enough that a relatively weak oxidant with a potential in the range of $+30$ to $-100\,\text{mV}$ versus Ag/AgCl would be sufficient to efficiently mediate electron transfer to the enzyme. The advantage of a low driving potential is that interference from redox-active species such as ascorbate is substantially reduced. Metallopolymers in which a poly(4-vinyl pyridine) or poly(N-vinyl imidazole) backbone contains covalently bound osmium polypyridyl complexes such as $[\text{Os(bpy)}_2\text{Cl}]^+$ or $[\text{Os}(4,4'\text{-dimethoxy-2,2' bipyridine})_2\text{Cl}]^+$ have proven to be particularly useful in this regard. The redox potential for the bipyridyl complex is approximately $+300\,\text{mV}$ but shifts to $+35\,\text{mV}$ on going to the dimethoxy derivative. These metallopolymers exist as hydrogels, thus providing an ideal microenvironment within which the enzyme can react. As discussed in detail above in Chapter 4, the rate of charge transport through these materials is often limited by either charge-compensating counterion transport or the movement of the redox centers through segmental motions of the polymer backbone. These processes tend to be rather slow and freely diffusing oxygen can compete in 'turning over' the immobilized enzyme. Some success in reducing oxygen dependence has been achieved by employing membranes which will impede oxygen diffusion into the film so that it does not reach the enzyme.

Fundamental investigations of the dynamics of electron transfer between mediators and enzymes indicate that their reactivities do not follow simple outer-sphere electron transfer mechanisms. Mikkelsen and co-workers [54] have developed the theory for intramolecular electron transfer involving the multiple attachment of ferrocene derivatives to GOx. The highest rate of intramolecular electron transfer was observed for a ferrocene carboxylic acid, i.e. $0.9\,\text{s}^{-1}$. If the rate of reaction with oxygen with the enzyme is assumed to be $2 \times 10^6\,\text{M}^{-1}\text{s}^{-1}$, then the mediated enzyme regeneration rate would have to be $5 \times 10^3\,\text{s}^{-1}$ if the sensor is placed in an air saturated solution ($[\text{O}_2] = 240\,\mu\text{M}$). Consistent with the current response being sensitive to the oxygen concentration, the experimental rates observed are significantly lower than this value.

5.3 Nanoparticles and Self-Assembled Monolayers

The foregoing discussion has focused on self-assembled monolayers formed on essentially flat electrode surfaces whose areas are vastly larger than those occupied by a single adsorbate. This field has now achieved a significant level of sophistication in terms of their structural characterization as well as their rational design for specific functions, e.g. chemically modulated switches. Although somewhat outside the scope of this book, another important area that exploits the unique properties of self-assembled monolayers is monolayer-protected metal clusters or nanoparticles.

Monolayer-protected clusters (MPCs), e.g. $Au_{55}[PPh_3]_{12}Cl_6$, have been extensively studied since the mid 1990s. However, the field was revolutionized by the seminal work of Schiffrin and co-workers [55] which demonstrated that large stable dodecane-thiol-protected gold nanoclusters could be created. The core diameters were somewhat polydisperse, ranging from 1–3 nm with a main population of 2.0–2.5 nm. The synthetic approach is highly flexible in that a range of thiols can be used and these protected metal nanoparticles are soluble in common organic solvents. Moreover, by varying the mole ratio of RSH to Au it is possible to control the sizes of the metal clusters. These protected nanoclusters can be processed like routine chemical compounds in the dry state and open up the real possibility of size–function studies which exploit the quantum chemical properties of these unusual structures.

While these MPCs exploit a similar surface modification chemistry to that of classical SAMS, these three-dimensional structures differ from the related two-dimensional systems in the following ways. First, the concentration of defect sites is typically significantly higher for the MPCs. Secondly, the radius of curvature is dramatically different from that encountered when using a macro- or microelectrode. These factors lead to a significantly larger fraction of the gold atoms actually binding alkane thiolate ligands in the MPCs than for macroscopic gold surfaces, i.e. 0.62 versus 0.33 thiols per surface-gold atom.

Functional groupings can be introduced into MPC monolayers by using functionalized thiols. However, changes in the identity of the stabilizing adsorbate often cause the core size to change. In order to maintain a constant average core size within an investigation, ligand-place-exchange reactions can be used to incorporate functional groupings in a subsequent step. By using this approach, a wide variety of molecules with interesting electrochemical and photophysical properties can be incorporated within the assembly [56].

5.3.1 Conductivities of Single Clusters – Molecular Switching

Monolayers which contain functional groups that are capable of binding to gold clusters are an important approach to forming two-dimensional arrays of nanoparticles. For example, as illustrated in Figure 5.25, gold nanoparticles have been linked to a gold electrode through redox-active linkers [57]. These bound clusters can then be individually addressed by an STM tip and the current–voltage characteristics probed. Systems of this kind represent an important approach to the 'bottom-up' creation of reversible molecular switches that can be integrated into electronic circuits. The system illustrated in Figure 5.25 involves a linear 4,4'-bipyridyl, (bpy^{2+}), moiety which contains two terminal thiol functional groups. One of these is bound to an electrode surface, while the other is bound to a gold nanocluster of 6 nanometers diameter. The STM can then be used to record the electrical properties of individual clusters. Significantly, the redox state of the bpy molecules control electron transport between the gold contacts. When the molecule is in the $bpy^{+\bullet}$ state, i.e. when the linker is reduced, relatively large currents flow between the gold substrate and the nanoparticle. This current flow is consistent with resonant electron tunneling across the junction. In contrast, when an oxidizing potential is applied and the linker

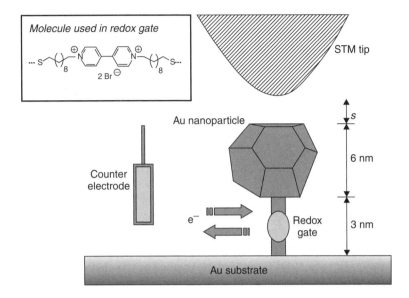

Figure 5.25 Schematic of the use of scanning tunneling microscopy to measure the conductivity of single gold clusters. Electrons can be injected into the redox gate by applying a suitable potential between the gold substrate and the counter electrode. The potential between the tip and the solution is controlled independently, with the gate consisting of up to 60 of the molecules shown in the inset. Reprinted with permission from D. I. Gittins, D. Bethell, D. J. Schiffrin and R. J. Nichols, 'A nanometre scale electronic switch consisting of a metal cluster and redox-addressable groups', *Nature (London)*, **408**, 67–69 (2000). Copyright (2000) Macmillan Magazines Limited

exists in the bpy^{2+} state, the tunneling current decreases dramatically. In this sense, Schiffrin and co-workers [57] have created an electrochemical switch whose state is determined by the potential needed to reduce the bridge by one electron. This switching potential depends intimately on the structure of the bridge, so allowing it to be synthetically tuned. However, real-world applications using redox-switchable integrated nanoelectronics must also consider the dynamics of the electron transfer processes. As discussed above in Section 5.2, the time-constants for electron transfer across tunneling junctions of this kind are often on the high nanosecond to low microsecond time-scale. These low switching rates, when compared to the GHz clock speeds achieved in conventional computer chips, coupled with their low gain, make this approach to molecular switching unlikely to become one of the 'engines' of a molecular electronics revolution. However, it is reasonable to expect that such systems will be exploited in low-gain devices such as sensors and in the design of new forms of computing logic and memory.

5.4 Electroanalytical Applications

5.4.1 Microarray Electrodes

Arrays are ensembles of microelectrodes that may consist of regularly, or irregularly, spaced assemblies of identical electrodes, ensembles of electrodes with identical shape but uneven dimensions, or disordered arrangements of irregularly shaped electrodes. Often, the individual elements are not independently addressable and all of the electrodes operate at the same applied potential. Under these conditions, the total response of the array depends on the relative size of the individual electrodes and the thickness of the diffusion layer that develops around each element during electrolysis. If the diffusion layers that develop at the individual electrodes merge, then a planar diffusion layer is created which extends over the entire surface of the array. Under such conditions, the array behaves like a large electrode whose surface area is given by the sum of the electrochemically active and inactive areas, while the background or double-layer charging current depends only on the dramatically smaller electrochemically active area.

Under the appropriate conditions, mixed SAMs composed of a long-chain alkane thiol and a short thiol form phase-separated domains of the short thiol with dimensions of the order of 10^4 nm^2. Given that the rate of heterogeneous electron transfer depends exponentially on the electron transfer distance, the small areas covered by the short alkane thiol represent 'hot-spots' at which redox reactions can proceed preferentially. For example, mixed SAMs of hexadecane thiol and 4-hydroxythiophenol give reproducible microarray behavior. Structures of this physical size remain an extreme challenge when using other approaches. Moreover, another advantage of this form of 'nanolithography' is that a degree of chemical selectivity can be achieved. For example, this hexadecane thiol/4-hydroxythiophenol mixed SAM shows a preferential response to neutral as opposed to charged analytes.

5.4.2　Selective Permeation

The ability to control the surface charge by terminating the interfacial supramolecular assemblies with charged functional groups can make the modified interface highly discriminating on the basis of electrostatic effects. Typically, the current is greatly enhanced at an ionic SAM for redox species of the opposite charge and strongly suppressed for a redox species of the same charge. For example, cationic dopamine can be detected in the presence of 100-fold higher concentration of anionic ascorbic acid at a gold electrode coated with ω-mercaptoalkanoic acid.

Crooks and co-workers [58] have demonstrated that pH influences the attraction of 4-aminothiophenol SAMs for charged inorganic and organic species through electrostatic interactions. Hence, when the solution pH is less than the pK_a of the monolayer, negatively charged species, e.g. anthraquinone-2,6-disulfonate, are adsorbed. In contrast, cationic species, such as $[Ru(NH_3)_6]^{3+}$, are repelled. A similar approach was also used by Cheng and Brajter-Toth [59] employing thioctic acid monolayers. These 'ion-gating' SAMs offer the advantage of selective responses whose target analyte can be switched simply by adjusting the pH of the solution or the potential applied to the electrode.

5.4.3　Preconcentration and Selective Binding

One of the most important approaches to generating an analyte-selective voltammetric response is to form a two-component SAM that includes a specific binding

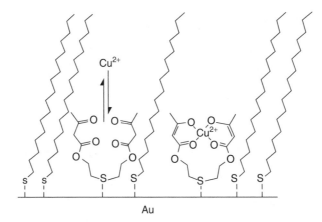

Figure 5.26 Schematic representation of a copper-selective 2,2′-thiobisethyl acetoacetate–*n*-octa-decylmercaptan SAM. Reprinted with permission from I. Rubinstein, S. Steinberg, Y. Tor, A. Shanzer and J. Sagiv, 'Ionic recognition and selective response in self-assembling monolayer membranes on electrodes', *Nature (London)*, **332**, 426–429 (1988). Copyright (1988) Macmillan Magazines Limited

site surrounded by alkane thiol. The latter insulates the electrode and blocks the direct redox reaction. As illustrated in Figure 5.26, Rubinstein *et al.* [60] were among the first to exploit this approach by co-immobilizing 2,2′-thiobisethyl acetoacetate (TBEA) and *n*-octadecyl mercaptan (OM) on gold. This approach exploits the known ability of acetoacetate-type ligands to selectively bind Cu(II), with the modified electrode detecting Cu(II) in the presence of other ions such as Fe(III). The use of a relatively long-chain alkane thiol efficiently blocks electron transfer away from the TBEA site. In addition, it exists in a liquid-like state and is therefore better able to accommodate the disorder caused by incorporating the relatively larger TBEA site. Significantly, the binding of Cu(II) depends strongly on the applied potential. Close to the potential of zero charge, no copper ion extraction was observed while binding was observed at both positive (field-assisted enolization) and negative (i.e. negative charge on TBEA) potentials. This system shows enhanced sensitivity when the OM was replaced with *n*-octadecyltrichlorosilane and trace amounts of Cu(II) (10^{-7} M), Pb(II) (10^{-5} M) and Zn(II) (10^{-9} M) can be detected.

5.4.4 SAM-Based Biosensors

Much of the research effort into electroanalytical applications of SAMs has focused on detecting organic and biochemical species. This dominance arises for three reasons. First, alkane thiol SAMs represent a hydrophobic environment which is highly amenable to extracting nonionic molecules. Secondly, the market for biomedical and environmental sensors continues to grow strongly, driven in part by legislation on anthropogenic emissions. Thirdly, the interactions between a SAM and an organic analyte can be tuned over a much wider range than is possible for hydrophilic species such as metal ions.

Enzyme-based biosensors continue to play important roles in this area because of the wide range of enzymes available and the selective responses that can be generated. As early as 1990, Bourdillon and Majda [61] demonstrated that a successful glucose sensor could be fabricated by immobilizing glucose oxidase in the head-group region of a bilayer consisting of *n*-octadecyltrichlorosilane (OTS) and a positively charged ferrocene amphiphile. Significantly, the lateral mobility of the ferrocene mediator is sufficiently high so that it can efficiently shuttle electrons between the electrode and the enzyme.

5.4.5 Kinetic Separation of Amperometric Sensor Responses

Electroanalytical chemistry is dominated by methods, e.g. differential pulse voltammetry and adsorptive stripping analysis, in which the response of the target analyte is distinguished from redox-active interferences on the basis of different formal potentials. Since the width of the electrochemical response for any species is a sizeable fraction of the potential scale, relying on the potential axis alone to generate a selective response provides only a very limited ability to resolve an analyte's response from that of an interfering species. In contrast, if one can use the time axis as well as the potential axis, separating the analyte's response from that of an interfering species becomes considerably more likely [62]. This time-resolved approach has benefited significantly from the dramatic expansion in the range of time-scales that can be resolved and exploited in electrochemistry with the advent of high-speed instrumentation and microelectrodes. The ability to *instrumentally* control the driving force, and hence the rate, of a reaction represents a distinct advantage of electrochemistry over other analytical approaches, e.g. spectroscopy. In hardly any other use of time-resolved analysis does one have the power to vary the kinetics in a simple, instrumentally based manner so as to optimize the separation achieved on the time axis.

The principle of the approach has been demonstrated by using two-component monolayers assembled using adriamycin and quinizarin [63]. This system challenges traditional electroanalytical approaches since the formal potentials of both redox couples in the mixture are identical, and the potential axis cannot be used to generate a selective response. As illustrated in Figure 5.27, the cyclic voltammetry observed for 30 μm radius mercury microelectrodes immersed in 5 μM solutions of adriamycin and quinizarin, respectively, where the supporting electrolyte is 1.0 M $HClO_4$, are similar to that theoretically predicted for a $2e^-/2H^+$ redox reaction, i.e. the FWHM and ΔE_p values are 50 ± 5 and 10 ± 5 mV, compared to the theoretical values of 45.3 and 0 mV, respectively. The adsorption thermodynamics follow the Langmuir isotherm over the concentration range from 2×10^{-8} to 2×10^{-5} M. Limiting surface coverages, Γ_s, of $1.1 \pm 0.1 \times 10^{-10}$ and $1.3 \pm 0.1 \times 10^{-10}$ mol cm^{-2}, and adsorption energy parameters, β, of $4.5 \pm 0.3 \times 10^5$ and $6.1 \pm 0.5 \times 10^5$ M^{-1}, are observed for adriamycin and quinizarin monolayers, respectively. Figure 5.28 illustrates the chronoamperometry data collected on the microsecond time-scale. The semi-log plots are linear, thus indicating that the rates of heterogeneous electron transfer to individual quinones within the assembly are all similar. Potential-dependent studies reveal that the standard heterogeneous electron transfer rate constants, k^0, are $3.1 \pm 0.2 \times 10^4$ and $1.0 \pm 0.1 \times 10^3$ s^{-1} (at pH 3.5) for adriamycin and quinizarin, respectively.

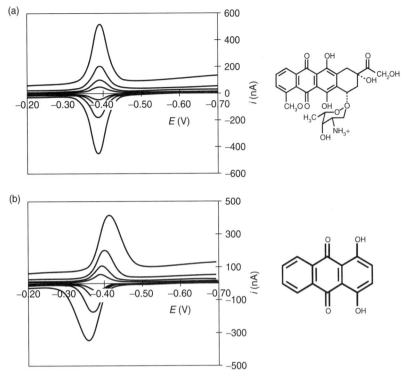

Figure 5.27   Cyclic voltammograms obtained for 30 μm radius mercury microelectrodes immersed in 5 μM solutions of (a) adriamycin and (b) quinizarin. The scan rates are, from top to bottom, 50, 20, 10 and 5 V s^{-1}; the supporting electrolyte is 1.0 M HClO$_4$. The cathodic currents are shown as 'up', while the anodic currents are shown as 'down'; the initial potential is −0.700 V. From R. J. Forster, *Analyst*, **121**, 733–741 (1996). Reproduced by permission of The Royal Society of Chemistry

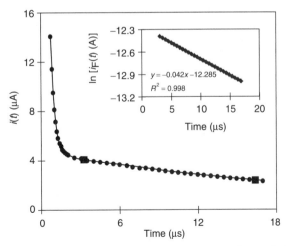

Figure 5.28   Current response for a 10 μm radius mercury microelectrode immersed in a 5 μM solution of adriamycin, following a potential step from −0.700 to −0.350 V; the supporting electrolyte is 1.0 M perchlorate at a pH of 4.5. The inset shows the semi-log plot for data between the marks on the current–time transient, with the time axis being referenced to the leading edge of the potential step. From R. J. Forster, *Analyst*, **121**, 733–741 (1996). Reproduced by permission of The Royal Society of Chemistry

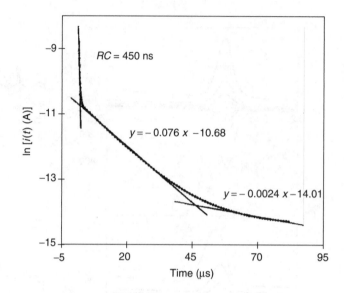

Figure 5.29   Log current versus time response for a 30 μm radius mercury microelectrode immersed in a solution containing 10 μM adriamycin and 10 μM quinizarin, following a potential step where the overpotential was 0.050 V; the supporting electrolyte is 1.0 M HClO₄. The time axis is referenced with respect to the leading edge of the potential step. From R. J. Forster, *Analyst*, **121**, 733–741 (1996). Reproduced by permission of The Royal Society of Chemistry

The formal potentials of adriamycin and quinizarin are almost identical. Therefore, binary monolayers, formed by simultaneous adsorption of both anthraquinones, exhibit only a single voltammetric peak. In these circumstances, traditional electroanalytical techniques cannot be used to determine the surface coverages of the individual species. However, as illustrated in Figure 5.29, the large difference in rate constant for the oxidation of the two anthraquinones can be exploited to temporally resolve the charge associated with oxidizing each adsorbate. The only requirement of this approach is that the interfacial kinetics of the individual components should be sufficiently different that two single exponential decays are observed.

5.5   Light-Addressable Assemblies

Photochemical and photophysical processes are the basis of many of the significant technological advances of the past 50 years. The advent of laser technology and the development of optical media have revolutionized the way in which information is stored and retrieved. This trend towards the development of optical devices is set to continue and will require that well-defined light-addressable molecular systems are developed for integration into circuitry or devices. An important element in these developments will be *interfacial photochemistry*. The latter is implicit in biological systems in which membranes separate donor and acceptor sites. Photovoltaic technology is also inherently interfacial. In the following sections, we will discuss the basic spectroscopy, reactivity and photochemistry of ISAs and explore how these complex, light-addressable, chemical systems can be exploited in functioning devices ranging from molecular switches to photocatalysts.

5.6 Surface–Photoactive Substrate Interactions

The immobilization of a photoactive species on a solid substrate may impact the photophysics of that material in a number of ways, as follows.

(a) *Close packing*. In many ISAs, the individual building blocks are often in close contact with one another, which may result in quenching of the electronically excited states created by irradiating the assembly. This behavior is expected to be particularly important in close-packed self-assembled monolayers. One method of reducing this source of quenching is to 'dilute' the luminescent species by coadsorption with a photochemically 'innocent' species.

(b) *Restricted mobility*. Reduced degrees of rotational or vibrational freedom may decrease non-radiative pathways and increase luminescence quantum yield. In luminescent films, this reduced mobility will be reflected in fluorescence anisotropy, which will increase with decreasing mobility.

(c) *Inaccessibility*. Solvent, ions and, in particular, oxygen may have limited access to the immobilized species, particularly in close-packed films or monolayers. This physical inaccessibility may prevent quenchers such as molecular oxygen from permeating into the film, hence causing increased luminescent lifetimes to be observed when compared with the same species dissolved in solution.

(d) *Local dielectric properties*. The microenvironment of the immobilized species, e.g. the local dielectric constant, may differ significantly when compared with homogeneous media. These changes can alter both radiative and non-radiative decay pathways, thus causing changes in the wavelengths of both absorption and emission, and the excited-state lifetimes of the immobilized species.

(e) *Direct surface interactions*. The nature of the surface, e.g. metal, semiconductor or glass, and the strength of the adsorbate–surface coupling will influence the extent to which the photophysical properties of the ISA differ from those found for the building blocks in solution. Direct electron transfer from a photochemically excited state may occur in the case of semiconductor or metal surfaces, often resulting in complete quenching of the excited state. Relatively inert surfaces such as silica may be employed to study the intermolecular behavior of surface-immobilized species in the absence of such effects.

   In the case of adsorbates which do not undergo direct photoinduced interfacial electron transfer, the surface may quench the excited states through energy transfer. Such behavior is particularly prevalent for metal surfaces as a result of their high density of states. However, some studies suggest that energy transfer to the metal continuum of states is strongly distance-dependent. A d^{-3} relationship between quenching efficiency and fluorophore metal distance is predicted, where d is the distance between the metal surface and the fluorophore [64]. In alkyl-linked chromophores, between five and ten atoms is sufficient to eliminate through-bond quenching of the excited chromophore.

(f) *Desorption*. In the case of adsorbed species, photoexcitation may lead to desorption, and particularly in weakly bound species or with the use of high-power irradiation, this will lead to increased non-radiative decay.

5.7 Photoactive Self-Assembled Monolayers

The primary focus in research into SAMs over the past two decades has been on developing systems which are capable of electrochemically induced reaction. This approach has provided significant new insights into those factors which control interfacial electron transfer. Aside from photovoltaic semiconductor and zeolite chemistry, interfacial photoinduced studies have primarily focused on Langmuir–Blodgett (LB) films. Although the Langmuir–Blodgett method results in excellent levels of predefined molecular organization, the resulting mono- and multilayers are physisorbed onto their solid substrates and are therefore only weakly bound. Their consequent lack of stability is a serious barrier to any long-term applications of LB films in molecular devices.

However, self-assembled and chemically adsorbed monolayers exhibit high degrees of self-organization, as well as excellent long-term stability. The introduction of photoactive components into the SAM is one of the most versatile ways of generating a molecular device. Light may be employed to initiate events that can subsequently be detected by electrochemical methods or may result in the generation of photocurrents. Light may also be utilized to initiate events such as isomerization, proton transfer or photoregulated binding, in order to create a bistable device, i.e. a *molecular switch*. A typical photoactive supramolecular species for SAM formation consists of a terminal group for strong surface adsorption, commonly a thiol, or a pyridine group which exhibits particularly strong interactions with gold, as well as silver and platinum substrates. A bridging unit, which will efficiently mediate electronic communication between the metal and photoactive unit, is also required. This bridge should be of sufficient length, whereby in the case of assemblies where a long lifetime is required from the photoactive group, as in the case of an optical signal output, the photoactive unit–metal surface interactions are kept to a minimum. Surfaces, particularly those of metals, may deactivate molecular excited states very efficiently through energy transfer to a continuum of electronic levels (see Chapter 2). Therefore, in contrast to electrochemical systems, for photonic applications a balance must be struck between effective electronic communication between the remote fluorophores and the surface and the need to prevent photochemical deactivation.

In the following discussion, some of the main approaches to electrooptical devices based on photoactive SAMs and various examples of coupled photon and electron transfer reactions which lead to measurable or usable output are discussed.

5.8 Photocurrent Generation at Modified Metal Electrodes

The vast majority of work on supramolecular systems as sensitizers for photocurrent generation has focused on large-band-gap semiconductors, such as TiO_2, with

Chapter 6 being devoted entirely to this subject. However, photoactive SAMs on conducting electrodes such as gold, silver and carbon are undergoing increasing investigation for photocurrent generation. These platforms allow electrochemical and photophysical measurements to be coupled, while the use of well-defined surfaces facilitate the creation of more organized assemblies.

An important consideration in the design of photovoltaic devices is the coupling of the supramolecular assemblies and the substrate surface. In semiconductor assemblies, terminal groups, such as carboxylate and phosphonate, have been exploited. However, as described earlier in Chapter 4, the use of metal electrodes in conjunction with terminal groups such as pyridine and thiols leads to exceptionally strong chemisorption and significant long-range order. For example, as illustrated in Figure 5.30, Yamada *et al.* [65] have linked a ruthenium bipyridine center to a methyl viologen via an alkyl (C7) chain. The MV^{2+} unit is then linked via a further C6 alkyl unit to a surface active disulfide terminal. These materials self-assemble on semitransparent vacuum-deposited Au/glass slides. Cyclic voltammetry of the SAM exhibited peak currents which were proportional to the scan rates, thus indicating immobilization.

Photoelectrochemical experiments in which the monolayer was in contact with a solution containing a 'sacrificial' donor, e.g. triethanolamine (TEOA), produced a photocurrent which increased with increasing cathodic potential. The photocurrent action spectrum corresponded to the absorbance spectra of $[RuC_7VC_6S]_2$, confirming that the bpy-based triplet metal-to-ligand change-transfer (^{3}MLCT) excited state was the origin of this photocurrent. As illustrated in Figure 5.31, photoexcitation of the ruthenium moiety results in electron transfer to the viologen species, MV^{2+}, hence resulting in the formation of a $Ru(III)$–MV^+ pair. Electron transfer from MV^+ to the electrode surface regenerates the MV^{2+} site and the $Ru(III)$ is reduced by the TEOA.

Fullerene-based 'supermolecules' have received significant interest as a result of their capability of reversibly accepting up to six electrons, and their participation in photoinduced electron transfer reactions. Electron transfer to fullerenes is accompanied by a very small reorganization energy. This small barrier frequently makes their reduction reactions fast, while their recombination reactions may be significantly slower because of their multi-step character. These two features lead to long-lived charge-separated states that can be used for energy production or for driving chemical reactions.

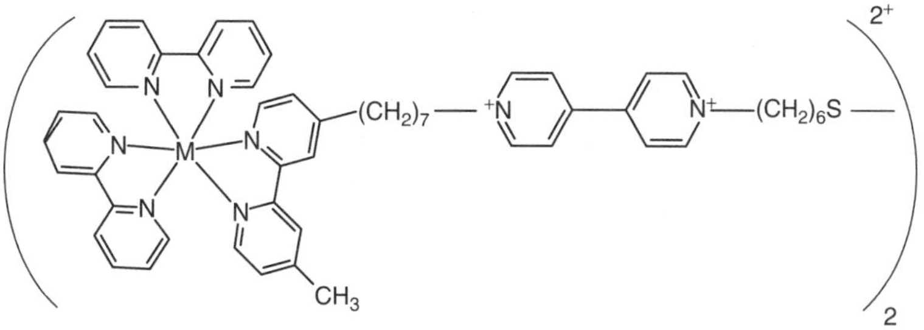

Figure 5.30 Molecular structure of the disulfide-linked ruthenium polypyridyl methyl viologen dyad, $[RuC_7VC_6S]_2$ [65]

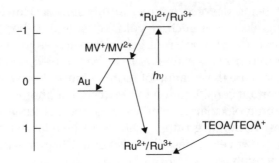

Figure 5.31 Proposed photocurrent reaction cycle for [RuC$_7$VC$_6$S]$_2$: MV, methyl viologen; TEOA, triethanolamine

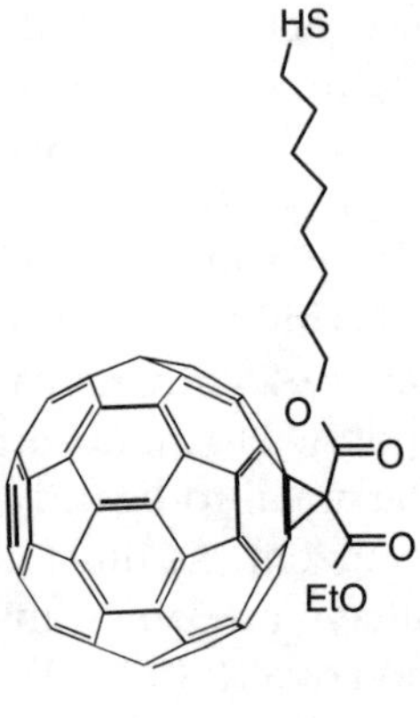

Figure 5.32 Molecular structure of the surface active thiol-terminated fullerene, reported by Neusch and co-workers [66]

Figure 5.32 shows the structure of a thiol-terminated fullerene studied by Neusch and co-workers [66]. This group reported the photocurrent generated by SAMs of this material on gold electrodes, with and without an ion-selective polyurethane membrane overlayer. Both systems show anodic photocurrents with high quantum yields and action spectra corresponding to C$_{60}$ absorption. No sacrificial donor was included and water was concluded to behave as the donor. Under optimal conditions, in deoxygenated solution and at a bias of 1.2 V, the quantum yield for the uncoated monolayer was 31 %, compared with 25 % for the polyurethane-coated system. However, the polyurethane-coated monolayer exhibited greater long-term stability and greater photoelectric response under non-biased conditions. The long-term stabilities of bare monolayers may represent a limitation to their application in molecular devices, although this work suggests that protection with appropriate films may offer solutions to this problem without any significant loss of efficiency.

In the next example, a mixed SAM is discussed which aims to utilize photoinduced energy and electron transfer processes to create a photocurrent in an approach which is reminiscent of the natural photosynthetic process. Figure 5.33 illustrates the molecular structures of the components of interest, i.e. the molecular triad ferrocene–porphyrin–fullerene (Fc–P–C$_{60}$) and a boron dipyrrin thiol (BoDy) [67]. Mixed monolayers were generated by coadsorption onto vacuum-deposited gold

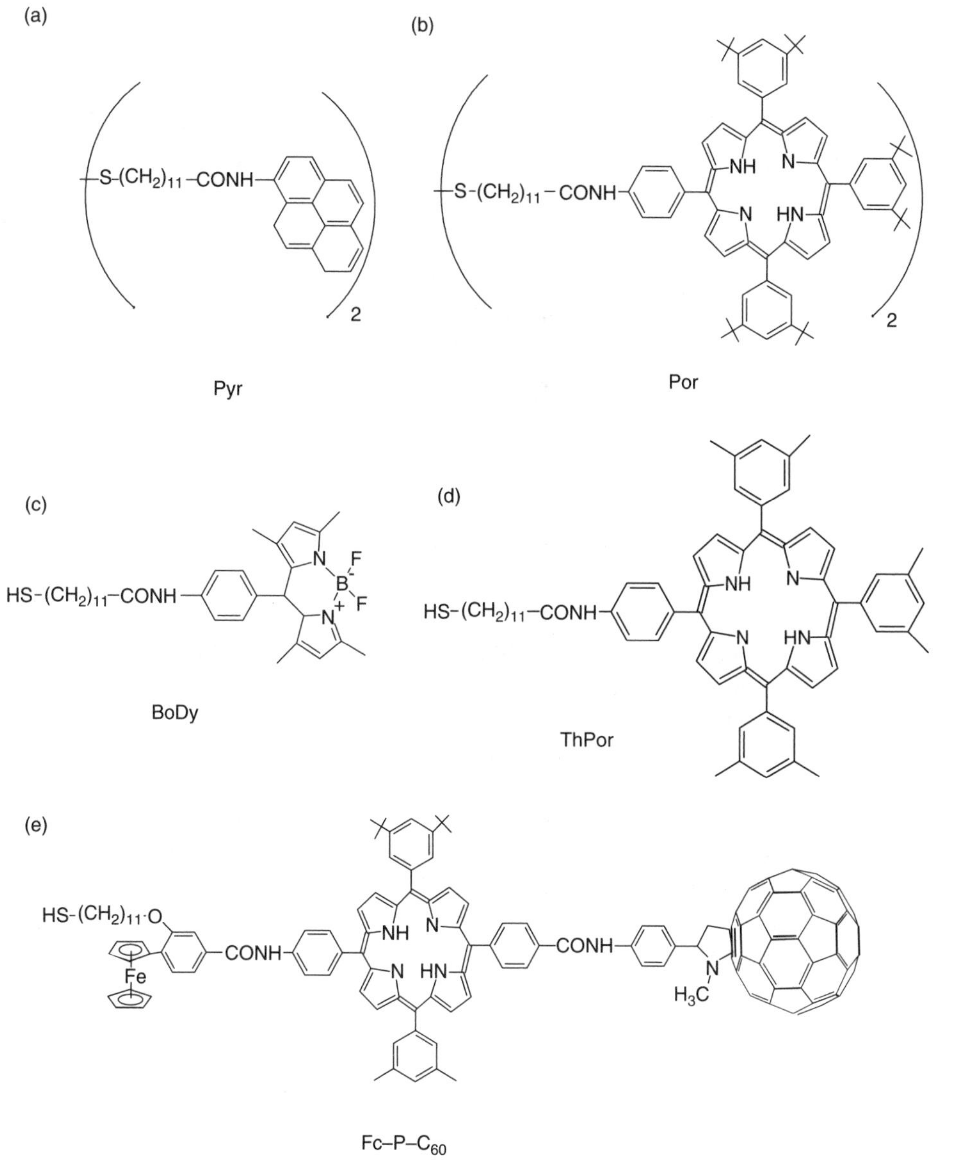

Figure 5.33 Structures of disulfide-terminated (a) pyrene (Pyr) and (b) porphyrin (Por), and thiol-terminated (c) boron–dipyrin (BoDy), (d) porphyrin (ThPor), and (e) the ferrocene–porphyrin–fullerene triad (Fc–P–C_{60})

(Au(111)). For optical studies, mica or glass substrates were used, while gold/titanium deposited onto a silicon (100) wafer was used as a substrate for time-resolved studies.

The photophysical and electrochemical properties of the components of this assembly are chosen to create a vectorial energy and electron transfer cascade away from the electrode surface. Figure 5.34 shows the energetics of such an assembly. It is worth noting that in this photocurrent-generating assembly the Au electrode

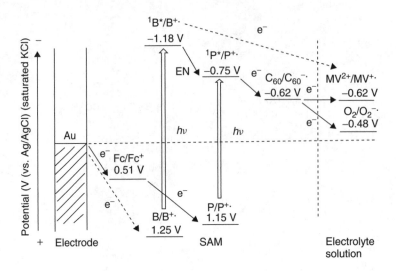

Figure 5.34 Energy level diagram showing the processes leading to cathodic photocurrent generation in Fe–P–C_{60} monolayer reported by Imahori

acts as an electron donor. This is contrary to what is observed for modified TiO_2 surfaces but is similar to what is found for p-type semiconductors such as NiO (see Chapter 6).

The experimental data obtained in this study are in agreement with the scheme proposed in Figure 5.34. The behavior of the Fc–P–C_{60} dyad under irradiation is considered first in isolation. On photoexcitation of the porphyrin moiety of Fc–P–C_{60}, photoinduced electron transfer to the fullerene takes place. Reduction of the resulting oxidized porphyrin by the ferrocene moiety yields the charge separated state, Fc^+–P–$C_{60}^{-•}$. In a photoelectrochemical cell, the $C_{60}^{-•}$ moiety is reoxidized by an electrolyte-based acceptor, such as MV^{2+} or O_2. Provided that an appropriate potential is applied, Fc^+ is reduced at the electrode. The photocurrent produced is, therefore, dependent on the potential applied to the gold electrode. This observation is promising, and 438 nm irradiation of this immobilized triad by itself yields a quantum yield for photocurrent production at about 20 %. However, the application of this assembly as a solar energy storage device is limited by the lack of absorption of the triad in the visible part of the spectrum [68].

The mixed SAM under discussion here has been designed to overcome this limitation and to increase the photocurrent efficiency in the visible part of the spectrum. The strong overlap of the emission spectrum of BoDy with the absorption of Fc–P–C_{60} in the 500 nm region is anticipated to facilitate efficient energy transfer between the two species. The photochemical action spectra of the mixed monolayer show contributions from both BoDy and Fc–P–C_{60}. The overall behavior of the assembly is therefore in agreement with the scheme shown in Figure 5.34. At 510 nm, a much increased quantum yield for photocurrent production of 50 % is obtained. This study shows that photocurrents can be created at metal electrodes. Relatively high photocurrent efficiencies are obtained; however, the fact that flat surfaces are used rather than the nanocrystalline approach taken for semiconductor substrates (see Chapter 6) means that the incident photocurrent efficiency (IPCE) values are substantially lower. For nanocrystalline systems, IPCE values of up to 10 % are obtained for the assembly discussed in this section, however, a value of about 1–2 % is reported on gold. This study is, nonetheless, a very elegant example of how with appropriate thermodynamic considerations, vectorial intramolecular energy and homo- and heterogeneous electron transfer may be achieved at modified interfaces.

5.9 Photoinduced Molecular Switching

One of the most promising applications of photochemically active self-assembled monolayers (SAMs) is in the development of optical switches. A potential molecular switch must possess some addressable, reversible response, which may be transducable as a binary signal. Typical switchable molecular responses that lead to switching behavior include *cis–trans* isomerization, protonation–deprotonation reactions, adsorbate reorientation, molecular shuttling, etc. An extensive range of elegant solution-phase supramolecular materials has been reported with switching moieties which are stimulated photochemically or electrochemically [69]. However, for practical molecular electronic applications spatial organization and electrochemical addressability are required.

The organization and study of photochemically switchable supramolecular SAMs is a rapidly developing area. In this present section, the principal approaches to achieving photoinduced switching by using an ISA are described. Figure 5.35 illustrates the first approach in which the structure of a surface-immobilized host, for example, an isomeric form, is photoswitched to a form that can more efficiently bind a solution-phase guest species which is redox-active. An amperometric signal can therefore be controlled at the interface. Figure 5.36 illustrates an alternative approach in which a photochemical switching at the SAM alters its packing density

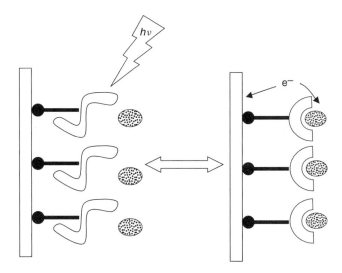

Figure 5.35 Schematic illustration of a photoswitch process in which the photoreaction, in this instance, photoisomerization, occurs at the SAM. This alters the ability of the monolayer species to recognize and interact with a redox-active solution-phase moiety

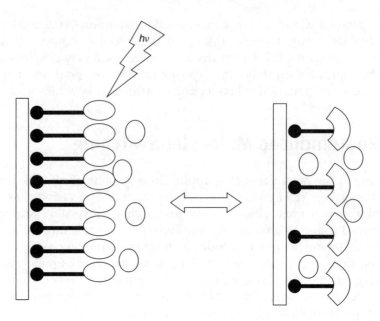

Figure 5.36 Schematic illustration of a photoswitch process in which the photoreaction occurs at the SAM, which alters the structure of the monolayer, thus permitting a redox solution-phase species to approach the electrode surface

in the monolayer. For example, photoisomerization may create a close-packed structure that efficiently blocks electron transfer to a solution-phase redox-active species which may be reversed to create a more porous structure that allows the target molecule to approach the electrode surface. Again, an amperometric response may be switched on or off by photochemically altering the interface. Figure 5.37 illustrates an example of switching in which light is used to modify the properties of the target molecule in solution rather than by changing the properties of the SAM. For example, the charge or structure of the solution species may be altered photochemically, which may then alter the ability of the monolayer to recognize and interact with this species, thereby changing the electrode response of this species.

As illustrated in the above figures, one of the most important photoswitchable processes in terms of persistence, reversibility and magnitude of effect is photoinduced isomerization. Willner and co-workers have conducted significant work in the area of photoswitchable and photoisomerizable SAMs [70]. An example of their work is shown in Figure 5.38 which illustrates an eosin monolayer self-assembled on a gold-coated quartz crystal [71]. Light-stimulated photoisomerization of the electron acceptors, *trans*-4,4'-bis(N-methylpyridinium)azobenzene (*trans*-4,4'-bNA) and *trans*-3,3'-bis(N-methylpyridinium)azobenzene (*trans*-3,3'-bNA) leads to reversible photoinduced binding to eosin. As shown in Figure 5.38(b), the approach works because irradiation at 355 nm creates the *cis*-isomer, which switches reversibly back to the *trans*-isomer when irradiated at $\lambda > 420$ nm. Spectrophotometric studies of the solution-phase interactions of *trans*-4,4'-bNA, and *trans*-3,3'-bNA and their *cis* analogues with eosin reveal the formation of eosin–bNA donor–acceptor complexes whose association constants depend on the form of the isomer of bNA.

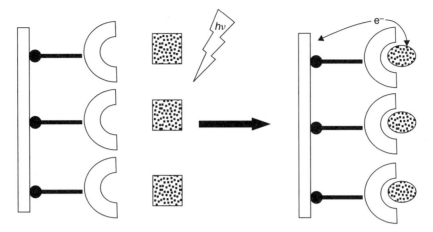

Figure 5.37 Schematic illustration of a photoswitch process in which the solution-phase species acts as the photoactive moiety. In this instance, photoisomerization alters the ability of the redox-active solution-phase moiety to recognize the monolayer and undergo a heterogeneous electron transfer process

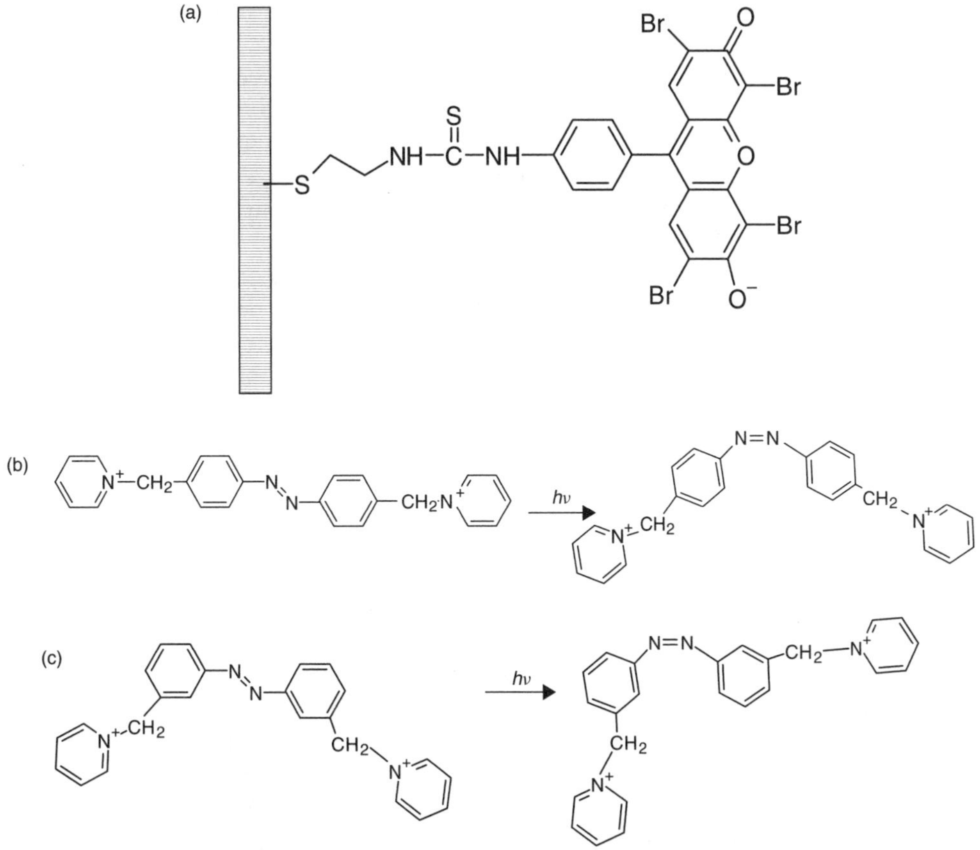

Figure 5.38 (a) Schematic of an eosin monolayer on a gold electrode, plus the molecular structures of (b) 4,4′-bis(*N*-methylpyridinium)azobenzene (4,4′-bNA) and (c) 3,3′-bis(*N*-methylpyridinium) azobenzene (3,3′-bNA), reported by Willner and co-workers [71]

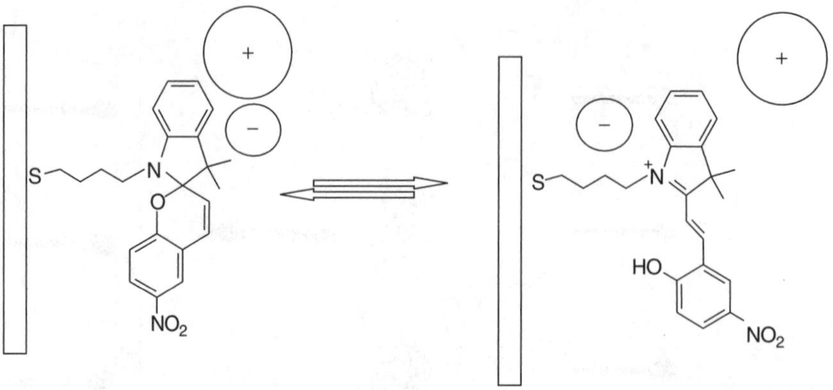

Figure 5.39   Schematic of a photoswitchable monolayer based on pyridine thiol/mercapto butyl-nitrospiropyran, reported by Doron and co-workers [72]

The *trans*-forms of the isomers, *trans*-4,4'-bNA and *trans*-3,3'-bNA, have binding affinities of $8.3 \times 10^3 \, M^{-1}$ and $3.8 \times 10^4 \, M^{-1}$, with Eosin respectively. In contrast, *cis*-4,4'-bNA, and *cis*-3,3'-bNA have binding affinities of $8.3 \times 10^3 \, M^{-1}$ and $3.8 \times 10^4 \, M^{-1}$, respectively.

Cyclic photoisomerization of the electron–acceptor between the *trans*- and *cis*-states permitted reversible piezoelectric transduction of the formation of the complexes with *trans*-4,4'-bNA, and *trans*-3,3'-bNA at the monolayer interface, and their dissociation upon photoisomerization to *cis*-4,4'-bNA and *cis*-3,3'-bNA.

A series of compounds which have received significant interest as photoisomerizable SAMs are the nitrospiropyrans. Figure 5.39 illustrates an example of a photoswitchable nitrospiropyran monolayer based on mercapto butylnitrospiropyran [72]. Optical excitation between 320 and 350 nm causes the nitrospiropyran monolayer to photoisomerize, yielding a positively charged nitromercyanine SAM which is protonated at pH 7. The nitrospiropyran monolayer may be restored by further irradiation at 495 nm. Monolayers of nitrospiropyran may therefore be employed for reversible photoinduced electrostatic discrimination between solution-phase species. Comparative studies on the interaction of negatively and positively charged electroactive substrates, 3,4-dihydroxyphenylacetic acid, dopamine and 3-hydroxytyramine, reveal that the amperometric response at the electrode, associated with the redox activity of these species, may be cycled between high and low levels by photochemically cycling the monolayer between neutral nitrospiropyran and cationic nitromercyanine. Dopamine exhibits similar but opposing photoswitchable amperometric responses, i.e. high at nitromercyanine and low at nitrospiropyran [72]. The anodic currents of the two substrates are comparable at the nitrospiropyran monolayer. However, as shown in Figure 5.40, at the nitrospiropyran monolayer the 3,4-dihydroxyphenylacetic acid exhibits a significant increase in amperometric signal when the monolayer is photoisomerized to the nitromercyanine state. Conversely, dopamine exhibits a high amperometric response at the nitrospiropyran monolayer which is significantly reduced at nitromercyanine. This is attributed to the electrostatic discrimination of the nitromercyanine layer. In the case of 3,4-dihydroxyphenylacetic acid, the anionic

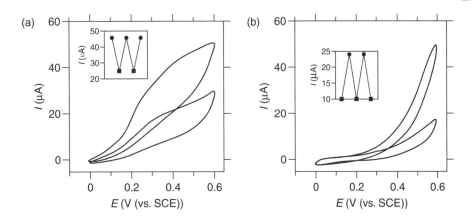

Figure 5.40 Cyclic voltammetry of (a) 3,4-hydroxyphenylacetic acid, and (b) dopamine at photoswitchable nitrospiropyran electrodes; the insets show the corresponding photoswitched reversible induced currents monitored at 470 mV. Reprinted with permission from A. N. Shipway and I. Willner, *Acc. Chem. Res.*, **34**, 6 (2001). Copyright (2001) American Chemical Society

substrate is electrostatically attracted to the monolayer, so providing efficient coupling to the electrode surface. However, the amperometric response of the cationic dopamine is reduced at nitromercyanine as a consequence of electrostatic repulsion between the two species. In this way, the monolayer can be optically addressed and amperometrically transduced, i.e. WRITE and READ functions have been achieved.

Photomodification of surface interactions may also be directed at a solution species, whose binding properties to a photoinactive SAM may be optically modified. A number of optoelectronic assemblies based on this principle have been developed. The cyclic polysugars, cyclodextrins, have proven particularly interesting in these applications as a result of their excellent binding properties, relative ease of synthetic functionalization and electrical inactivity which eliminates interference in signal transduction. Figure 5.41 illustrates the structure of a thiol-substituted cyclodextrin monolayer and the photoisomerization reaction of a bipyridinium–azobenzene diad reported by Willner and co-workers [73]. The association constants of both isomers with the β-CD cavity in solution were obtained spectrophotometrically. The *trans*-isomer was shown to have an affinity which was 10-fold that of the photogenerated *cis*-isomer. The bipyridinium–azobenzene guest possesses potentially two electroactive components, i.e. the bipyridine unit and the azo unit, both of whom may be reduced. The reduction potential of the latter moiety is pH-dependent and at a sufficiently high pH (10.5) the bipyridinium moiety is the only redox process occurring between -0.2 and 0.65 V. Therefore, the guest species possesses a binding affinity to the CD which may be photomodulated, and a redox chemistry which may be employed for generation of an optically induced amperometric signal. The thiol-terminated cyclodextrin was assembled on a gold wire electrode through *in situ* reaction of a surface bound thiol ester with β-aminocyclodextrin. The surface coverage of monolayer was estimated from coulometry of the thiol ester precursor and quartz crystal microbalance studies to be 3×10^{-10} mol cm^{-2}. Plots of the voltammetric peak current versus the scan rate are linear for the *trans*-isomer on the CD-modified electrode, thus confirming that the

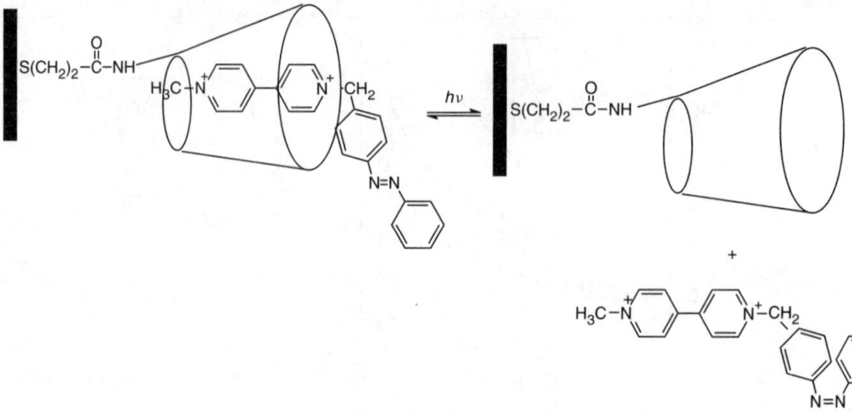

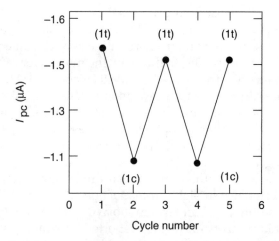

Figure 5.41 Optoelectronic functions of the cyclodextrin-monolayer-modified electrode and bipyridinium–azobenzene diad, as reported by Willner and co-workers [73]. Photoisomerization of the diad alters its ability to include in the cyclodextrin cavity and consequently 'write' to the electrode

Figure 5.42 Cyclic amperometric responses of a β-CD functionalized monolayer on photoswitching of the bipyridinium–azobenzene diad; currents are measured at a constant potential of −0.6 V. From M. Lahav, K. T. Ranjit, E. Katz and I. Willner, 'Photosimulated interactions of bipyridinium–azobenzene with a β-aminocyclodextrin monolayer-functionalized electrode: An optoelectronic assembly for the amperometric transduction of recorded optical signals', *Isr. J. Chem.*, **37**, 185–195 (1997). Reproduced by permission of Laser Pages

bipyridinium–azobenzene dyad is surface-confined, i.e. it is included in the β-CD. Photoswitching of the redox response was conducted by separately photolyzing the electrolyte solution containing the *trans*-isomer at 355 nm, before reintroducing the solution to the electrochemical cell. The *trans*-isomer could be reformed by removing the solution from the cell and photolyzing it at 375 nm. Figure 5.42 illustrates the cyclical amperometric response observed when the bipyridinium–azobenzene species is reversibly photoisomerized. An amperometric response is observed at the same potential for both isomers. However, the peak current for the *cis*-isomer is less than 50 % of that found for the *trans*-isomer because of its lower association constant. Quartz crystal microbalance studies further confirm the association of the bipyridinium–azobenzene species with the monolayer. Frequency changes of the β-CD functionalized crystal on exposure to solutions of the *cis*- and *trans*-isomers provide estimates of the surface coverage of 8.3×10^{-11} mol cm^{-2} and 3.6×10^{-11} mol cm^{-2} by these species, respectively. Studies of peak-to-peak separations of the redox waves of the *cis*- and *trans*-isomers, and application of Laviron's theory [74], yielded estimates of the heterogeneous electron transfer rates of the surface-confined bipyridinium units of 21 s^{-1} for the *trans*-isomer and 42 s^{-1} for the *cis*-isomer. The authors associate the enhanced electron transfer rate for the *cis*-isomer to the closer distance between the bipyridinium and the electrode surface as a consequence of its binding in the CD cavity.

A particularly interesting example of a photoswitch, where optically induced reaction occurs within the SAM, which alters the electrochemical response of a solution-based electroactive species, was reported by Mirkin and co-workers [75]. Photon gating is induced in a self-assembled monolayer via structural changes associated with *cis–trans* isomerization of the immobilized species. Figure 5.43(a) illustrates the structures of the photoisomers of the thiol-terminated azobenzene ferrocene (ABF) dyads reported, while Figure 5.43(b) illustrates the principle by which monolayers of these materials are considered to photoswitch. Heterogeneous electron transfer between solution-phase ferrocyanide and a gold electrode modified by this thiol-terminated azobenzene–ferrocene (ABF) dyad is mediated by the ferrocene moieties in the SAM. A two-component SAM was constructed by immersing gold wafer electrodes in solutions of *cis*-azobenzene (c-AB), followed by immersion in a solution of ABF, whereby exchange of c-AB for ABF occurs. Voltammetry indicates that 1 % (or 5.0×10^{-12} mol cm^2) of the film is comprised of ABF. Gold SAMs comprised of only c-AB effectively block heterogeneous electron transfer to solution-phase ferrocyanide. The latter can be oxidized at electrodes modified with mixed monolayers but the corresponding reduction currents are dramatically lower. This highly reproducible, irreversible response is associated with ferrocene-mediated rectification of electron transfer. Irradiation of the SAM with ambient light leads to *trans–cis* isomerization of AB yielding a normal electrochemical response for the ferrocyanide species. The resulting c-AB adsorbates have larger 'footprints', estimated on the basis of crystallography to be 1.5–2 times that of the associated *trans*-isomer. Isomerization is thought to create modifiable channels in the monolayer structure, so allowing the solution-phase ferrocyanide direct access to the electrode surface. Differential capacitance measurements, which provide a measure of film porosity, confirm a greater film porosity in the case of the *cis*-SAM. This work illustrates the impact of the mediating bridge in electron transfer reactions and how disruption of SAM structure via optical stimulation can be employed to modulate the rate of heterogeneous electron transfer.

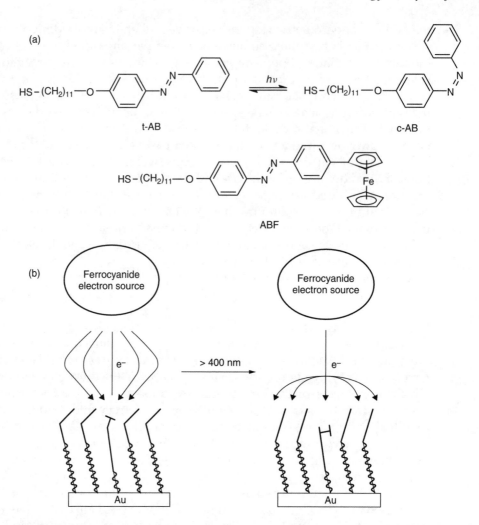

Figure 5.43 (a) Structures of the *cis-* and *trans-*thiol-terminated azobenzene–ferrocene dyads, and (b) illustration of the principle of operation of the photogated reaction in the SAM of these complexes, whereby photoinduced structural changes in the monolayer allow solution-phase ferrocyanide ions access to the electrode surface. Reprinted with permission from D. G. Water, D. J. Campbell and C. A. Mirkin, *J. Phys. Chem., B*, **103**, 402 (1999). Copyright (1999) American Chemical Society

5.10 Luminescent Films

Luminescence has proven a useful probe of structure and dynamics in a broad range of heterogeneous media, from zeolites to micelles to biomaterials. The sensitivity of this process to its environment, and its ease of detection, make it a highly versatile analytical tool. To date, luminescence as a probe of solid–liquid interfacial processes in SAMs is still relatively limited. However, with the development of fluorescence-based analytical methods of increasing spatial and temporal resolution, it is likely to be used increasingly in answering fundamental questions regarding monolayer behavior.

As discussed previously, immobilization of a luminescent species onto a metallic or semiconductor surface frequently results in surface-induced quenching of radiative processes. These quenching processes can be reduced by extending the distance between chromophore and surface, by employing long surface linkers, or by employing innocent platforms such as fused silica or silica oxide surfaces. There are circumstances, however, when an immobilized species on a metal surface may exhibit a higher emission intensity as a result of surface plasmon enhanced luminescence. For example, Ishida and Majima employed surface plasmon enhanced fluorescence spectroscopy to investigate the exchange reaction of a self-assembled monolayer with fluorescent porphryine disulfides [76]. Figure 5.44(a) illustrates the

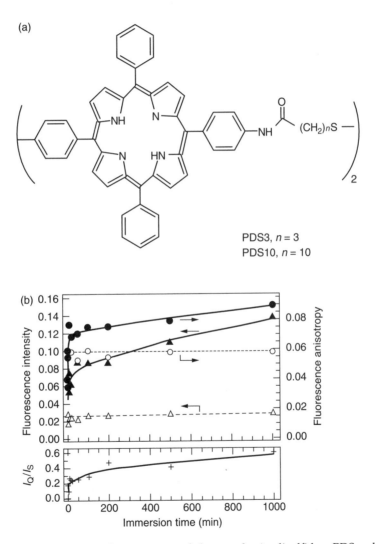

Figure 5.44 (a) General molecular structure of the porphyrin disulfides, PDS*n*, described by Ishida and Majima [76]. (b) Changes in the surface plasmon enhanced fluorescence spectra for the exchange reaction of a decane thiol SAM with a 50 mmol dm^{-3} 1,2-dichloroethane solution of PDS10 ($\lambda_{em} = 725$ nm; $\lambda_{ex} = 425$ nm). From A. Ishida and T. Majima, *J. Chem. Soc., Chem. Commun.*, 1299–1300 (1999). Reproduced by permission of The Royal Society of Chemistry

structure of the porphyrin disulfides (PDSn) reported in this work; two porphyrins were investigated with different chain lengths, i.e. PDS3, which contained three alkyl linkers and PDS10, which contained ten alkyl groups. Decane thiol SAMs were self-assembled on gold films deposited on right-angle prisms which were then immersed in a solution of the porphyrin disulfides. The fluorescent properties were measured by surface plasmon excitation, as a function of the exposure time of the film to porphyrin. Very weak fluorescence indicated that no exchange occurred in the case of the porphyrin disulfide PDS3 – this was attributed to the short chain length of this species which prevented surface binding. Figure 5.44(b) illustrates the changes in fluorescent properties occurring in the decane thiol film as a function of the immersion time in the porphyrin disulfide PDS10 solution. The fluorescence intensity increased significantly over time, indicating that this 'supermolecule' exchanges efficiently with the decane thiol SAM. The fluorescence anisotropy also increases, thus indicating reduced degrees of freedom of the porphyrin during exchange. This is associated with π–π stacking in the film and is further reflected in changes in the porphyrin Soret/Q bands emission intensity ratio which suggests J-aggregate-like formation at higher exchange levels. Exchanging the decane thiol SAMs with a Zn porphyrin analogue of PDS10 results in fluorescence from this species which is fully quenched by subsequently exposing the SAM to PDS10.

This quenching is attributed to energy transfer from the Zn porphyrin to PDS10, suggesting that the porphyrin adsorption is site-specific, i.e. the molecules tend to cluster on the surface, possibly as a consequence of π-stacking interactions between surface-bound porphyrin.

Beyond providing insight into exchange processes, fluorescent monolayers may also prove useful in sensor technology. An interesting example of a fluorescent monolayer with potential sensing application was reported by Russell and co-workers [77]. This group also reported on the application of an optical waveguide as a means of exciting detectable fluorescence in a monolayer. Figure 5.45(a) illustrates the structures of the phthalocyanine monolayer molecules, Pc1 and Pc2, employed in this study. The monolayers were constructed on glass slides onto which layers of chromium and gold were vapor deposited. The films were analyzed by reflection–adsorption infrared (RAIR) spectroscopy which suggested that monolayers of Pc1 and Pc2 were oriented differently on the metallic surface. As illustrated in Figure 5.45(b), fluorescent measurements were conducted whereby the glass slide behaved as a waveguide and allowed illumination of the film along its longitudinal. Both SAMs were observed to emit, with a fluorescence λ_{max} at approximately 800 nm, which was red-shifted compared to the solution. Self-quenching appeared to be limited in the monolayer and this was attributed to steric inhibition of overlaps of the Pc macrocycles due to the bulky substituents on the rings. Pc1 exhibited a greater luminescence intensity than Pc2; this was attributed to the longer thiol-terminated linker to the metal surface in Pc1, estimated to be between 11 and 12 Å, which is sufficient to reduce surface quenching. By employing the flow cell shown in Figure 5.45(b), the fluorescence intensities of monolayers of Pc1 were investigated in the presence of NO_2 gas. Using this approach, quantifiable alterations in the fluorescence signal were obtained down to 10 ppm with no interference from CO_2 or CO. However, the film's stability and response time were limited. Stern–Volmer analysis was applied to obtain a lower-limit estimate of the

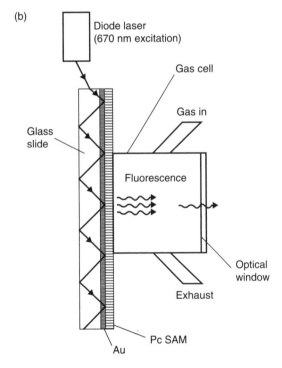

Figure 5.45 (a) Molecular structures of the thiol-terminated phthalocyanine compounds reported by Russell and co-workers [77]. (b) Schematic of the experimental configuration used for evanescent wave excitation and emission detection at Pc SAMs, including a gas flow-through cell. Reprinted with permission from T. R. E. Simpson, D. J. Revell, M. J. Cook and D. A. Russell, *Langmuir*, **13**, 460 (1997). Copyright (1997) American Chemical Society

quenching rate of 7.44×10^5 mg^{-1} dm^3 s^{-1}. These results suggest that fluorescent SAMs may become useful as sensor materials if problems with film ruggedness are overcome. Surface-enhanced fluorescence is likely to become increasingly important for developments in this field.

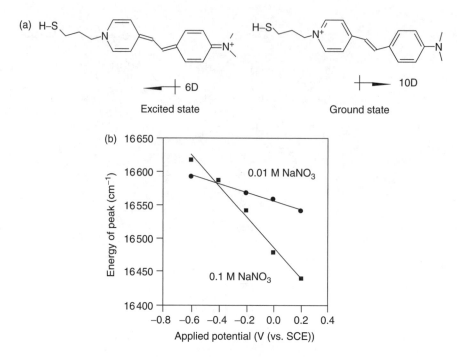

Figure 5.46 (a) Structures of the surface active fluorescent hemicyanine probe reported by Pope and Buttry [78]. The dipole moments for the ground and excited states of this species are also indicated. (b) Plots of emission peak energy versus applied potential for a dye monolayer, at an excitation wavelength of 476.5 nm, with a laser power of 3 mV at the sample surface. The continuous lines indicate the linear least-squares regression of the data. Reprinted from *J. Electroanal. Chem.*, **498**, J. M. Pope and D. A. Buttry, 'Measurements of the potential dependence of electric field magnitudes at an electrode using fluorescent probes in a self-assembled monolayer', 75–86, Copyright (2001), with permission of Elsevier Science

The sensitivity of a fluorophore to its environment, i.e. solvatochromism, means that fluorescence spectroscopy can be used to probe the local microenvironment of an adsorbate. Pope and Buttry exploited the fluorescent thiol-terminated hemicyanine dye illustrated in Figure 5.46(a) to interrogate the potential dependence of the electric field strength in an electric double-layer at a metal–organic thin film electrolyte interface [78]. The dye molecules were coadsorbed with alkane thiols onto roughened Ag or smooth Au electrodes. Two types of diluent were investigated, namely *n*-alkyl thiols (where $n = 12$ or 3). In the former case, this generated a configuration in which the dye is embedded in long-chain groups, forcing it to maintain a configuration which is coaxial with the alkyl chains. In the latter case, the dye is external to the alkyl thiols. The monolayer structure, especially the dye orientation, was probed by using RA-FTIR spectroscopy and molecular modeling. These measurements indicate that the long axis of the dye molecule, and by implication, its charge-transfer axis, is perpendicular to the plane of the electrode surface. The use of *in situ* emission spectroscopy of the fluorescence on roughened gold substrates revealed the highest luminescence intensity of the emission of the dye species at 580 nm, which was attributed to surface plasmon effects.

Fluorescent spectra were collected while the SAMs were held at potentials of between +0.2 and −0.6 V. Reproducible blue shifts in λ_{max} were observed as the applied potential becomes more negative. While reproducible changes in luminescent intensity were also observed, the potential dependence of these changes is complex. Figure 5.46(b) shows plots of the emission peak energy of the monolayer as a function of the applied potential in 0.01 and 0.1 M $NaNO_3$. The smaller slope of the 0.01 versus the 0.1 M $NaNO_3$ plot is consistent with weaker electric fields in the lower-ionic-strength medium. The plots intersect at −0.4 V, suggesting the field is independent of ionic strength at this potential. The authors interpret this as the potential at which the dye molecules experience no electric field and attribute this to E_{pzc}, the potential of zero charge. The plots are approximately linear and their slopes could be used to calculate the dependence of the electric field on the applied potential, i.e. dE/dE_{app}.

The principle behind this investigation is electrochromism or Stark-effect spectroscopy. The electronic transition energy of the adsorbed chromophore is perturbed by the electric field at the electric double layer. This is due to interactions of the molecular dipole moment, in the ground and excited states, with the interfacial electric field induced by the applied potential. The change in transition frequency $\Delta\nu$, is related to the change in the interfacial electric field, ΔE, according to the following:

$$h\Delta\nu = -\Delta\mu\Delta E\cos\theta - \tfrac{1}{2}\Delta\alpha_{ge}\Delta E^2\cos\theta \qquad (5.10)$$

where $\Delta\mu$ is the total change in dipole moment between the ground and excited states, θ is the angle between the electric field vectors, which is estimated to be 1, and $\Delta\alpha_{ge}$ is the change in polarizability between the molecular ground and excited states.

By employing these parameters, the authors calculated dE/dE_{app} to be 4×10^4 and 3×10^5 V cm^{-1} V^{-1} dye–C3 alkyl and dye–C12 alkyl SAMs, respectively. These values refer to the change in electric field experienced by the dye species per volt change in applied potential. In each instance, these values suggest that the dye molecules experience an electric field within the SAM which is more than an order of magnitude weaker that predicted by models such as the Helmholtz and Gouy–Chapman double-layer theory. In the case of the dye–C12 SAM, the authors attribute this discrepancy to the penetration of ions from the supporting electrolyte into the SAM. For the dye–C3 alkyl assembly, the dye lies external to the alkyl chains and is thought to orient parallel to the electrode, so reducing the effective electric field. This work is an excellent example of the powerful insight that fluorescence spectroscopy can provide into fundamental interfacial properties.

5.11 Photoinduced Processes in Bio-SAMs

An important conclusion of supramolecular and biomolecular chemistry is the importance of non-covalent interactions in facilitating electron transfer. Significant study and debate has been dedicated to the role of hydrogen bridges, electrostatic interactions and π-stacking in promoting long-range electron transfer. Consequently, the coupling of biomaterials to photoactive SAMs offers the possibility of achieving a significant insight into the electron and energy transfer properties of biological systems.

One such study, undertaken by Morita *et al.* [79], investigated photocurrent generation in the helical peptide SAMs illustrated in Figure 5.47. The photocurrent could be switched from anodic to cathodic by changing the sacrificial donor–acceptor species.

Peptides have long been identified as excellent mediators of electron transfer. This has been attributed to a number of their characteristics, including their π and σ non-bonding orbitals, which are thought to provide effective mediation pathways, and the electric fields generated by the α-helical segments, which are thought to accelerate electron transfer. Self-assembled monolayers provide a useful means of addressing such bridge-mediated issues in electron transfer and Morita *et al.* have reported on the study of disulfide-terminated self-assembled monolayers on gold, comprising α-helical chains linked to photoactive *N*-ethylcarbazolyl (Ecz) appendages [79]. The helical content of the various peptide chains was obtained from circular dichroism spectra as Ecz–A12–SS (45.5 %), SS–A12–Ecz (49.2 %) and Ecz–A20–SS (81.8 %), confirming that helical structures are stabilized in the longer peptides. Cyclic voltammetry, EQCM and AC impedance measurements indicated that the tridecapeptide SAM with N-terminal binding to gold was more densely packed than the C-terminal bound monolayers. Fourier-transform infrared reflection–absorption spectroscopy (FTIR-RAS) yielded helix tilt angles of approximately 40° for the monolayers on gold. Photocurrent generation was investigated in photoelectrochemical cells, employing either an electron acceptor, methyl viologen, or electron donors, such as EDTA or triethanolamine. The photocurrent action spectrum corresponded to the absorption spectrum of the Ecz moiety, thus confirming it to be the source of the photoexcitation (see Table 5.2).

The extent of electronic coupling mediated by the peptide bridges was determined by using semi-classical electron transfer theory (as outlined earlier in Chapter 2), and estimates of the tunneling parameter, β, were made for the SAMs of Ecz–A12–SS as $0.58 \, \text{Å}^{-1}$, SS–A12–Ecz as $0.60 \, \text{Å}^{-1}$ and Ecz–A20–SS as $0.46 \, \text{Å}^{-1}$. These β-parameters were significantly lower than that found for electron tunneling through alkanes, i.e. $1.0 \, \text{Å}^{-1}$, hence making the electron transfer through peptides much less sensitive to changes in electron transfer distance.

The impact of the peptide dipole on the electron transfer was investigated by studying the potential dependence of the anodic photocurrent in the presence of an uncharged donor, i.e. TEOA. Electron transfer was observed to be accelerated in SAMs in which the helix dipole is directed towards the electrode

Table 5.2 Percentage quantum efficiencies of the cathodic and anodic photocurrent generation in aqueous solutions of MV^{2+}, TEOA and EDTA.

Peptide system	MV^{2+}	TEOA	EDTA
Ecz–A12–SS	0.43	0.19	0.17
SS–A12–Ecz	0.33	0.45	0.42
Ecz–C11–SS	0.19	0.10	0.04
Ecz–A20–SS	0.04	0.07	0.03

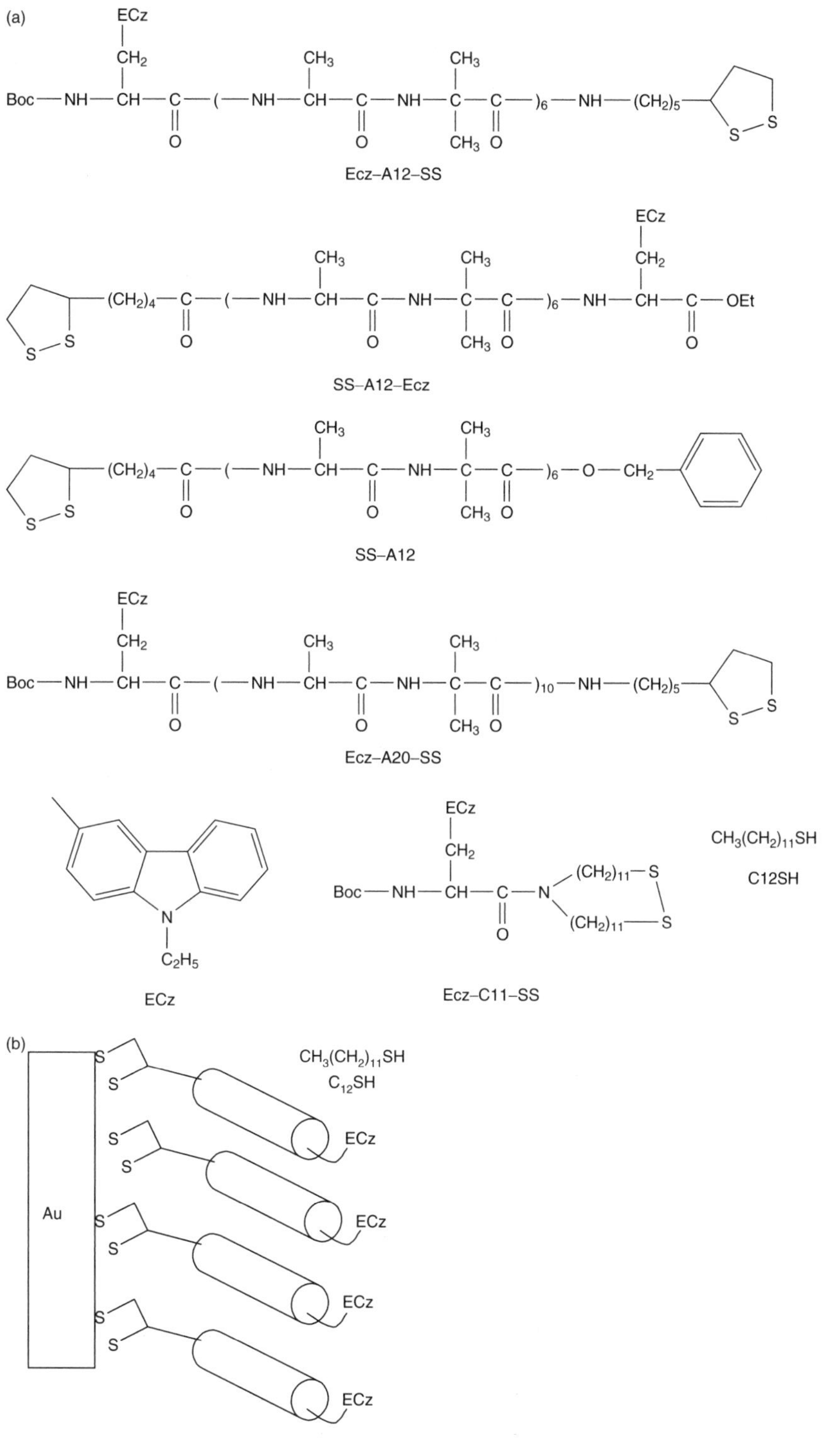

Figure 5.47 (a) Molecular structures of the disulfide-terminated α-helical chains containing *N*-ethylcarbazolyl (Ecz) reported by Morita *et al.* [79]. (b) Schematic of a typical arrangement of SAMs containing such helical chains at a gold substrate

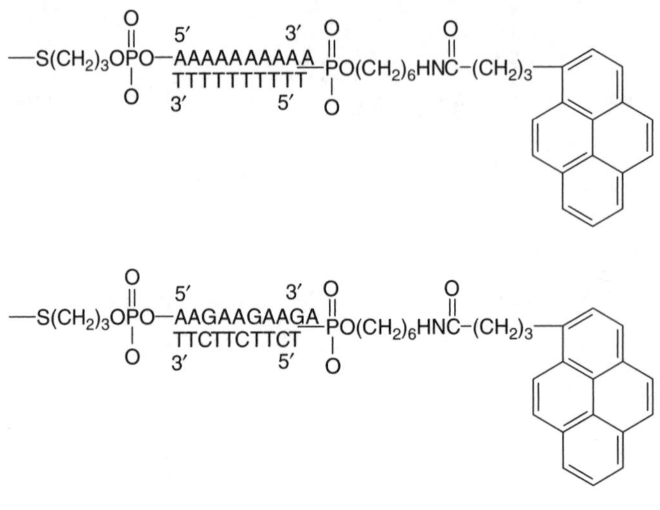

Figure 5.48 Molecular structures of two of the photoactive pyrene end-labeled oligonucleotides reported by Reese and Fox [80]

surface, thus illustrating the importance of helical dipoles in peptide-mediated electron transfer.

Previous work on the mediation of electron transfer in biological molecules suggests that π-stacking promotes long-range transfer; furthermore, the nature of the intervening base-pairs appears to play an important role in the rate and efficiency of electron transport through DNA. For example, guanine appears to play a promotional role, possibly as a consequence of its hole-acceptor ability, while thymidine plays a less supportive role. By employing specific base sequences or DNA fragments as bridges linking photo- or electrochemically active species, it may be possible to control the electron transfer rate and direction. Figure 5.48 illustrates the oligonucleotides and their associated pyrene end-labels reported by Reese and Fox [80]. Monolayers were formed stepwise on a gold surface by first immobilizing the oligonucleotide, which was then labeled by using complementary binding. Immobilization of the oligonucleotides was confirmed from grazing-angle reflectance FTIR spectroscopy and cyclic voltammetry of $K_2[Fe(CN)_6]$ solution at the surface-modified electrodes. Surface modification with the oligonucleotide caused significant attenuation of the current response, associated with electrostatic blocking of the ferrocyanide. This blocking was almost complete on incorporation of the end-label, which further limits access to the electrode surface. Emission spectroscopy confirmed the incorporation of pyrene into the monolayer in both SAMs.

Photocurrent generation was studied in deaerated aqueous solutions of methyl viologen, MV^{2+}, exciting at 346 nm to correspond to the pyrene absorbance. Anodic photocurrent was observed in both SAMs, which was eliminated in the absence of MV^{2+} or in the presence of hydroquinone. Figure 5.49 illustrates the time courses of the dark-current response and light-induced response for the mono-layers on gold. The polarity of the photocurrent was opposite to that expected for an oligonucleotide-mediated reduction of the pyrene excited state. Direct pho-toinjection from excited pyrene would be expected to increase in the presence of hydroquinone, as it would re-reduce the pyrene, whereas there is no photocurrent

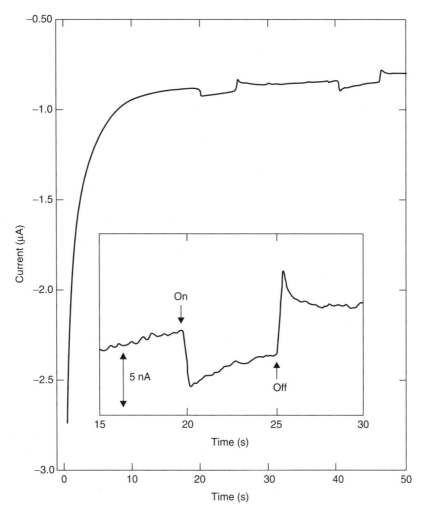

Figure 5.49 Time course for the limiting current at gold modified with the oligonucleotide 2/4 in methylviologen (MV^{+2}) solution (0.1 mM in 0.1 NaCl) at an applied potential of 0 mV (vs. Ag/AgCl). The inset shows the time domain in which the electrode is alternatively exposed to and shielded from 346 nm illumination; the arrows indicate when the shutter is open or closed. From R. S. Reese and M. A. Fox, 'Spectral and cyclic voltammetric characterization of self-assembled monolayers on gold of pyrene end-labeled oligonucleotide duplexes', *Can. J. Chem.*, **77**, 1077–1084 (1999), with permission from The National Research Council of Canada

under such conditions. The authors attribute the anomalous photocurrent polarity to alterations occurring in the electrode SAM double layer in the presence of MV^{2+}.

5.12 Photoinduced Electron and Energy Transfer in SAMs

In principle, thermal and photoinduced electron transfers should exhibit the same dependencies on bridge nature and length, as well as electrode materials. Many of the studies on electrochemically stimulated heterogeneous electron transfers on monolayers apply equally well to photoinduced charge injection, since on a simple level the difference between the ground-state and photoinduced electron transfer is derived from the increased driving force in the photoinduced reaction.

5.12.1   Distance Dependence of Photoinduced Electron and Energy Transfer

Electrochemical studies on SAMs have proven invaluable in elucidating the impact of various molecular parameters such as bridge structure, molecular orientation or the distance between the electroactive species and electrode surface. As described above in Section 5.2.1, the kinetics of heterogeneous electron transfer have been studied as a function of bond length for many systems. Similarly, the impact of bridge structure and inter-site distances have been studied for various supramolecular donor–acceptor systems undergoing photoinduced electron transfer in solution. In both types of study, electron transfer is observed to increase as the distance between the donor and acceptor decreases. As discussed earlier in Chapter 2, the functional relationship between the donor–acceptor distance and the electron transfer rate depends on the mechanism of electron transfer, which in turn depends on the electronic nature of the bridge.

Direct measurement of photoinduced heterogeneous electron transfer kinetics have not yet been reported for SAMs on metal electrodes. However, as described later in Chapter 7, ultrafast spectroscopic techniques have been applied to the study of these processes in large-band-gap semiconductors. The distance dependence of photoinduced electron transfer of ISAs has been investigated by using indirect approaches, e.g. by investigating the photocurrent quantum yield. However, many of these results are ambiguous because heterogeneous electron transfer may occur between the electrode and solution-phase reactants. Notwithstanding these challenges, there have been a significant number of carefully executed studies investigating the influence of bridge length on photoinduced reactions in SAMs. Figure 5.50 illustrates the complexes examined in one such study by Yamada and co-workers [81]. Here, two ruthenium tris(2,2'-bipyridine) viologen-linked disulfides, $[RuC_nVC_6S]_2$, with alkyl spacers containing $n = 3$ and $n = 7$ methylenes separating the viologen and ruthenium centers, and two ruthenium disulfide complexes, $[RuC_mS]_2$, in which the ruthenium and disulfide units were separated by $m = 13$ and $m = 17$ methylene groups, were synthesized and immobilized on gold surfaces.

Although quantum yields were not reported, luminescence studies on solutions of these complexes reveal the following relationships between the emission intensities: $[RuC_{17}S]_2 \approx [RuC_{13}S]_2 > [RuC_7VC_6S]_2 \gg [RuC_3VC_6S]_2$. These data indicate that there is no luminescence quenching in $[RuC_{13}S]_2$, but significant quenching in $[RuC_nVC_6S]_2$, which is increased in the complex with the shorter distance between the viologen and photoactive ruthenium center. Self-assembled monolayers of these materials on gold electrodes were examined by cyclic voltammetry and quartz crystal microbalance studies. These investigations reveal that the surface coverage is less than that expected for a dense monolayer in each case. Photocurrent measurements were conducted on SAMs of all four complexes in the presence of triethanolamine, which behaved as a sacrificial donor. An anodic photocurrent was

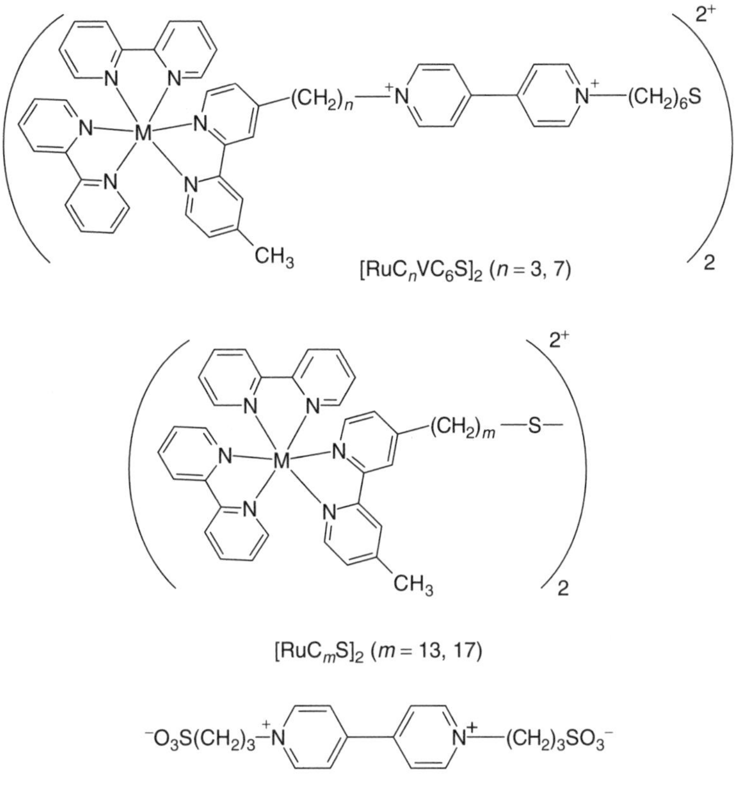

Figure 5.50 Molecular structures of the disulfide-linked ruthenium polypyridyl methylviologen diads, $[RuC_nVC_6S]_2$, and ruthenium disulfide complexes, $[RuC_mS]_2$ (of variable chain lengths n and m, respectively), plus the structure of the electron acceptor 4ZV, reported by Yamada and co-workers [81]

observed, with the photocurrent action spectra corresponding to the absorbance spectra of the complex, thus confirming the Ru to bpy metal-to-ligand charge transfer (MLCT) to be the origin of the photocurrent. As Figure 5.51 illustrates, anodic photocurrent was observed to be substantially larger for $[RuC_7VC_6S]_2$ than $[RuC_3VC_6S]_2$, with little photocurrent being observed for the $[RuC_mS]_2$ complexes. The proposed reaction for this process is as follows:

$$Bpy-Ru^{2+}-V^{2+} + h\nu \longrightarrow Bpy^{-\bullet}-Ru^{3+}-V^{+} \longrightarrow Bpy^{-\bullet}-Ru^{3+}-V^{+} + TEOA + e$$

$$\longrightarrow Bpy-Ru^{2+}-V^{2+} \quad (5.11)$$

Excitation of the complexes leads to photoinduced electron transfer from the excited ruthenium polypyridyl site to the viologen acceptor. The Ru^{2+} site is restored through electron transfer from the TEOA or back-electron transfer from the bipyridine, while the viologen is oxidized by the electrode, thus generating the photocurrent. As illustrated in Figure 5.51, this mechanism is supported by experiments in which the electron acceptor 4ZV (see Figure 5.50) reduced the

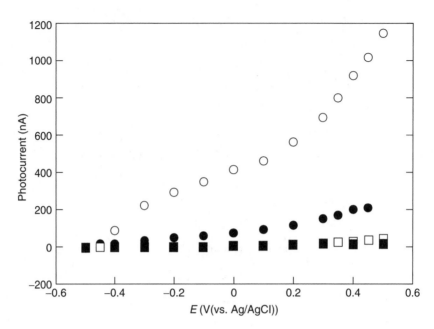

Figure 5.51   Applied potential dependence on the photocurrent intensities of RuC_7VC_6S/Au (○), RuC_3VC_6S/Au (●), $RuC_{17}S/Au$ (□) and $RuC_{13}S/Au$ (■) electrodes: λ_{ex}, 470 nm; [TEOA] = 5 × 10^{-2} M; [NaClO$_4$] = 0.1 M. Reprinted from *Thin Solid Films*, **350**, Y. Koide, N. Terasaki, T. Akiyama and S. Yamada, 'Effects of spacer-chain length on the photoelectrochemical responses of mono-layer assemblies with ruthenium tris(2,2'-bipyridine)–viologen linked disulfides', 223–227, Copyright (1999), with permission from Elsevier Science

photocurrent to approximately zero for RuC_7VC_6S/Au but bore considerably less influence on [RuC$_3$VC$_6$S]$_2$. This dependence of chain length on the ability of external viologen, 4ZV, to interfere with photocurrent efficiency is attributed to the faster forward- and back-electron transfers observed in the shorter-chain-length species.

Uosaki and co-workers have reported on the surface-immobilized porphyrin–ferrocene–thiol linked molecules shown in Figure 5.52(a) [82]. SAMs of these materials were observed to yield very efficient photocurrent generation at gold in cells containing methylviologen (MV^{2+}) as an electron acceptor. The mechanism of this process is illustrated in Figure 5.52(b). Photoexcitation leads to electron transfer from the porphyrin to the MV^{2+}, followed by alkylferrocene-mediated heterogeneous electron transfer between the electrode and photooxidized porphyrin [82]. Alkyl chains containing six, eight and eleven carbons were investigated. The authors observed that the longer the alkyl chain between porphyrin and ferrocene, then the larger the photocurrent. Quantum efficiencies of 4.0, 11 and 12% were observed, respectively, for the three SAMs at an electrode potential of −0.2 V, with the corresponding photoenergy conversion efficiencies being 0.7, 1.8 and 2% for the complexes containing alkyl chains lengths of six, eight and eleven carbons, respectively.

These results suggest that the effective inhibition of energy transfer and reverse electron transfer underlies the effect of bond length on photoinduced electron transfer at the gold electrodes modified with the SAMs of these molecules.

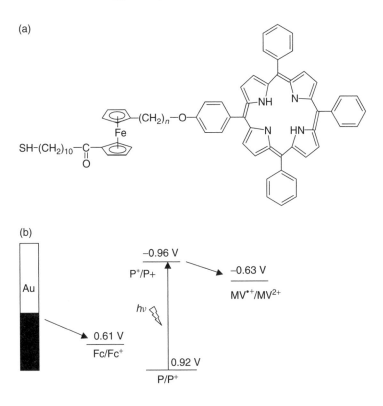

Figure 5.52 General molecular structure of the surface active porphyrin–ferrocene–thiol supramolecular complexes reported by Uosaki and co-workers [82]. (b) Energy diagram illustrating the mechanism behind photocurrent generation in SAMs of such complexes at cathodic electrode potentials of the Fc/Fc$^+$ couple: P, porphyrin; Fc, ferrocene; MV^{2+}, methyl viologen

5.12.2 Photoinduced Energy Transfer

Photoinduced energy transfer in SAMs may occur via concentration quenching in densely packed monolayers or through Dexter or Förster mechanisms in mixed donor and acceptor self-assembled monolayers. Alternatively, a single supramolecular species containing both the energy donor and acceptor may be bound to a surface. In order for photoinduced energy transfer to be observable, surface quenching must be limited. This problem may be reduced by increasing the distance between the photoactive species and the surface. Although the majority of studies on energy transfer in immobilized species have been conducted on Langmuir–Blodgett films, a number of studies have been conducted on energy transfer in mixed self-assembled monolayer systems.

Imahori *et al.* reported on the energy transfer in mixed self-assembled monolayers of pyrene and porphyrin [83]. The molecular structures of the disulfide-terminated chromophores are illustrated in Figure 5.53. Porphyrin and pyrene dimers were coadsorbed onto Au(111) mica substrates in different ratios. The ratios of pyrene to porphyrin were experimentally determined from absorption spectroscopy and the pyrene dimer was observed to adsorb preferentially, possibly due to its strong

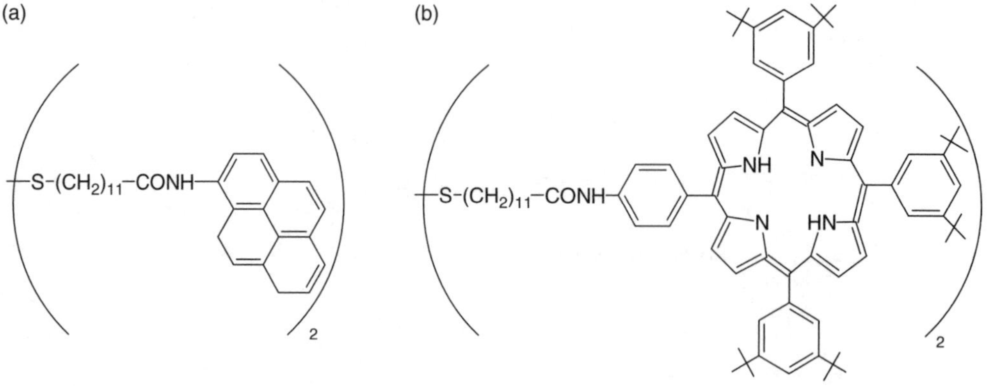

(a)

(b)

Figure 5.53 Molecular structures of the (a) pyrene and (b) porphyrin monolayer components in the mixed self-assembled monolayers reported by Imahori *et al.* [83]

$\pi - \pi$ stacking interactions. Cyclic voltammetry indicated that pyrene (Py) and porphyrin (Por) individually form densely packed monolayers. Adsorption of Py and Por onto gold led to a decrease in the fluorescent lifetime. The lifetime of Py decreased biexponentially from 7.4 and 3.2 ns in dichloromethane to 23 ps on gold, while the Por lifetime decreased from 8.1 ns in solution to 40 ps on gold. These decreases are attributed to efficient energy transfer to the surface states. Both the lifetimes of Py and Por decreased in mixed monolayers with increasing Por:Py ratios. The decrease in excited state lifetime is thought to originate from efficient energy transfer (>62 %) from the singlet excited pyrene moiety to the porphyrin, followed by energy migration among the porphyrin moieties in the densely packed monolayer. In conclusion, intermolecular energy transfer within the mixed monolayer has to compete with energy transfer to the surface.

5.12.3 Monolayer Mobility and Substrate Roughness

With the exception of single crystals and mercury, the substrate on which a monolayer is formed is not atomically flat. In addition, the adsorbate may be mobile on the surface. Therefore, an important question is how molecular motion and substrate morphology impact on the behavior of the SAM. These issues have been addressed to some extent on electrochemical systems but considerably less so on SAMs containing photoactive moieties. This situation is surprising given that fluorescence anisotropy measurements can provide an effective insight into both the distributions of molecular orientations caused by surface roughness and surface diffusion. However, there have been some insightful studies reported in the literature. For example, Figure 5.54 illustrates the molecular structures of a range of oligobis-phosphonates reported by Horne and Blanchard [84]. These investigations probed the effects of monolayer concentration and substrate roughness on the motional dynamics and lifetimes of zirconium phosphate monolayers deposited on oxidized Si (100) and fused silica surfaces. Four monolayer types were studied, one containing BDP and the other QDP, on both fused silica and oxidized Si (100) substrates.

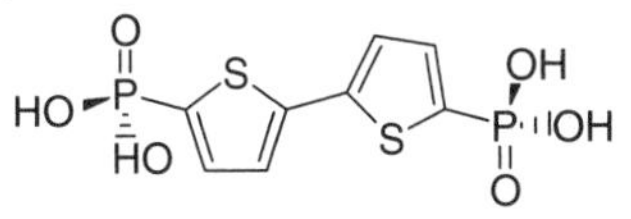

Figure 5.54 Molecular structures of the photoactive thiopheneoligobisphosphonates, BDP and QDP, and the diluent alkylphosphonates, DDBPA and HBPA, described in the work of Horne and Blanchard [85]

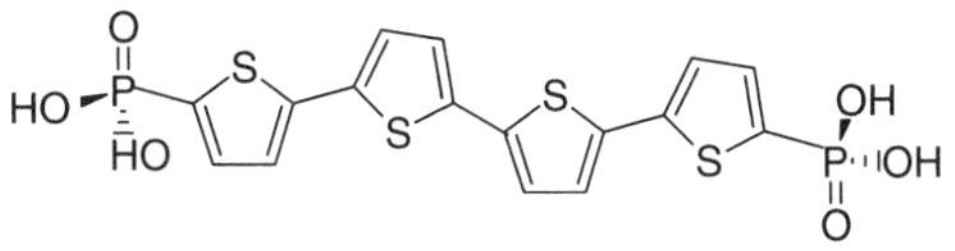

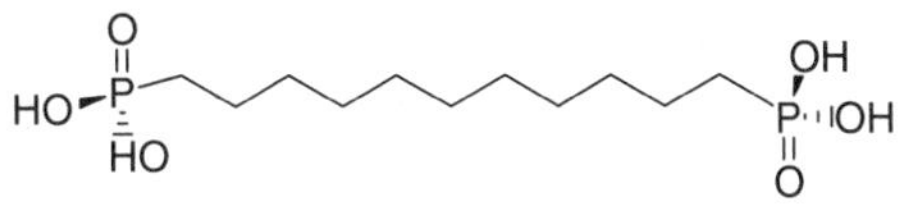

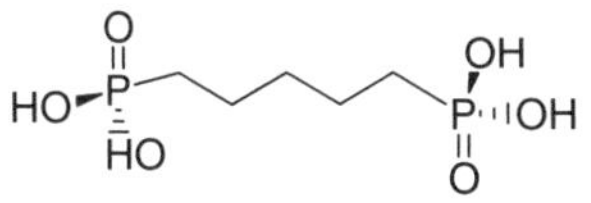

These thiophene oligobisphosphonates are the photochemically active components, while DDBPA and HBPA are inactive components employed for diluting the active species on the QDP- and BDP-functionalized surfaces, respectively (see Figure 5.54). Both the oxidized Si (100) and fused silica substrates possess the same OH binding sites. However, the surface roughness of the substrates are different. The fluorescent decays of the monolayers are biexponential, comprising a short-lived component of ca. 200 ps and a long-lived component of ca. 1200 ps. The fluorescent lifetimes are independent of both substrate identity and chromophore concentration within the monolayer. However, the relative fractions of long- and short-lived components in the lifetimes are sensitive to the chromophore concentration. Time-resolved fluorescence anisotropy studies were conducted in order to elucidate the chromophore mobility within the monolayers. The anisotropy parameters for the monolayers on each substrate are shown in Table 5.3. The decay of the anisotropy function over

Table 5.3 Induced orientational anisotropy data for various monolayer systems[a].

Monolayer system	$R(0)$	t_{MR} (ps)	$R(\infty)$	θ (degrees)
BDP on silica	–	–	0.06 ± 0.01	35 ± 2
QDP on silica	–	–	0.07 ± 0.05	39 ± 7
BDP on silicon	0.40 ± 0.15	393 ± 119	0.18 ± 0.08	19 ± 11
QDP on silicon	0.32 ± 0.06	479 ± 153	0.13 ± 0.06	32 ± 7

[a]The angles between the transition moments, δ, are 31 and 0° for BDP and QDP, respectively (determined from Equation (5.12)).

time, $R(t)$, was modeled according to the following:

$$R(t) = R(\infty) + [R(0) - R(\infty)]\exp(-t/\tau_{MR}) \qquad (5.12)$$

where τ_{MR} is the motional relaxation time constant, $R(\infty)$ is the steady-state anisotropy, and $R(0)$, the anisotropy at time zero, is related to the angle between the chromophore absorption and emission transition moments, δ, according to the following:

$$R(0) = 2/5P_2(\cos\delta) \qquad (5.13)$$

where P_2 is the second-order Legendre polynomial.

For BDP, $R(0)$ is 0.24, corresponding to an angle of 31°, while for QDP $R(0)$ is 0.40, i.e. corresponding to an angle of 0°.

Because the chromophores are tethered to the surface, rotational motion of the chromophore is confined to a conical volume. In contrast, for chromophores in solution, $R(\infty) = 0$, since the chromophores attain random orientations at infinite time. In monolayers, $|R(\infty)| > 0$, since they are motionally restricted, and $R(\infty)$ is related to the average tilt angle of the chromophores according to the following relationship:

$$R(\infty) = 2/5P_2(\cos\theta_{ex})P_2(\cos\theta_{em}) < P_2(\cos\theta) > 2 \qquad (5.14)$$

The angles θ_{ex} and θ_{em} are the transition moment angles for excitation and emission respectively with respect to the center of the motional cone axis, with $\theta_{ex} = 0°$ and $\theta_{em} = \delta$ in this instance. The average tilt angles of the chromophores within the monolayers are provided in Table 5.3. Fluorescence anisotropy studies on BDP and QDP monolayers on silica reveal no decay of $R(t)$, implying there is no chromophore motion on the time-scale being investigated.

Figure 5.55 illustrates (a) the polarized emission intensities of a 1% monolayer of QDP on silicon, and (b) the decay of this anisotropy. A similar behavior is also reported for BDP. It is clear that unlike the silica substrate, chromophore motion does occur on silicon. The possibility of a large amplitude rotation of the chromophore is excluded on the basis that the observed behavior is concentration-independent and also independent of the nature of the chromophore. Therefore, it is uninfluenced by the 'footprint' of the molecule, which is contrary to what would be anticipated if large amplitude precessional motions were occurring. On the basis of computational models, chromophore motion is attributed to an intramolecular thiophene ring rotation, which is not expected to be strongly influenced by a hindered environment. The difference in behavior on silicon and silica is attributed to the differences in surface roughness of the two substrates. AFM studies of the surfaces reveals that on the 50 Å horizontal length-scale, the variations in surface height are ±2 Å on silica, compared with ±10 Å on silicon. Furthermore, silicon exhibits a lower density of active hydroxyl groups. Both surface characteristics should allow for greater motional freedom of adsorbates on silicon.

Electrochemical studies on SAMs have proven remarkably effective in elucidating the basic mechanisms underlying electron transfer processes. Solution-phase studies have for many years illustrated the power of photochemistry and spectroscopy in elucidating photoinduced reaction mechanisms. The marriage of these two experimental approaches promises to bring new and exciting insights into the behavior

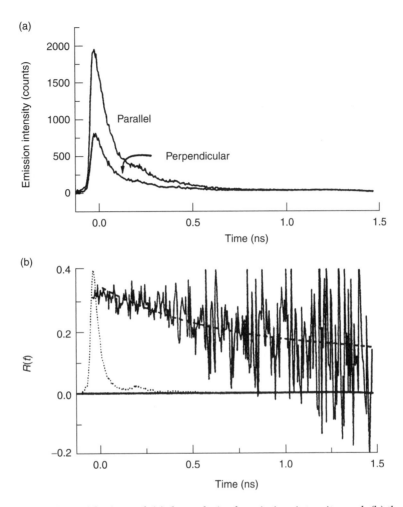

Figure 5.55 Variation with time of (a) the polarized emission intensity and (b) the induced orientational anisotropy function for a 1% QDP monolayer on silicon. Reprinted with permission from J. C. Horne and G. J. Blanchard, *J. Am. Chem. Soc.*, **120**, 6336 (1998). Copyright (1998) American Chemical Society

and mechanisms of interfacial species since, although numerous elegant photoactive SAMs have been developed, photoinduced electron transfer remains relatively poorly explored for spontaneously adsorbed and self-assembled monolayers. Many fundamental questions remain to be answered and significantly greater research effort in this field appears justified since future molecular-scale devices are most likely going to require surface organization.

5.13 Multilayer Assemblies

It is evident that the ability to direct and control energy and electron transfer processes in functionalized monolayers via optical and electrical processes is continuously evolving. However, greater complexity and extensive three-dimensional organization will inevitably be required for applications such as high-density information storage. Multilayer synthesis at surfaces makes it possible to fabricate donor and acceptor species, layer by layer, through self-assembly. The number of reports on such species for optoelectronic applications are increasing. In Section 5.2.2 above, we discussed the value of fullerene in photoinduced electron transfer and photocurrent generation as a component of a supramolecular SAM. Ikeda *et al.* recently reported on the efficient photocurrent generation at self-assembled multilayers of a fullerene–calix[3]arene inclusion complex and an anionic porphyrin polymer [85]. Figure 5.56 shows the self-assembled multilayer structure and the individual component molecules from this report. The multilayer was constructed by stepwise self-assembly of the molecular layers onto indium–tin oxide (ITO). Three multilayer electrodes were examined, i.e. one containing just layers '1' and '2', the 3-mercaptoethanesulfonate and hexacationic homooxacalix[3]arene–[60]fullerene (2:1) complex, with the second and third systems containing these layers ('1' and '2'), as well as a further (third) layer containing the polymer-bound porphyrin shown in Figure 5.56, either in its native state, '3', or as its Zn porphyrin analogue, '4'.

Electronic spectroscopy studies of these multilayer electrodes support the structures whereby red-shifts and peak broadening of the porphyrin Soret band indicates aggregation similar to that of porphyrins on ITO. Surface concentrations of the porphyrin moieties are estimated to be 7.6×10^{-11} and 5.5×10^{-11} mol cm^{-2} for '3' (the native porphyrin) and '4' (the Zn porphyrin), respectively. Photocurrent measurements on the multilayers were performed by employing ascorbic acid as the sacrificial donor, and stable anodic photocurrents were observed. The photocurrent action spectra show considerably greater photocurrent densities in the multilayers containing the porphyrins and also indicate that the latter are acting as the photosensitizing units. Thermodynamics and bias voltage data suggest that the photocurrent flows from porphyrin to fullerene to electrode, with the porphyrin being restored via its electron exchange with ascorbic acid. The photocurrent quantum yields are 14 % for the porphyrin and 21 % for the Zn porphyrin multilayers, with these values being among the highest quantum yields recorded to date.

Quantum dots are a curious particulate form of matter, lying between the classical concepts of molecules and bulk materials. Molecular clusters of metals and semiconductors are termed as quantum dots when they exist in a size regime where the particle size is of the order of the wavelength of an electron. This leads to behavioral trends which are best described by quantum mechanics. Figure 5.57 illustrates a comparison of the electronic states of bulk matter, in this case a semiconductor, a quantum dot, or nanocluster, and a single molecule. In bulk materials, the electron states converge to generate bands, where in the case of metals these are conductance and valence bands separated by a band-gap in semiconductors. In nanoparticles, spatial confinement leads to the formation of discrete electronic energy levels which are analogous to molecular orbitals in molecules, with the electrons being localized in these discrete states. As a result of their quantum confinement, many nanoparticle materials exhibit interesting optical properties, for example, absorbances which depend on dot size, and also luminescence. Long-lived excited states may be generated in quantum dots after photon absorption. Excitation leads to the formation of a hole–electron pair, or *exciton*, which is spatially confined to a three-dimensional quantum well in which valence and conduction bands are quantized. The energy

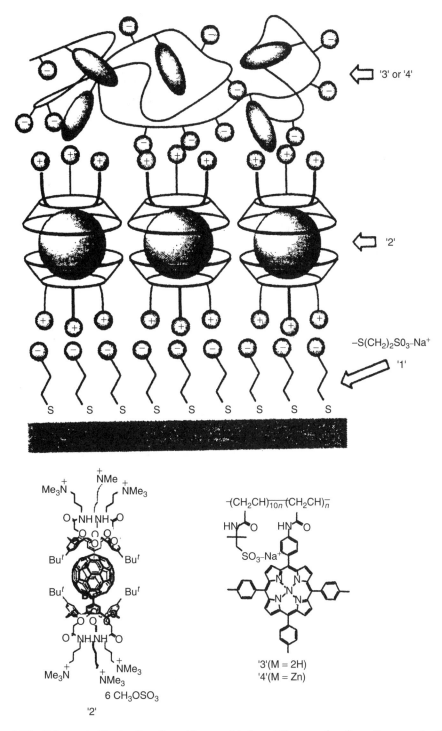

Figure 5.56 Schematic illustrating the self-assembled multilayers of sodium 3-mercaptoethane-sulfonate (layer '1'), hexacationic homooxacalix[3]arene–fullerene (2:1) complex (layer '2'), with either the anionic porphyrin polymer ('3') or its zinc porphyrin analogue ('4') forming the third layer. Reprinted with permission from A. Ikeda, T. Hatano, S. Shinkai, T. Akiyama and S. Yamada, *J. Am. Chem. Soc.*, **123**, 4855 (2001). Copyright (2001) American Chemical Society

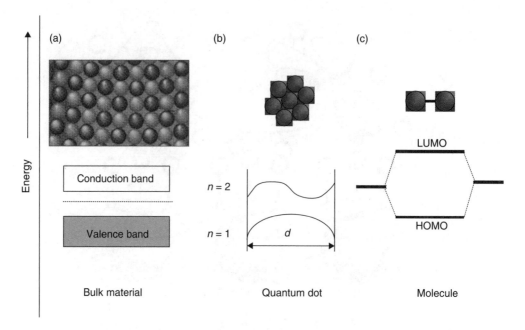

Figure 5.57 Illustration of the relationship between particle size and energy levels for (a) a bulk material possessing energy bands, (b) a nanoparticle with quantum-confined energy levels, and (c) a single molecule with molecular orbitals

separation between these bands is dependent on the size of the dot, which therefore dictates the optical properties, including luminescence quantum yield, wavelength and lifetime.

The photochemical properties of nanoparticulate CdS, which is one of the most studied of the quantum dot materials, are well known. Particular advantages in working with such materials in potential optoelectronic or photoelectrochemical devices lies in the control permitted over spectroscopic properties based on the size of the particles. A potential difficulty in using nanoparticles in mono- and multilayers lies in their tendency to aggregate, thereby losing their quantum-confined properties. Uosaki and co-workers have reported on multilayer CdS nanoparticles, self-assembled on gold with alkane thiols [86]. The binding and structures of these multilayers are illustrated in Figure 5.58. Alternate layers of 1,6-hexane dithiol or 1,10-decane dithiol and CdS nanoparticles were assembled on gold which had been vapor deposited on glass slides. This was achieved by immersing the slide, alternatively, in solutions of dithiol and a dispersion of CdS nanoparticles. The layer-on-layer structure of the dithiol SAM and monolayer of CdS nanoparticles was confirmed by Fourier-transform infrared reflection–absorption spectroscopy, which showed significant differences in the mid-IR region between the Au–dithiol SAM and Au–dithiol–CdS multilayers, and X-ray photoelectron spectroscopy (XPS). Clear evidence for the presence of CdS in the Au–dithiol–CdS multilayer was observed from the appearance in the photoelectron spectra of Cd $3d_{5/2}$ and Cd $3d_{3/2}$ peaks at 405.7 and 412.5 eV, respectively. Inductively coupled plasma-mass spectrometry (ICP-MS) measurements were conducted to determine the average surface concentration of cadmium atoms. Based on these measurements,

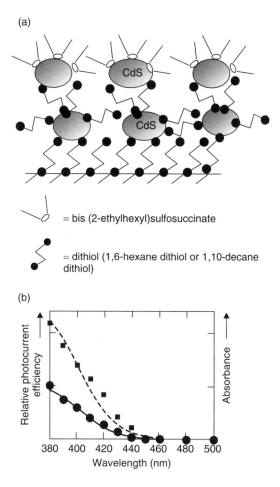

Figure 5.58 (a) Schematic illustration of the binding in a Au–dithiol–CdS–dithiol–CdS multilayer, where the CdS nanoparticles are formed from reversed micelles onto a modified gold surface. (b) Action spectra of composite films and absorption spectra of the CdS nanoparticle dispersion before being immobilized: continuous and dashed lines show the action spectra on different scales; ● and ■ represent the relative photocurrent efficiencies of Au–dithiol–CdS and Au–dithiol–CdS–dithiol–CdS, respectively. Reprinted with permission from O. Nakanishi, B. Ohtani and K. Uoskai, *J. Phys. Chem., B*, **102**, 1571 (1998). Copyright (1998) American Chemical Society

the Au–dithiol–CdS system was calculated to contain 3×10^{12} particles cm^{-1}, compared with 5.4×10^{12} particles cm^{-1} for an Au–dithiol–CdS–dithiol–CdS multilayer where the dithiol was hexane dithiol, and 6×10^{12} particles cm^{-1} where the dithiol was decane dithiol.

When holding the composite films at 0 V (vs. Ag/AgCl) under illumination ($\lambda \le$ 450 nm), an anodic photocurrent was observed in the presence of a triethanolamine donor whose relative efficiency was dependent on the excitation wavelength. There was good agreement between the absorption spectra of a dispersion of CdS nanoparticles and the photocurrent action spectra. This indicated that the CdS particles maintained their integrity in the multilayer without any problems of aggregation.

The mechanism proposed involved photoexcitation of the CdS particle which generates an electron–hole pair; the hole then oxidizes the solution triethanolamine while the electron is injected into the gold substrate to produce the anodic photocurrent. The relative photocurrent efficiency of the Au–dithiol–CdS system was twice that of the Au–dithiol–CdS–dithiol–CdS multilayer, which further confirms the role of the CdS in creating the photocurrent. This work also demonstrates the advantages of employing multilayers for photocurrent generation. The absorption cross-section increases as a consequence of using a multilayer, whereby a higher concentration of photoactive species can be accommodated at the interface when compared with a monolayer.

Luminescent multilayer systems have been reported extensively in Langmuir–Blodgett studies; however, reports of intrinsically luminescent self-assembled multilayers are far rarer. Bard has reported on an intriguing luminescent multilayer comprised of alternate layers of 1,6-hexane dithiols and Cu ions (see Figure 5.59),

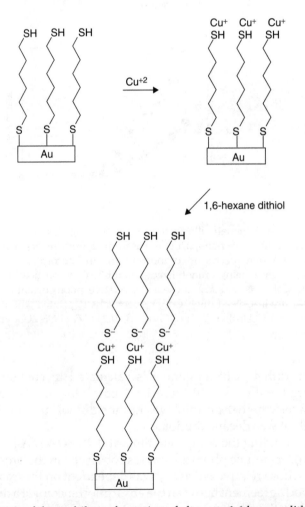

Figure 5.59 Schematic of the multilayer formation of alternate 1,6-hexane dithiol molecules and Cu(II) ions [87]

which were assembled on Au-sputtered films on glass slides [87]. These films showed strongly Stokes-shifted emission ($\lambda_{ex} = 400$ nm), centered at approximately 680 nm. Luminescence would suggest that the copper exists in the multilayer in the Cu(I) state, while the copper ions are deposited in the multilayer in the 2+ state. XPS studies indicated that the multilayers contained Cu(I) and this was confirmed from cyclic voltammetry. As Cu(I) is unlikely to bind to two thiolate ligands, it was concluded that the oxidation of the thiolates by Cu(II) leads to the disulfide–Cu(I) structure shown in Figure 5.59.

Solid luminescent materials, with identical emission λ_{max} values, were also obtained on the reaction of 1,6-hexane dithiol with cupric perchlorate. The origins of the luminescence in both the film and solid appear to be the same, although this is not yet clearly understood. It is likely that they are associated with Cu–Cu interactions in what is essentially a polynuclear structure.

These multilayers may offer significant advantages over monolayers in terms of luminescent intensity, as it becomes possible to maintain luminescent centers at a significant distance from the electrode surface. Furthermore, layers of luminophores may be deposited, thus increasing their concentrations in the films. Both properties may lead to significantly increased luminescent intensities over those of monolayers. These are important considerations if such materials are to prove useful in miniaturized optoelectronic applications.

5.13.1 Photoinduced Charge Separation in Multilayers

As discussed earlier, a disadvantage of the use of organic polyelectrolytes as components for the self-assembly of multilayer films is the substantial interdigitation of the layers. Intermingling of different layers is a significant barrier to developing devices based on efficient charge separation. To address this issue, Mallouk and co-workers have adapted the electrostatic self-assembly methods by introducing inorganic components [88]. With these mixed inorganic–organic multilayers, effective charge separation can be obtained, as shown in the next two examples. The inorganic polyelectrolytes used are α-Zr(HPO$_4$)$_2$.H$_2$O (α-ZrP) or KTiNbO$_5$. These materials are polyanionic sheets of well-defined layer thickness and it was anticipated that with such layers interpenetration of alternating anion/cation layers can be prevented. Organic–inorganic multilayers were grown by sequential adsorption reactions, as outlined in Figure 5.60. The substrate (glass or silica) is primed with an aminoalkylsilane derivative, which at neutral pH contains protonated amino groups. When this modified surface is dipped into a solution containing α-ZrP, anionic sheets of the polyanion are adsorbed as a monolayer. In turn, this anionic surface can absorb a monolayer of polycations, such as polyallylamine (PAH) as shown in step 2 of Figure 5.60. Steps one and two can be repeated many times to produce a well-defined multilayer assembly. AFM and fluorescence studies indicate that the layer grows in a stepwise fashion with little interpenetration of the PAH layers. The ability of these multilayer assemblies to achieve long-lived charge separation has been investigated by using the ruthenium polypyridyl/viologen system shown in Figure 5.61. This assembly consists of an amine-modified silica particle with an α-ZrP/polymer-bound viologen/α-ZrP/polymer-bound [Ru(bpy)$_3$]$^{2+}$

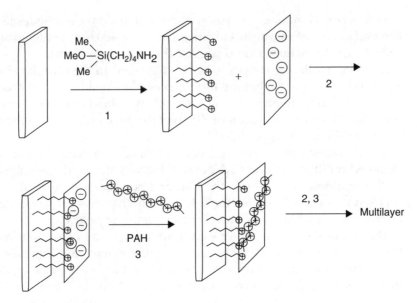

Figure 5.60 Schematic representation of the sequential steps taken for the formation of multilayers based on electrostatic self-assembly using cationic polymers and anionic α-ZrP sheets (see text for further details). Reprinted from *Coord. Chem. Rev.*, **185–186**, D. M. Kaschak, S. A. Johnson, C. C. Waraksa, J. Pogue and T. E. Mallouk, 'Artificial photosynthesis in lamellar assemblies of metal poly(pyridyl) complexes and metalloporphyrins', 403–416, Copyright (1999), with permission from Elsevier Science

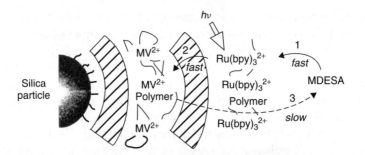

Figure 5.61 Schematic representation of a $[Ru(bpy)_3]^{2+}/\alpha$-ZrP viologen structure on silica, plus the sequence of fast (1, 2) and slow (3) electron transfer steps that follow photoexcitation of the photoactive ruthenium-containing polymer; MDESA, *p*-methoxyaniline diethylsulfonate. Reprinted from *Coord. Chem. Rev.*, **185–186**, D. M. Kaschak, S. A. Johnson, C. C. Waraksa, J. Pogue and T. E. Mallouk, 'Artificial photosynthesis in lamellar assemblies of metal poly(pyridyl) complexes and metalloporphyrins', 403–416, Copyright (1999), with permission from Elsevier Science

multilayer system. Upon photoexcitation of the ruthenium complex, a metal-to-ligand charge-transfer (MLCT) state is created. In the absence of viologen, a triplet MLCT emission is observed, but with viologen in solution efficient quenching of this emission by electron transfer to the electron acceptor viologen is observed. With the immobilized Ru–viologen assembly shown in Figure 5.61, no quenching of the ruthenium excited state is observed. This observation suggests that electron transfer through the α-ZrP barrier is inefficient. This is not unexpected since the modest

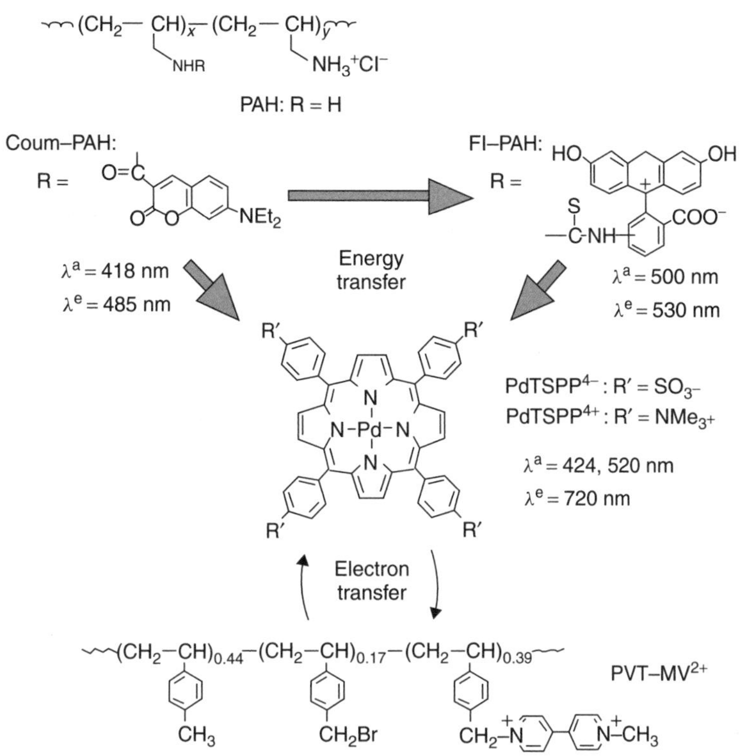

Figure 5.62 Energy and electron transfer pathways for the antenna polymers Coum–PAH and Fl–PAH, the photosensitizers PdTAPP^{4+} and PdTSPP^{4-}, and the electron acceptor PVT–MV^{2+}. Reprinted with permission from D. M. Kaschak, J. T. Lean, C. C. Waraska, G. B. Saupe, H. Usame and T. E. Mallouk, *J. Am. Chem. Soc.*, **121**, 3435 (1999). Copyright (1999) American Chemical Society

driving force of about 0.28 V may not be enough to drive electron transfer across the 8 Å thick inorganic sheet. However, a reduction of the viologen is observed in the presence of *p*-methoxyaniline diethylsulfonate (MDESA) in solution. The presence of this electron donor results in the formation of the reduced [Ru(bpy)$_3$]$^{+}$ species. This compound is a much better electron donor than the excited [Ru(bpy)$_3$]$^{2+*}$ species and as a result electron donation to the viologen sites is very fast ($\leq$50 ns) and the charge-separated state is long-lived (τ is about 40 µs) since the back-reaction must occur via the ruthenium polymer layer and the α-ZrP sheet.

This approach has also been applied to the design of artificial photosynthetic devices. Natural photosynthesis is based on intermolecular energy and electron transfer processes with charge separation that ultimately converts light into chemical energy. In nature, the light absorption is carried out by 'light-harvesting' compounds, which pass on their energy to the active center, a porphyrin dimer, where charge separation takes place. Energy transfer between the light-harvesting centers and the reaction center takes place via overlap of the donor emission and the acceptor absorption spectra, i.e. energy transfer proceeds according to the Förster mechanism. Charge separation at the active center is obtained by a subtle arrangement of electron donors and acceptors. Many attempts have been made to devise artificial systems which mimic this behavior. The most important and difficult step

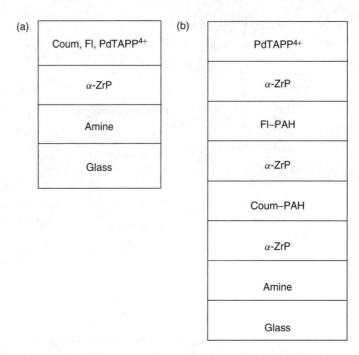

Figure 5.63 Two examples of multilayer systems where the energy transfer components are (a) coadsorbed, and (b) sequentially adsorbed. Reprinted with permission from D. M. Kaschak, J. T. Lean, C. C. Waraska, G. B. Saupe, H. Usame and T. E. Mallouk, *J. Am. Chem. Soc.*, **121**, 3435 (1999). Copyright (1999) American Chemical Society

is to achieve long-lived charge separation. In order to cover as much as the visible light spectrum as possible, several types of light-absorbing chromophores have been proposed. As shown in Figure 5.63, Mallouk and co-workers [88,89] have used a combination of inorganic and organic ionic sheets to build an artificial 'leaf'. The assemblies are based on the use of inorganic cationic separators and anionic organic polymers containing the light-harvesting compounds, fluorescein, coumarin and porphyrins. As Figure 5.62 shows, the electronic properties of the components involved allow for efficient energy transfer to occur. The coumarin polymer has its absorption maximum at 412 nm and emits at 485 nm. This emission coincides with both the absorption spectrum of the fluorescein polymer (λ_{max}, 500 nm) and with the Q-band of the porphyrins (λ_{max}, 520 nm). The emission of the fluorescein polymer is again coincident with the porphyrin absorption. The porphyrins show an additional absorption feature at 424 nm. Therefore, this assembly is well set up for light absorption over a wide range of the visible spectrum (between 380 and 500 nm) and should also be capable of setting up an efficient energy cascade. The PVT–MV^{2+} layer does not absorb in the visible region. The viologen groups are therefore not participating in the energy cascade, but are acting as electron acceptors to the porphyrin sensitizer. Because of the short excited state lifetimes of the coumarin and fluorescein components, energy transfer must be fast. Introducing a palladium metal center into the porphyrin increases the efficiency of inter-system crossing to the long-lived triplet emitting state. Long-lived charge separation is only obtained when the electron acceptors and donors are separated in different layers.

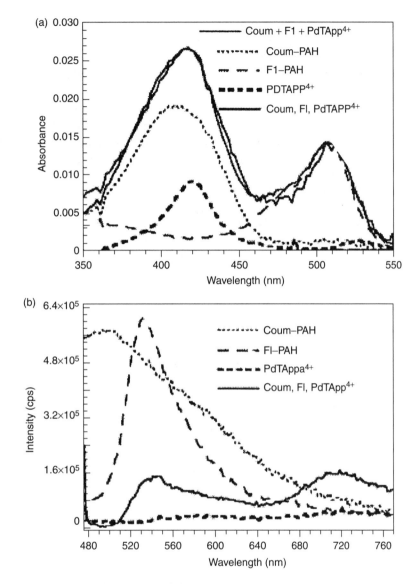

Figure 5.64 (a) UV–vis adsorption, and (b) steady-state emission ($\lambda_{ex} = 450$ nm) spectra of individual chromophore layers and of the coadsorbed triad (Figure 5.63a) (Coum–PAH, Fl–PAH and PdTAPP^{4+}) on glass/amine α-ZrP. Reprinted with permission from D. M. Kaschak, J. T. Lean, C. C. Waraska, G. B. Saupe, H. Usame and T. E. Mallouk, *J. Am. Chem. Soc.*, **121**, 3435 (1999). Copyright (1999) American Chemical Society

In order to investigate the effectiveness of energy transfer as a function of the layer assembly, both intralayer and interlayer multilayers were prepared.

Figure 5.63 shows the multilayers discussed in this section, while Figure 5.64 and Figure 5.65 show the emission spectra for an assembly containing coadsorbed (Figure 5.63a) or sequentially adsorbed (Figure 5.63b) energy transfer components and for the individual components. By using ellipsometry and UV–vis spectroscopy, the thickness of the layers was obtained accurately. The α-ZrP/dye–PAH bilayers are typically 22 Å thick, while the PdTSPP^{4+}/PAH layers have a thickness of about 9 Å. Each dye–PAH layer adds 10–12 Å to the total film thickness, while PAH deposited at pH 5 has a thickness of 6 Å, with each α-ZrP layer adding 10–12 A. The thickness observed for the PAH layer suggests that when coated at pH 5, the polymer chains are partially coiled.

For the multilayers shown in Figure 5.63, the absorption spectra for the assemblies can be obtained by the direct addition of the individual components. Figures 5.64 and 5.65 show, however, that the emission spectra are not additive, with the spectra obtained providing direct evidence for the presence of efficient energy transfer. Upon excitation at 450 nm, there is almost complete quenching of the coumarin emission at 485 nm, substantial quenching of the fluorescein emission and a greatly enhanced porphyrin emission. A comparison of Figures 5.64 and 5.65 shows that the increase of the porphyrin emission is strongest for the coadsorbed layer where intra layer energy transfer is taking place. The higher quantum yields obtained for energy transfer in assemblies such as those shown

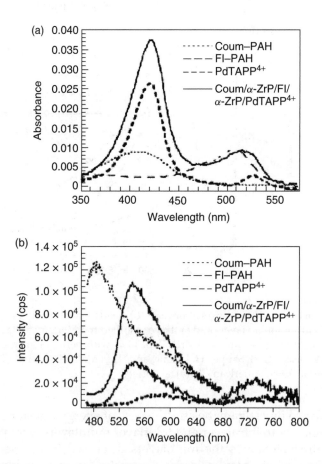

Figure 5.65   (a) UV–vis adsorption, and (b) steady-state emission ($\lambda_{ex} = 450$ nm) spectra of individual chromophore layers and the multilayer triad shown in Figure 5.63(b). Reprinted with permission from D. M. Kaschak, J. T. Lean, C. C. Waraska, G. B. Saupe, H. Usame and T. E. Mallouk, *J. Am. Chem. Soc.*, **121**, 3435 (1999). Copyright (1999) American Chemical Society

in Figure 5.63(a), where the partners are not separated by an inorganic layer, are consistent with the closer separation between the different components. Therefore, to achieve efficient energy transfer separation of the various components is clearly not advantageous.

These results show that a layer-by-layer assembly based on electrostatic inter-actions provides an attractive route to the organization of assemblies capable of complex energy and electron processes. The non-penetrating inorganic layers facilitate efficient charge separation and by a careful choice of flat-band poten-tials can be utilized to further enhance charge separation. On the other hand, the interpenetration of organic polyelectrolyte layers can be used to optimize energy transfer processes. In the systems discussed there is no active interaction at the organic–inorganic interface. Studies of the efficiency of charge injection in photo-voltaic cells based on ruthenium dyes on nanocrystalline TiO_2 surfaces have shown that the presence of a covalent bond between the solid substrate and the photoactive dye greatly enhances the charge injection. The application of such techniques here can be expected to further enhance the capabilities of multilayer assemblies to act as artificial photosynthetic models.

5.14 Electrochemistry of Thin Redox–Active Polymer films

The physical structure of spontaneously self-assembled monolayers is often highly ordered on the molecular length-scale. In contrast, the primary structures of poly-mers tend to be rather disordered. However, as discussed earlier in Chapter 4, the secondary structure may be well-defined because of self-associations between the monomer units or between the monomer units and solution. The 'persistence length' of this secondary structure often strongly depends on the concentration of the active groups and the nature of the contacting solvent, as well as the tempera-ture. As discussed in Chapter 4, polymer coatings can be obtained on a wide range of substrates and the physical properties and compositions of the materials can be tuned to a particular application. In the following, the structural aspects of polymer films will be addressed. It is clear that a fully molecular-scale organization such as that found in hydrogen-bonded crystals or in self-assembled monolayers cannot be expected. However, understanding and manipulating the molecular structures of these layers to achieve a particular electrochemical or photophysical behavior are important objectives. Of particular interest is the manner in which the dynamics of electron and energy transfer can be modulated by external factors, such as a contacting electrolyte.

One area where the relationship between the structure of the polymer matrix and the physical processes of the thin layer has been studied in detail is that of electrodes modified with polymer films. The polymer materials investigated in these studies include both conducting and redox polymers. Such investigations have been driven by the many potential applications for these materials. Conducting polymers have been applied in sensors, electrolytic capacitors, batteries, magnetic storage devices, electrostatic loudspeakers and artificial muscles. On the other hand, the development of electrodes coated with redox polymers have been used extensively to develop electrochemical sensors and biosensors. In this discussion,

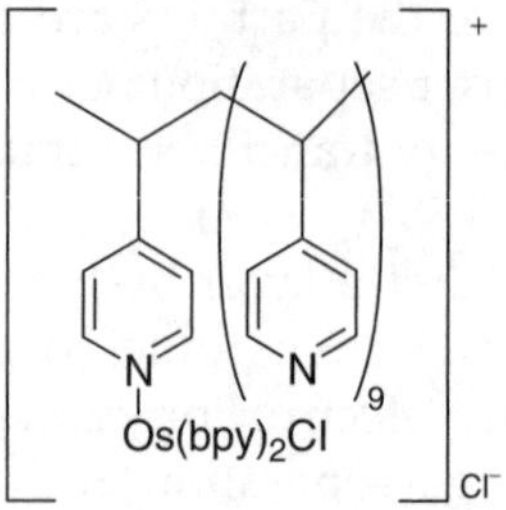

Figure 5.66   Molecular structure of the redox polymer [Os(bpy)$_2$(PVP)$_{10}$Cl]Cl

the emphasis will be placed on investigations of how the polymer structure effects electron transfer within the ISA and across the film–solution interface, and will be based on data obtained from both electrochemical quartz microbalance and neutron reflectivity studies.

The example considered is the redox polymer, [Os(bpy)$_2$(PVP)$_{10}$Cl]Cl, where PVP is poly(4-vinylpyridine) and '10' signifies the ratio of pyridine monomer units to metal centers. Figure 5.66 illustrates the structure of this metallopolymer. As discussed previously in Chapter 4, thin films of this material on electrode surfaces can be prepared by solvent evaporation or spin-coating. The voltammetric properties of the polymer-modified electrodes made by using this material are well-defined and are consistent with electrochemically reversible processes [90,91]. The redox properties of these polymers are based on the presence of the pendent redox-active groups, typically those associated with the Os$^{(II/III)}$ couple, since the polymer backbone is not redox-active. In sensing applications, the redox-active site, the osmium complex in this present example, acts as a mediator between a redox-active substrate in solution and the electrode. In this way, such redox-active layers can be used as electrocatalysts, thus giving them widespread use in biosensors.

The electrochemical properties of these redox polymers are quite different from those observed for conducting polymers. They are conducting only over a limited potential range, with the redox centers being essentially independent from each other. Since the osmium polymers under discussion here are hydrophilic, they form hydrogels which have an open structure which interacts strongly with the contacting electrolyte. The three-dimensional organization of the layers depends strongly on external influences, so making it difficult to obtain a detailed picture of their structure. The latter depends on variables such as layer thickness, the nature of solvents, electrolytes and temperature. Figure 5.67 illustrates the close-to-ideal voltammetry observed for these systems. Such voltammograms can be used to study the structure of the polymer layer by identifying the nature of the charge-transport process through it.

5.14.1   Homogeneous Charge Transport

Three processes can control the rate of homogeneous charge transport through a redox-active polymer film, i.e. electron self-exchange between redox-active centers,

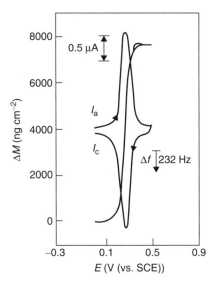

Figure 5.67 Cyclic voltammogram and frequency potential plot for $[Os(bpy)_2(PVP)_{10}Cl]Cl$ in 0.1 M *p*-toluene sulfonic acid. The potential scan rate was $1\,mV\,s^{-1}$, with a surface coverage of $2.0 \times 10^{-8}\,mol\,cm^{-2}$. Δf is frequency change. Reprinted from *J. Electroanal. Chem.*, **389**, A. P. Clarke, J. G. Vos, A. R. Hillman and A. Glidle, 'Overall redox switching characteristics of osmium-containing poly(4-vinylpyridine) films immersed in aqueous *p*-toluene sulfonic acid', 129–140, Copyright (1995), with permission from Elsevier Science

ion movement to maintain electroneutrality, or the polymer movement required to bring together redox-active centers so that electron self-exchange can occur.

Which of these processes will control the charge transport for a particular film will depend on a combination of measurement conditions, the nature of the polymer backbone and the concentration (loading) of the redox-active centers in the polymer matrix. Detailed investigations are needed to identify which particular process controls the behavior of a polymer film. These involve studies on the dependence of the charge transport rate on variables such as the nature of the electrolyte and the temperature. The fastest charge transport rates are expected when the process is controlled by electron hopping. In this case, the matrix does not play a part in the charge transport and the self-exchange between the redox centers is rate-limiting. In such situations, charge transport rates of the order of 10^{-8}–$10^{-6}\,cm^2\,s^{-1}$ are obtained. However, for most redox polymers the nature of the polymer film will be rate-determining. Important factors are the permeability of the polymer matrix, the amount of swelling, and the distance between redox centers. The charge-transport parameters observed for the osmium polymers discussed in the following are therefore significantly reduced from the 'ideal' electron self-exchange controlled rate.

For investigating the nature of the charge-transport process, thermodynamic studies are particularly useful. In these studies, the charge transport rate, D_{CT}, is measured over a range of temperatures. Its temperature-dependence can be described by using the Arrhenius equation, as follows:

$$D_{CT} = D_{CT}^0 \exp(-E_a/RT) \tag{5.15}$$

The entropy change, ΔS, can be estimated from the Eyring equation, as follows:

$$D^0_{CT} = e\delta^2(k_B T/h)\exp(\Delta s/R) \tag{5.16}$$

where E_a is the activation energy, δ is the mean separation between the redox centers, k_B is the Boltzmann constant and h is the Planck constant. Of particular interest for the determination of the rate-limiting factors are the values obtained for the activation energy and the entropy change. A negative entropy and a small activation energy are indicative for self-exchange or for the ion movement being rate-limiting. When polymer movement is the controlling factor, a large positive entropy and a large activation energy are expected.

Tables 5.4 and 5.5 provides the homogeneous charge transport diffusion coefficients, D_{CT}, for osmium polymers with different loadings in $HClO_4$ and H_2SO_4. Further information about the nature of these processes can be obtained by determining the thermodynamic parameters. These parameters are also summarized

Table 5.4 Effects of redox-site loading and sulfuric acid concentration on the charge-transport properties as obtained by cyclic voltammetry of $[Os(bpy)_2(PVP)_n Cl]Cl$ films.

Metal loading, n	Concentration (M)	D_{CT} (293 K) ($\times 10^{11}$ cm^2 s^{-1})	E_a (kJ mol^{-1})	$\Delta S^{\ddagger}$ (J mol^{-1} K^{-1})
5	0.1	32	20	−103
	1.0	108	25	23
10	0.1	25	11	−147
	1.0	167	116	221
15	0.1	17	21	−123
	1.0	78	90	121
20	0.1	21	22	−123
	1.0	75	80	82
25	0.1	210	65	36
	1.0	116	75	65

Table 5.5 Effects of redox-site loading and perchloric acid concentration on the charge-transport properties as obtained from cyclic voltammetry of $[Os(bpy)_2(PVP)_n Cl]Cl$ films.

Metal loading, n	Concentration (M)	D_{CT} (293 K) ($\times 10^{11}$ cm^2 s^{-1})	E_a (kJ mol^{-1})	$\Delta S^{\ddagger}$ (J mol^{-1} K^{-1})
5	0.1	26	49	−9
	1.0	9	42	−39
10	0.1	5	224[a], 35[b]	548[a], −106[b]
	1.0	0.5	92	92
15	0.1	11	163	333
	1.0	5	151	295
20	0.1	11	160	320
	1.0	4	154	291
25	0.1	12	158	304
	1.0	4	157	298

[a] At $T > 285$ K.
[b] At $T < 285$ K.

in these tables and were obtained by carrying out CV experiments between 277 and 311 K.

At first instance, the data given in these tables are somewhat surprising. If the availability of charge-compensating counterions limited the rate of diffusional charge transport through the film (D_{CT}), then this parameter should increase with the electrolyte concentration. However, in perchloric acid and for low loadings in sulfuric acid, the opposite is observed. Secondly, we would expect charge transport to be higher for redox polymers with a higher osmium loading since this would decrease the distance between the redox-active groups. Again, this is not the case, especially in sulfuric acid where the highest charge transport rates are observed for the lower loadings. The data presented in Tables 5.4 and 5.5 show that the charge transport rate depends strongly on the nature of the electrolyte. These observations indicate that there are substantial differences in the polymeric structures of the films between electrolytes and that these structures also depend on the loading of the polymer backbone. Such differences are also reflected in the activation parameters. Negative entropy and a small activation energy identify electron hopping or ion movement, while polymer movement is expected to show a large positive entropy and a large activation energy. These considerations indicate that in both H_2SO_4 and $HClO_4$, for high loadings, electron hopping or ion movement processes are limiting, while from $n = 15$ to $n = 25$ the movement of the polymer chains becomes rate-limiting for 1.0 M electrolytes. These data suggest that the structure changes significantly between different electrolytes and different loadings. Perchloric acid is known to cross-link poly(vinyl pyridine), resulting in a closed structure for which the access of counterions needed to maintain electroneutrality is hindered. Sulfuric acid, on the other hand, creates an open swollen structure where counterions have free access.

5.14.2 Electrochemical Quartz Crystal Microbalance Studies

The strong dependence of the layer structure on the nature of the contacting electrolyte has been further investigated by using the electrochemical quartz crystal microbalance (EQCM). As discussed above in Chapter 3, this technique is based on the measurement of the frequency with which a coated quartz crystal vibrates, and this frequency can then be related to the mass of this crystal provided that the material attached to the surface is rigid. In this way, the changes that occur in thin films as a result of redox processes can be monitored.

Two representative pictures for an electrode coated with [Os(bpy)$_2$(PVP)$_{10}$Cl]Cl at different *p*-toluene sulfonic acid concentrations are shown in Figures 5.67 and 5.68 [92]. In Figure 5.67 (see above), where the electrolyte is 0.1 M *p*-toluene sulfonic acid, the S-shaped curve shows that the mass increases during oxidation of the metal centers within the film from Os(II) to Os(III). However, in sharp contrast Figure 5.68 shows that the same oxidation process in a 0.01 M solution of the same acid shows an apparent decrease in mass. An increase in the mass is expected upon oxidation, since an additional anion is needed to maintain electroneutrality. The mass change observed in Figure 5.67 suggests ingress of one toluene sulfonate counterion and about 17 molecules of water.

Against this background the 'mass loss' calculated from the frequency change observed in 0.01 M electrolyte is very surprising. A detailed analysis of the data

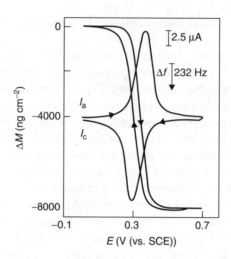

Figure 5.68 Cyclic voltammogram and frequency potential plot for [Os(bpy)$_2$(PVP)$_{10}$Cl]Cl in 0.01 M *p*-toluene sulfonic acid. The potential scan rate was 5 mV s^{-1}, with a surface coverage of 2.0×10^{-8} mol cm^{-2}. Reprinted from *J. Electroanal. Chem.*, **389**, A. P. Clarke, J. G. Vos, A. R. Hillman and A. Glidle, 'Overall redox switching characteristics of osmium-containing poly(4-vinylpyridine) films immersed in aqueous *p*-toluene sulfonic acid', 129–140, Copyright (1995), with permission from Elsevier Science

reveals that the frequency change is not related to a change in mass, but to a large change in the polymer structure. Strikingly, the polymer structure becomes extremely open and is no longer rigid in the concentration range 0.01–0.05 M. As a result, the Sauerbrey equation cannot be applied. These studies demonstrate that changes in polymer structure can occur as the result of relatively small changes in electrolyte concentration, polymer composition and layer thickness.

5.14.3 Interfacial Electrocatalysis

Polymer-modified electrodes have shown considerable utility as redox catalysts. In many cases, modified electrode surfaces show an improved electrochemical behavior towards redox species in solution, thus allowing them to be oxidized or reduced at less extreme potentials. In this manner, overpotentials can be eliminated and more selective determination of target molecules can be achieved. In this discussion, a mediated reduction process will be considered, although similar considerations can be used to discuss mediated oxidation processes. This mediation process between a surface-bound redox couple A/B and a solution-based species Y can be describes by the following:

$$A + e \longrightarrow B \tag{5.17}$$

$$B + Y \overset{k}{\longrightarrow} A + Z \tag{5.18}$$

The redox potential of the surface-bound redox couple is an important factor in achieving electrocatalysis, or more correctly, mediated reduction of the analyte. In

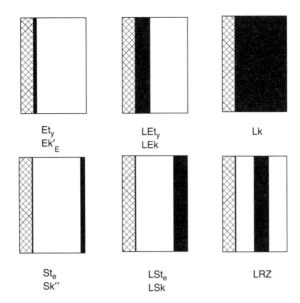

$$
\begin{array}{ccc}
\text{Et}_y & \text{LEt}_y & \text{Lk} \\
\text{Ek}'_E & \text{LEk} & \\
\end{array}
$$

$$
\begin{array}{ccc}
\text{St}_e & \text{LSt}_e & \text{LRZ} \\
\text{Sk}'' & \text{LSk} & \\
\end{array}
$$

Figure 5.69 The different reaction layers that are possible in the analysis of the mediation process in modified electrodes, together with the notations which relate to the rate-determining steps in each case (see text for further details). Hatched region is the electrode. The black section the reaction layer

order for B to be able to catalyze the reduction of Y, the redox potential of the A/B couple should be less positive than of the solution couple Y/Z.

However, the interplay between electrolyte and polymer layer needs to be considered when optimizing the performances of polymer-modified electrodes. Structural factors will influence the interfacial ion transport and this will have a direct effect on the mechanism and location within the polymer layer of the mediation process. The following discussion will show that the nature of the mediation process can be changed dramatically by changing the electrolyte, from a situation where an electrochemical sensor with good sensitivity is obtained, to a situation where the sensitivity obtained is not much better than that observed for the bare electrode.

Albery and Hillman have derived a detailed theoretical model which allows the kinetic parameters that drive the mediation reaction to be determined by using rotating-disk voltammetry [93]. The parameters that are considered in this analysis are the diffusion rate of the electroactive species, D_y, the electron transfer rate, D_e, and the rate of the catalytic reaction, k. The mechanism of the mediation reaction will depend on structural factors such as film morphology and film thickness which may influence the rate-determining step. The analysis developed describes the mediation process in terms of reaction layers. A representation of these reaction layers is shown in Figure 5.69, together with labels which illustrate the rate-determining steps in this particular case. In order to determine the case most appropriate for a particular reaction, a diagnostic scheme was developed. This approach is illustrated in Figure 5.70 and is based on the analysis of rotating-disk measurements. In the following discussion, this schematic approach will be used to identify the nature of the mediation process for electrodes modified with the redox polymer $[Os(bpy)_2(PVP)_{10}Cl]Cl$.

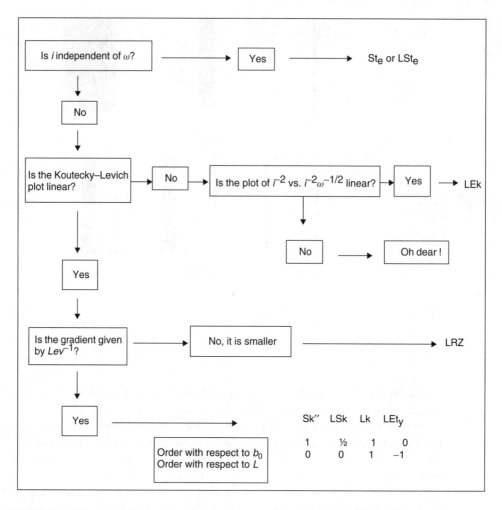

Figure 5.70 Diagnostic scheme used for determination of the mechanism for mediation and position of the reaction layer in modified electrodes. From W. J. Albery and A. R. Hillman, *RSC Annu. Rep.*, **78**, 377 (1981). Reproduced by permission of The Royal Society of Chemistry

Mediation of Fe(III) by an electrode modified with [Os(bpy)$_2$(PVP)$_{10}$Cl]Cl

Given the formal potential of the osmium polymers (about 250 mV (vs. SCE)) [94], electrodes modified with these polymers can be use to mediate the reduction of Fe(III) as follows:

$$Fe(III) + Os(II) \longrightarrow Fe(II) + Os(III) \tag{5.19}$$

This reaction can be utilized to determine Fe(III) at potentials where it is not redox-active at an unmodified glassy carbon electrode. The mediation of this reaction in two different electrolytes, i.e. 1 M perchloric acid and 0.1 M sulfuric acid, will be discussed. EQCM and neutron reflectivity measurements have shown that, while in H$_2$SO$_4$ a swollen structure is obtained, in perchloric acid a much more closed morphology is obtained which inhibits ion movement.

To analyze the mediation process, rotating-disk measurements were carried out. As discussed above in Chapter 3, in such an experiment an electrode is rotated at speeds of up to several thousands of rpm in the presence of a redox species in solution. By a variation of the potential, a situation is reached where the current becomes independent of the potential applied. For a bare electrode, this limiting current, i_{LEV}, is given by the Levich equation, as follows:

$$i_{LEV} = 0.62nAFD^{2/3}v^{-1/6}y\omega^{1/2} \tag{5.20}$$

where A is the electrode area, F the Faraday constant, D the diffusion rate of the substrate, v the viscosity of the solvent, and ω the rotation speed of the electrode. This technique allows for the control of substrate transport from the bulk solution to the electrode surface. The catalytic properties of the modifying layer can then be probed by comparing measurements made at bare and at modified electrodes. Figure 5.71 illustrates a typical example of rotating-disk electrode data obtained for a polymer-modified electrode in the presence of Fe(III). The modified electrode behavior cannot be represented by the simple Levich equation. Instead the limiting current is defined by the following:

$$1/i_{lim} = i/i_F + 1/i_{LEV} \tag{5.21}$$

$$1/i_{lim} = (0.62nFAk'_{ME}y)^{-1} + (0.62nAFD^{2/3}v^{-1/6}y\omega^{1/2})^{-1} \tag{5.22}$$

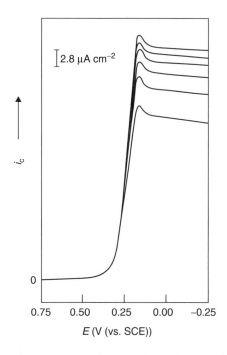

Figure 5.71 Rotating-disk voltammograms showing the reduction of a 0.2 m/mM $[Fe(H_2O)_6]^{3+}$ solution in 0.1 M H_2SO_4 at an electrode modified with $[Os(bpy)_2(PVP)_{10}Cl]^+$. The rotation rates, from top to bottom, are 500, 1000, 1500, 2000, 2500 and 3000 rpm, respectively; the surface coverage is 5×10^{-9} mol cm^{-2}. From R. J. Forster and J. G. Vos, *J. Chem. Soc., Faraday Trans.*, **87**, 1863–1867 (1991). Reproduced by permission of The Royal Society of Chemistry

where k'_{ME} is the electrochemical rate constant and y the concentration of the substrate. A plot of i^{-1} versus $\omega^{-1/2}$ yields a Koutecky–Levich plot, for which the intercept represents k'_{ME} and the gradient is the reciprocal of the Levich slope. The following discussion will be concentrating on determining the reaction layer, by using the approach outlined in Figure 5.70.

Mediation in H_2SO_4

Figure 5.71 above shows a typical rotating-disk voltammogram for the reduction of Fe(III) mediated by an electrode surface modified with $[Os(bpy)_2(PVP)_{10}Cl]Cl$. The figure clearly shows that mediation takes place when Os(III) is being reduced to Os(II) and is indicative of the mediation reaction shown in Equation (5.19). Figure 5.72 shows the Koutecky–Levich plots obtained from data such as those shown in Figure 5.71. Figure 5.70 will now be used to analyze these data. The first step in the diagnostic diagram considers whether the current in the rotating-disk voltammograms depends on the rotation rate, ω. For the modified electrode, i_{Lev} is proportional to ω so the limiting cases St_e and LSt_e can be eliminated. In addition, the Koutecky–Levich plots are linear so the LEk case is not operative. The slopes observed for the plots are constant and are identical, to within experimental error, to those obtained for a bare electrode. This eliminates the LRZ case. As a result of this analysis, the only cases to be considered at this stage are Sk″, LSk, Lk and LEt_y. These can be distinguished by examining the dependence of k'_{ME} on the

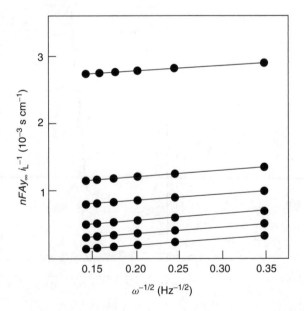

Figure 5.72 Typical Koutecky–Levich plots for the reduction of a 0.2 mM $[Fe(H_2O)_6]^{3+}$ solution in 0.1 M H_2SO_4 at platinum electrodes modified with layers of different surface coverages of $[Os(bpy)_2(PVP)_{10}Cl]^+$. From top to bottom, the surface coverages (in mol cm^{-2}) are 1.7×10^{-10}, 1.8×10^{-9}, 2.7×10^{-9}, 5.0×10^{-9}, 1.1×10^{-8}, and 'bare' platinum. From R. J. Forster and J. G. Vos, *J. Chem. Soc., Faraday Trans.*, **87**, 1863–1867 (1991). Reproduced by permission of The Royal Society of Chemistry

Table 5.6 Dependence of k_{ME} on the layer thickness L in different solvents where the $[Fe(H_2O)_6]^{3+}$ concentration is 0.2 mM.

		k_{ME} (cm s^{-1})	
L (nm)	Γ (mol cm^{-2})	0.1 M H$_2$SO$_4$	1.0 M HClO$_4$
150	1.0×10^{-8}	6.3×10^{-3}	2.2×10^{-4}
70	4.9×10^{-9}	2.8×10^{-3}	2.2×10^{-4}
38	2.7×10^{-9}	1.5×10^{-3}	2.1×10^{-4}
25	1.8×10^{-9}	0.9×10^{-3}	2.0×10^{-4}
10	7.0×10^{-10}	0.3×10^{-3}	2.0×10^{-4}

layer thickness, L, and the concentration of the redox active species, b_0, in the film. Table 5.6 shows that the dependence of the electrochemical rate constant on the layer thickness is linear over the thickness range studied. This observation, together with the fact that the dependence of k'_{ME} on b_0 is also linear, established that in 0.1 M H$_2$SO$_4$, the Lk mechanism provides an appropriate description of the system. In this case, the catalytic reaction occurs throughout the polymer layer and is limited by the rate of the catalytic reaction between the immobilized redox centers and the analyte. Interestingly, for thicker layers ($>10^{-7}$ mol cm^{-2}) the mechanism switches to the LSk case which corresponds to total catalysis where the rate of diffusion of the substrate from the solution to the electrode controls the mediation process. This behavior indicates that the mediation process is not limited by electron transport, t_e or substrate diffusion, t_Y. Such a situation is indicative of an open, swollen structure, which allows rapid access of ions and substrate.

Mediation in 1 M HClO$_4$

The limiting current of the rotating-disk voltammograms also depends on the rotation rate in HClO$_4$. In addition, as shown in Figure 5.73, the Koutecky–Levich plots are linear and have a slope similar to that observed for a bare electrode. Moreover, this figure suggests that k'_{ME} depends, at best, only weakly on the layer thickness and this is confirmed by the data shown in Table 5.6. This result suggests that the limiting case in this electrolyte is either Sk″ or LSk. These two cases can be distinguished by their dependence on b_0. A detailed analysis of the electrochemical rate constant as a function of the concentration of the redox-active centers in the film again shows a linear dependence on the active-site concentration which identifies the limiting case in perchloric acid as Sk″. In this case, the mediated reaction takes place in a layer close to the polymer–electrolyte interface and is controlled by the second-order rate constant for the reaction between the surface-bound redox couple and the substrate in solution.

Control of substrate partitioning by chemical cross-linking

The above example shows how the interfacial processes occurring at thin polymer films can be manipulated by a change in the contacting electrolyte. This section

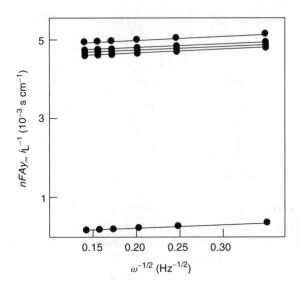

Figure 5.73 Typical Koutecky–Levich plots for the reduction of a $[Fe(H_2O)_6]^{3+}$ solution in 1.0 M $HClO_4$ at platinum electrodes modified with layers of different surface coverages of $[Os(bpy)_2(PVP)_{10}Cl]^+$. From top to bottom, the surface coverages (in mol cm^{-2}) are 1.7×10^{-10}, 1.8×10^{-9}, 2.7×10^{-9}, 5.0×10^{-9}, 1.1×10^{-8}, and 'bare' platinum. From R. J. Forster and J. G. Vos, *J. Chem. Soc., Faraday Trans.*, **87**, 1863–1867 (1991). Reproduced by permission of The Royal Society of Chemistry

shows that it is possible to control the partitioning of solution reagents into a polymer film by cross-linking the polymer [95]. As outlined above, in H_2SO_4 the mediation of the reduction of Fe(III) by the osmium polymer takes place by the Lk mechanism in which the whole layer is active. This mechanism points to an efficient partitioning of the substrate into the polymer layer. Surprisingly, a different rate-determining step is observed when the same polymer is cross-linked with 1,2-dibromedecane. Analysis of the rotating-disk voltammetry data indicates that in the same electrolyte the mediation takes place according to the kinetic regime Sk″. Under these circumstances, the mediation process takes place in a very thin region close to the polymer–electrolyte interface. The behavior observed is independent of the amount of cross-linking over the range 1–20 %. Importantly, the cross-linking process does not significantly affect the rate of charge transport through the films. This observation suggests that while the ion movement and electron transport are not changed, a lack of partitioning of the substrate into the layer is responsible for the behavior observed. This result is non-trivial. In the electrolyte used, for both cross-linked and non-cross-linked materials, all of the unlabeled pyridine groups will be protonated and the polymer is polycationic. Therefore, electrostatic repulsion is unlikely to be the cause of the change of mechanism. The lack of partitioning of $[Fe(H_2O)_6]^{3+}$ into the film is more than likely explained by steric reasons. This is further confirmed by the fact that when *p*-dibromobenzene is used as a cross-linker the kinetic regime changes to Lk, hence indicating the participation of the whole layer in the mediation process.

The osmium-containing polymers can also be used for the mediated reduction of NO_2^- to produce NO^+ [96]. Rotating-disk voltammetry shows that for

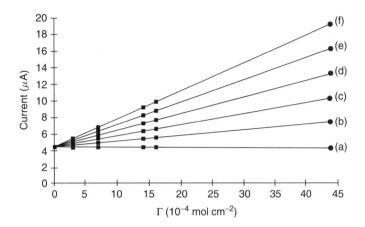

Figure 5.74 Electrode current versus film thickness, Γ, for the reduction of (a) 2.0×10^{-4} M Fe(III) with (b) 2.0×10^{-3}, (c) 4.0×10^{-3}, (d) 6.0×10^{-3}, (e) 8.0×10^{-3} and (f) 1.0×10^{-2} M NO_2^- in 0.1 M H_2SO_4; the electrode rotation speed is 500 rpm. Reprinted with permission from A. P. Doherty and J. G. Vos, *Anal. Chem.*, **65**, 3424 (1993). Copyright (1993) American Chemical Society

the 1,10-dibromodecane cross-linked polymer, mediated reduction takes place according to the Lk mechanism, i.e. mediation takes place throughout the layer. In contrast, Fe(III) reduction occurs exclusively at the film–solution interface. This difference in the location of the redox processes between Fe(III) and nitrite can be used to determine both species at the same time. This objective can be achieved by using a series of electrodes, e.g. in an electrode array, which are coated with layers of different thickness. Figure 5.74 shows a plot of the limiting currents as a function of layer thickness, obtained from rotating-disk voltammetry for a solution containing Fe(III) and varying amounts of NO_2^-. From these plots, the Fe(III) concentration of the solution can be obtained from the intercept, while the slopes of the lines yield the nitrite concentration.

The observation that individual redox process can be isolated within the same film is an important one as it allows for the simultaneous determination of normally interfering analytes. This approach could be extended to many other systems and would greatly increase the amount of analytical information that can be gained from solutions which contain several redox-active species.

5.15 Conclusions and Future Directions

Optimism.... The ability to control the rate of electron and energy transfer across molecular bridges impacts diverse areas ranging from the development of molecular electronics to understanding biosystems. This chapter has illustrated some of the key developments in this important field. Many of the case studies presented have provided significant new insight into those factors which control electron and energy transfer both across surface–ISA interfaces and within the assemblies themselves. The field has developed from using relatively coarse variables such as distance to modulate electron transfer rates into more subtle systems in which the electronic structure of the bridge can be reversibly changed in response to the local microenvironment, e.g. through a protonation reaction. In addition, the theoretical descriptions have not only grown in sophistication but they have become vitally more accessible to experimental testing and verification.

5.15.1   Challenges for the Next Decade

Molecular optoelectronic devices

The key principles underpinning the field of molecular electronic and photonic devices have been demonstrated. However, many of the systems proposed lack the order, stability or synthetic flexibility to be useful for real-world devices. The actual quality of the films used in some literature reports is not adequately addressed. Information about the ISA structure at both the molecular and microscopic length-scales, as well as its mechanical properties, is as vital as the detailed studies of their optoelectronic properties. However, it is equally important that the focus of the field does not shrink to the point of obsession about creating specific devices or to focus too strongly on commercially oriented goals. Rather, by achieving a more complete understanding of the forces which control the assembly of ISAs, as well as those which dictate electron and energy transfer, the applications are likely to flow naturally.

Combining electrochemistry and photochemistry

The electrochemical and spectroscopic communities often appear to be divided by differences in language that often prevent the free flow of information about common issues. Unifying electrochemical and photophysical approaches to probing electron transfer processes is likely to be one of the challenges that will be met in the coming decades. Increasingly, electrochemistry and photochemistry are being combined for probing and controlling interfacial supramolecular devices.

Quantum-chemical models

The next decade will see a shift away from phenomenological models towards microscopic (atomic-level) descriptions of the behavior of interfacial supramolecular assemblies. These novel approaches will continue to benefit from the interplay between quantum chemistry and modern theories of charge transfer. It is now almost 25 years since the cluster model of the metal surface was first used to describe the structure and composition of the electrical double layer. Now quantum chemical calculations of model *supermolecules*, such as *metal cluster plus reactants*, promise to provide a truly elementary insight into heterogeneous electron transfer at redox-active monolayers. In particular, *Ab initio* calculations using the computationally less demanding density functional theory coupled with the Kohn–Stam formalism would appear to be an especially attractive approach to exploring the fundamentals of heterogeneous electron transfer.

Models of interfacial photochemistry

Over the past two decades, a wealth of supramolecular systems with interesting photophysical properties have been created. These systems help complex, often biological, systems to be more fully understood. However, these studies have primarily been conducted in solution where long-range molecular order is typically absent. It is clear that in order to achieve the next level of organization, i.e. long-range order, platforms or frameworks will be required. This next step, interfacial supramolecular photochemistry, is a field still in its relative infancy but has promise to generate systems that can be interrogated and controlled by using an optoelectronic input that may address the substrate as well as the assembly. We have outlined some interesting examples of supramolecular interfacial assemblies which are both electro- and photoactive. However, the majority of such reports focus on the structure and functioning of these species, and therefore a significant number of issues remain to be explored, for example, the impact of substrate identity on the photochemical properties of the interfacial supermolecule. Significant exploration in the late 1970s and early 1980s was undertaken on the interfacial behavior of excited-state molecules physisorbed onto metal surfaces. The same level of investigation has not yet been dedicated to chemisorbed photoactive supramolecular assemblies. For example, it appears that approximately five alkyl carbons is sufficient to limit the interfacial quenching of an excited state by a metal. Furthermore, it is known that parameters such as the density of states of the metal can influence the coupling strength between the adsorbate and metal and therefore the rate of electron transfer in heterogeneous reactions. Detailed exploration of how interfacial energy transfer is influenced by such factors remains largely unexplored. Such studies will provide insight into the means of optimizing interfacial interactions, or even avoiding them.

Photoactive SAMs and multilayers containing 'biobridges'

Optoelectronic applications will require photoactive SAMs capable of intermolecular or heterogeneous electron transfer. In order to maintain its photochemical integrity at an interface, a photoactive moiety must be maintained at some distance from the electrode surface to prevent quenching, but must also be coupled strongly enough to support electron transfer. Bridges based on biomaterials such as DNA and proteins, or indeed other bridges based on comparable bonding patterns, may have these properties. Weak intermolecular forces, such as π-stacking and hydrogen bonding, have been shown to support the strong coupling of redox sites in biomaterials over long distances. SAMs and multilayers bridged through such bonds are likely to prove very valuable in future optoelectronic applications.

References

[1]　Forster, R.J. (2000). Ultrafast electrochemical techniques, in *Encyclopedia of Analytical Chemistry*, Vol. 11, R. Meyers (Ed.), Wiley, Chichester, UK, pp. 10142–10171.

[2]　Chidsey, C.E.D. (1991). *Science*, **251**, 919.

[3] Acevedo, D. and Abruña, H.D. (1991). *J. Phys. Chem.*, **95**, 9590.

[4] Finklea, H.O. and Hanshaw, D.D. (1992). *J. Am. Chem. Soc.*, **114**, 3173.

[5] Finklea, H.O., Liu, L., Ravenscroft, M.S. and Punturi, S. (1996). *J. Phys. Chem.*, **100**, 18852.

[6] Weber, K., Hockett, L. and Creager, S. (1997). *J. Phys. Chem., B*, **101**, 8286.

[7] Smalley, J.F., Feldberg, S.W., Chidsey, C.E.D., Linford, R., Newton, M.R. and Liu, Y.-P. (1995). *J. Phys. Chem.*, **99**, 13141.

[8] Feng, Z.Q., Imabayashi, S., Kakiuichi, T. and Niki, K. (1997). *J. Chem. Soc., Faraday Trans.*, **93**, 1367.

[9] Song, S., Clark, R.A., Bowden, E.F. and Tarlov, M.J. (1993). *J. Phys. Chem.*, **97**, 6564.

[10] Tarlov, M.J. and Bowden, E.F. (1991). *J. Am. Chem. Soc.*, **113**, 1847.

[11] Forster, R.J. (1996). *Inorg. Chem.*, **35**, 3394.

[12] Walsh, D.A., Keyes, T.E., Hogan, C.F. and Forster, R.J. (2001). *J. Phys. Chem., A*, **105**, 2792.

[13] Gosavi, S. and Marcus, R.A. (2000). *J. Phys. Chem., B*, **104**, 2067.

[14] Forster, R.J., Loughman, P.J. and Keyes, T.E. (2000). *J. Am. Chem. Soc.*, **122**, 11948.

[15] Brunschwig, B.S. and Sutin, N. (1999). *Coord. Chem. Rev.*, **187**, 233.

[16] Ishida, T., Mizutani, W., Akiba, U., Umemura, K., Inoue, A., Choi, N., Fujihira, M. and Tokumoto, H. (1999). *J. Phys. Chem., B*, **103**, 1686.

[17] Weber, K.S. and Creager, S.E. (1998). *J. Electroanal. Chem.*, **458**, 17.

[18] Forster, R.J., Loughman, P.J., Figgemeier, E., Lees, A.C., Hjelm, J. and Vos, J. (2000). *Langmuir*, **16**, 7871.

[19] Bagchi, G. (1989). *Ann. Rev. Chem.*, **40**, 115.

[20] Sutin, N. (1982). *Acc. Chem. Res.*, **15**, 275.

[21] Barr, S.W., Guyer, K.L., Li, T.T.-T., Liu, H.Y. and Weaver, M.J. (1984). *J. Electrochem. Soc.*, **131**, 1626.

[22] Sutin, N. and Brunschwig, B.S. (1982). Electron transfer in weakly interacting systems, in *Inorganic Reaction Mechanisms*, ACS Symposium Series, Vol. 198, American Chemical Society, Washington, DC, pp. 105–135.

[23] Forster, R.J., Vos, J.G. and Keyes, T.E. (1998). *Analyst*, **123**, 1905.

[24] Forster, R.J. and O'Kelly, J.P. (1996). *J. Phys. Chem.*, **100**, 3695.

[25] Finklea, H.O. and Hanshew, D.D. (1992). *J. Am. Chem. Soc.*, **114**, 3173.

[26] Marcus, R.A. (1956). *J. Chem. Phys.*, **24**, 966.

[27] Marcus, R.A. (1963). *J. Phys. Chem.*, **67**, 853.

[28] Schmickler, W.J. (1977). *J. Electroanal. Chem.*, **82**, 65.

[29] Bockris, J.O'M. and Khan, S.U.M. (1979). *Quantum Electrochemistry*, Plenum Press, New York, Ch. 8, pp. 235–288.

[30] Chidsey, C.E.D. (1991). *Science*, **251**, 919.

[31] Weber, K. and Creager, S.E. (1994). *Anal. Chem.*, **66**, 3164.

[32] Tender, L., Carter, M.T. and Murray, R.W. (1994). *Anal. Chem.*, **66**, 3173.

[33] Forster, R.K. and Keyes, T.E. (2001). *J. Phys. Chem., B*, **105**, 8829.

[34] McConnell, H.M. (1961). *J. Chem. Phys.*, **35**, 508.

[35] Laviron, E. (1983). *J. Electroanal. Chem.*, **146**, 15.

[36] He, P., Crooks, R.M. and Faulkner, L.R. (1990). *J. Phys. Chem.*, **94**, 1135.

[37] Forster, R.J. and O'Kelley, J.P. (2001). *J. Electroanal. Chem.*, **498**, 127.

[38] O'Hanlon, D. and Forster, R.J. (2000). *Langmuir*, **16**, 702.

[39] Ward, M.D. (1997). *Chem. Soc. Rev.*, **26**, 365.

[40] Juris, A., Balzani, V., Barigelletti, F., Campagna, S., Belser, P. and Von Zelewsky, A. (1988). *Coord. Chem. Rev.*, **84**, 85.

[41] Jones, W.E. and Foxe, M.A. (1994). *J. Phys. Chem.*, **98**, 5095.

[42] Pflug, J.S. and Faulkner, L.R. (1980). *J. Am. Chem. Soc.*, **102**, 6144.

[43] Pflug, J.S., Faulkner, L.R. and Seitz, W.R. (1983). *J. Am. Chem. Soc.*, **105**, 4890.

[44] Kuhn, H.J. (1970). *Chem. Phys.*, **53**, 101.

[45] Chance, R.R., Prock, A. and Silbey, R. (1978). *Adv. Chem. Phys.*, **37**, 1.

[46] Forster, R.J. and Keyes, T.E. (1998). *J. Phys. Chem., B*, **102**, 10004.

[47] Liu, Z., Hashimoto, K. and Fujishima, A. (1990). *Nature*, **347**, 658.

[48] Doron, A., Portnoy, M., Lion-Dagan, M., Katz, E. and Willner, I. (1996). *J. Am. Chem. Soc.*, **118**, 8937.

[49] Willner, I., Pardo-Yissar, V., Katz, E. and Ranjit, K.T. (2001). *J. Electroanal. Chem.*, **497**, 172.

[50] Song, S., Clark, R.A., Bowden, E.F. and Tarlov, M.J. (1993). *J. Phys. Chem.*, **97**, 6564.

[51] Van Dyke, B.R., Saltman, P. and Armstrong, F.A. (1996). *J. Am. Chem. Soc.*, **118**, 3490.

[52] Willner, I., Heleg-Shabtai, V., Blonder, R., Büchmann, A.F. and Heller, A. (1996). *J. Am. Chem. Soc.*, **118**, 10321.

[53] Rajagopolan, C.R., Aoki, A. and Heller, A. (1996). *J. Phys. Chem.*, **100**, 3719.

[54] Badia, A., Carlini, R., Fernandez, A., Battaglini, F., Mikkelsen, S.R. and English, A.M. (1993). *J. Am. Chem. Soc.*, **115**, 7053.

[55] Brust, M., Walker, M., Bethell, D., Schiffrin, D.J. and Whyman, R. (1994). *J. Chem. Soc., Chem. Commun.*, 801.

[56] Miles, D.T. and Murray, R.W. (2001). *Anal. Chem.*, **73**, 921.

[57] Gittins, D.I., Bethell, D., Schriffin, D.J. and Nichols, R.J. (2000). *Nature*, **408**, 67.

[58] Sun, L., Johnson, B., Wade, T. and Crooks, R.M. (1990). *J. Phys. Chem.*, **94**, 8869.

[59] Cheng, Q. and Brajter-Toth, A. (1992). *Anal. Chem.*, **64**, 1998.

[60] Rubinstein, I., Steinberg, S., Tor, Y., Shanzer, A. and Sagiv, J. (1988). *Nature*, **332**, 426.

[61] Bourdillon, C. and Majda, M. (1990). *J. Am. Chem. Soc.*, **112**, 1795.

[62] Forster, R.J. and Faulkner, L.R. (1995). *Anal. Chem.*, **67**, 1232.

[63] Forster, R.J. (1996). *Analyst*, **121**, 733.

[64] Chance, R.R., Prock, A. and Silbey, R. (1975). *J. Chem. Phys.*, **62**, 2245.

[65] Yamada, S., Koide, Y. and Matsuo, T. (1997). *J. Electroanal. Chem.*, **426**, 23.

[66] Enger, O., Neusch, F., Fibbioli, M., Echegoyen, L., Pretsch, E. and Diederich, F. (2000). *J. Mater. Chem.*, **10**, 223.

[67] Imahori, H., Norieda, H., Yamada, Y., Nishimura, Y., Yamazaki, I., Sakata, Y. and Fukuzumi, S. (2001). *J. Am. Chem. Soc.*, **123**, 101.

[68] Imahori, H., Yamada, H., Nishimura, Y., Yamazaki, I., Sakata, Y. and Fukuzumi, S. (2000). *J. Phys. Chem., B*, **104**, 2099.

[69] Balzani, V., Credi, A., Raymo, F.M. and Stoddart, J.F. (2000). *Angew. Chem. Int. Ed. Engl.*, **39**, 3349.

[70] Shipway, A.N. and Willner, I. (2001). *Acc. Chem. Res.*, **34**, 6.

[71] Ranjit, K.T., Marx-Tibbon, S., Ben-Dov, I., Willner, B. and Willner, I. (1996). *Isr. J. Chem.*, **36**, 407.

[72] Doron, A., Katz, E., Tao, G. and Willner, I. (1997). *Langmuir*, **13**, 1783.

[73] Lahav, M., Ranjit, K.T., Katz, E. and Willner, I. (1997). *Isr. J. Chem.*, **37**, 185.

[74] Laviron, E.J. (1990). *J. Electroanal. Chem.*, **101**, 9.

[75] Water, D.G., Campbell, D.J. and Mirkin, C.A. (1999). *J. Phys. Chem., B*, **103**, 402.

[76] Ishida, A. and Majima, T. (1999). *J. Chem. Soc., Chem. Commun.*, 1299.

[77] Simpson, T.R.E., Revell, D.J., Cook, M.J. and Russell, D.A. (1997). *Langmuir*, **13**, 460.

[78] Pope, J.M. and Buttry, D.A. (2001). *J. Electroanal. Chem.*, **498**, 75.

[79] Morita, T., Kimura, S. and Kobayashi, S. (2000). *J. Am. Chem. Soc.*, **122**, 2850.

[80] Reese, R.S. and Fox, M.A. (1999). *Can. J. Chem.*, **77**, 1077.

[81] Koide, Y., Terasaki, N., Akiyama, T. and Yamada, S. (1999). *Thin Solid Films*, **350**, 223.

[82] Kondo, T., Kanai, T., Iso-O, K. and Uoskai, K. (1999). *Z. Phys. Chem.*, **212**, S23.

[83] Imahori, H., Nishimura, Y., Noreda, H., Karita, H., Yamazaki, I., Sakata, Y. and Fukuzumi, S. (2000). *J. Chem. Soc., Chem. Commun*, 661.

[84] Horne, J.C. and Blanchard, G.J. (1998). *J. Am. Chem. Soc.*, **120**, 6336.

[85] Ikeda, A., Hatano, T., Shinkai, S., Akiyama, T. and Yamada, S. (2001). *J. Am. Chem. Soc.*, **123**, 4855.

[86] Nakanishi, O., Ohtani, B. and Uoskai, K. (1998). *J. Phys. Chem., B*, **102**, 1571.

[87] Brust, M., Blass, P.M. and Bard, A.J. (1997). *Langmuir*, **13**, 5602.

[88] Kaschak, D.M., Johnson, S.A., Waraksa, C.C., Pogue, J. and Mallouk, T.E. (1999). *Coord. Chem. Rev.*, **185–186**, 403.

[89] Kaschak, D.M., Lean, J.T., Waraksa, C.C., Saupe, G.B., Usami, H. and Mallouk, T.E. (1999). *J. Am. Chem. Soc.*, **121**, 3435.

[90] Forster, R.J., Vos, J.G. and Lyons, M.E.G. (1991). *J. Chem. Soc., Faraday Trans.*, **87**, 3761.

[91] Forster, R.J. and Vos, J.G. (1991). *J. Electroanal. Chem.*, **314**, 135.

[92] Clarke, A.P., Vos, J.G., Hillman, A.R. and Glidle, A. (1995). *J. Electroanal. Chem.*, **389**, 129.

[93] Albery, W.J. and Hillman, A.R. (1981). *RSC Annu. Rep.*, **78**, 377.

[94] Forster, R.J. and Vos, J.G. (1991). *J. Chem. Soc., Faraday Trans.*, **87**, 1863.

[95] Doherty, A.P. and Vos, J.G. (1993). *Anal. Chem.*, **65**, 3424.

[96] Doherty, A.P. and Vos, J.G. (1992). *J. Chem. Soc., Faraday Trans.*, **88**, 2903.

6 Interfacial Electron Transfer Processes at Modified Semiconductor Surfaces

Because of the presence of a well-defined energy gap between the conduction and the valence band, semiconductors are ideally suited for investigation of the interfacial interactions between immobilized molecular components and solid substrates. In this chapter, interfacial assemblies based on nanocrystalline TiO_2 modified with metal polypyridyl complexes will be specifically considered. It will be shown that efficient interaction can be obtained between a molecular component and the semiconductor substrate by a matching of their electronic and electrochemical properties. The nature of the interfacial interaction between the two components will be discussed in detail. The application of such assemblies as solar cells will also be considered. The photophysical processes observed for interfacial triads, consisting of nanocrystalline TiO_2 surfaces modified with molecular dyads, will be discussed. Of particular interest in this discussion is how the interaction between the semiconductor surface and the immobilized molecular components modifies the photophysical pathways normally observed for these compounds in solution.

6.1 Introduction

One of the driving forces behind the current interest in supramolecular chemistry is the potential application of supramolecular systems as nano-scale molecular devices. However, there are a number of criteria than need to be fulfilled before real-life devices can become a reality. These include organization and addressability, and in the case of light-driven devices, long-lived charge separation. In solution, these issues can only be addressed to a limited extent and this realization has led to the view that the incorporation of active solid components in supramolecular devices would be beneficial. With a solid substrate, molecular components can be arranged in a well-defined manner, for example, as monolayers, through adsorption or covalent attachment. The hybrid, interfacial supramolecular assembly obtained in this manner can also be addressed and the information stored can be read by using spectroscopic or electrochemical techniques. In addition, a well-chosen solid substrate can actively interact with the immobilized molecular components. This active participation may, for example, lead to improved charge separation. There is, therefore, a great interest in studying molecular components attached to active solid supports. Nanocrystalline semiconductors are particularly attractive examples of the latter. Transparent nanocrystalline semiconductor electrodes have been investigated in great detail since they form the basis of potential practical applications, such as solar cells, photo- and electrochromic windows and lithium intercalation batteries. Nanostructured semiconductor films have also been applied as antibacterial coatings and for the light-assisted degradation of organic pollutants. As a result, metal oxides such as TiO_2, ZnO, NiO, SnO_2 and others have been modified with molecular components and the nature of the light-driven processes of these assemblies has been studied in great detail. This interest has been driven to a large extent by the proven ability of TiO_2 surfaces, modified with a range of organic and inorganic molecular components, to act as efficient photovoltaic cells. In a series of detailed studies, Grätzel and co-workers and other groups have shown that efficient solar energy devices can be assembled from nanocrystalline TiO_2 surfaces modified with ruthenium polypyridyl complexes. With these components, overall efficiencies of over 10 % can be obtained with relatively low light intensities. The surprising observation with these devices is that injection of electrons into nanocrystalline TiO_2 surfaces is very fast (in the sub-picosecond range), while the recombination process is several orders of magnitude slower, thus resulting in long-lived charge separation. In this chapter, the interaction between molecular components and solid substrates will be presented. The discussion will concentrate on light-driven processes and focus mainly on modified TiO_2 surfaces. Issues discussed include practical applications such as solar cells and electrochromic devices, as well as the fundamentals of the interactions between molecular components and solid substrates. In the literature, the immobilization of many different types of molecular components on solid substrates has been reported. In this discussion, ruthenium-polypyridyl-type complexes and some rhodium- and osmium-based analogues will be discussed as examples. Several good reviews dealing with the issues discussed in this chapter have recently appeared [1–3].

6.2 Structural and Electronic Features of Nanocrystalline TiO₂ Surfaces

One of the fundamental issues in interfacial supramolecular assemblies is how the solid substrate interacts with the molecular components and how the photophysical and electrochemical behaviors of the molecular components are affected by their interaction with this substrate. In order to assess this interaction, as well as to be able to devise methods in which the properties of the assembly can be altered by modifications of the substrate, it is necessary to first consider the properties of the solid component. In this discussion, the fundamental properties of semiconductors and also the effect of particle size on these properties will be considered.

6.2.1 Electronic Properties of Bulk TiO₂

The electronic properties of solid materials are normally described in terms of the band gap. In this description, the valence band is the ground state and the

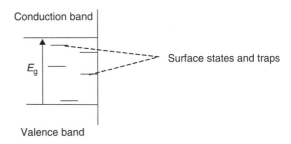

Figure 6.1 Illustration of the electronic structure for bulk semiconductor materials

conduction band the excited state. The band gap is then defined as the energy needed to promote an electron from the valence band, which acts as the highest occupied molecular orbital (HOMO) to the lowest unoccupied molecular orbital (LUMO), as shown in Figure 6.1. The nature and position of the valence and conduction bands control the properties of solid materials. For conductors, these bands overlap and as a result electrons can move freely from the valence to the conduction band. For insulators, the valence and conduction bands are separated by a considerable band gap, which thus prevents a thermally driven population of the conduction band at room temperature; these materials are therefore non-conducting. For semiconductors, an intermediate state is observed, where the gap between the valence and conduction bands is smaller than that observed for insulators and thermal energy may be sufficient to populate the conduction band from the valence band. The efficiency of this population may not be very large and as a result the conductivity of semiconductors is limited. This thermally driven population of the conduction band also means that the conductivity of semiconductors is temperature-dependent, i.e. the conductivity increases with increasing temperature. The extent of this increase depends on the magnitude of the band gap. As illustrated in Figure 6.2, population of the valence band can also be obtained by optical means and indeed the photoconductivity of semiconductors such as TiO$_2$ and others has been studied in great detail. As expected from their band structures, semiconductors absorb radiation only if their energies are equal to, or larger than, that of the band gap. As a result, the band gap of the material under investigation can be determined from the threshold wavelength of this radiation. A schematic diagram of the band-gap properties of a series of semiconductors is shown in Figure 6.2. Although the main factor that determines the electronic properties of semiconductors is the band gap, the role of lattice defects also needs to be addressed. In semiconductors, lattice defects are common and, as illustrated in Figure 6.1, their presence leads to the formation of sub-surface states and traps. Their presence is important since, in particular, surface states may be involved in light-induced processes and may affect charge injection, especially in nanoparticles.

The picture that emerges suggests that the electronic properties of solid-state semiconductors are similar to those of single molecules, where excitation of the materials also leads to the transfer of an electron from a HOMO to a LUMO state. However, there are important differences. These are related to the nature of the HOMO and LUMO in bulk materials and in single molecules. Quantum mechanics dictates that in single molecules molecular orbitals have discrete energies and are located on single molecules. In the solid state, this picture is no longer valid.

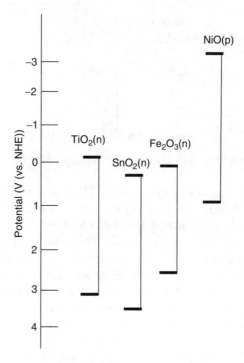

Figure 6.2 Band-gap positions of some selected semiconductor materials at pH 1

Individual atoms can no longer be identified and the number of HOMO and LUMO orbitals becomes very large, with the result that they merge into a continuum, i.e. the valence and conduction band. Molecular orbitals are no longer identified with particular molecules but are instead delocalized. As a result of this delocalization, the energy gap between the HOMO and LUMO orbitals is reduced as the number of bulk atoms increases, and in addition the difference between the ground state and the excited state for solid-state materials is smaller than that observed for the corresponding single molecules.

6.2.2 Electronic Properties of Nanoparticles

The question that arises is how the electronic properties of small, nano-sized particles can be described. Is there a limit at which the bulk description of valence and conduction bands is no longer valid and where a new model needs to be developed? Both theoretical calculations and experimental data suggest that the traditional bulk state band-gap model is no longer valid for small clusters. Calculations have shown that for Si clusters, bulk properties are not observed for clusters containing less than something like 10^5 atoms. For particles containing less that a thousand atoms, both the electronic and the physical structure are different from the bulk, while between 1000 and 100 000 atoms, the lattice structure is 'bulk-like', although the electronic structure differs considerably. In practical terms, it has been observed that the color of small particles, and therefore the energy of the band gap, depends strongly on the particle size.

The application of semiconductors as substrates in interfacial supramolecular assemblies has been dominated so far by films consisting of nanoparticles. In an attempt to understand the properties of such particles, and in particular, issues such as light-induced charge separation, their electronic properties will be discussed in some detail. Relevant issues in this respect are quantum effects, the size of the band gap, charge transport and band bending.

The band gap of a semiconductor represents the energy needed to produce an electron and a hole in a stable configuration, which minimizes their Coulomb interaction. Theoretical considerations show that [4] when two of these charge carriers approach they create a so-called *Wannier state*, which can be represented by a hydrogen-like Hamiltonian, as follows:

$$\hat{H} = (\hbar^2/2m_h \nabla_h^2) - (\hbar^2/2m_e)\nabla_e^2 - e^2/\varepsilon|r_e - r_h| \tag{6.1}$$

where m_h and m_e are the effective masses of the hole and electron, respectively, and ε is the dielectric constant. An important point is that the effective masses of the two charges in semiconductors are often much lower than the electron mass and these low values are an indication of the extent of delocalization. Furthermore, since the dielectric constant of semiconductors is typically between 5 and 12, the Coulomb attraction between the two particles is practically absent. This combination of small effective masses and weak Coulombic attraction results in an exciton wave function, which extends over a large region. The physical dimension of this exciton is defined by the Bohr radius r_B, which is defined as follows:

$$r_B = h^2\varepsilon_0\varepsilon/(e^2\pi m) \tag{6.2}$$

where ε is the dielectric constant and m is the effective mass of the charge carrier. However, this bulk picture breaks down for particles with a size that is similar to r_B. Since the dielectric constant and the effective mass of the charge carrier are properties of a particular material, this Bohr radius is substrate-dependent. For example, for CdS and TiO₂ values of 24 and 3 Å, respectively, are obtained. Therefore, the bulk description of the electronic properties of semiconductor particles is no longer valid for particles which are similar in size or smaller than the physical dimension of the wave function, which represents the lowest energy exciton. At this limit, the particle size is no longer sufficiently large enough to warrant the application of a model where wave functions are delocalized over many different molecules. At these sizes, electron–hole interactions are starting to dominate the physical behavior of the particles and a molecular rather than a bulk approach becomes more appropriate. As a result, the energy eigenvalues are no longer continuous and the valence and conduction band description needs to be replaced by discrete, localized states, the number of which will be determined by the number of units in the crystal. In addition, the delocalized molecular orbitals have to be replaced with wave functions, which are localized in space. Both the LUMO and HOMO energies are expected to change as a result of the increased localization of the system. For small semiconductor particles with a radius R, the band gap can be defined by the following:

$$E_g(R) = E_g(R = \infty) + h^2\pi^2/2R^2(1/m_e + 1/m_h) - 1.8e^2/\varepsilon R \tag{6.3}$$

where $E_g(R = \infty)$ is the band gap of the bulk semiconductor. This equation shows that for small particles the energy gap depends on the particle size. The first term represents the band gap in a macroscopic semiconductor, and the other two terms are indicative of decreasing particle size delocalization, which leads to the formation of localized molecular orbitals. Therefore, both are related to the size of the particle. The second term is related to the sum of the confinement energies for the electron and the hole as a result of quantum localization, while the third term is the Coulombic interaction energy. This last term predicts a decrease of the band gap with decreasing particle size. However, since this decrease depends on $1/R$, and the increase of the second term is proportional to $1/R^2$, the overall band gap for nanoparticles is expected to *increase* with *decreasing* particle size. This is illustrated in Figure 6.3. This decrease is not unexpected since as noted before delocalization in bulk semiconductors leads to a reduction of the HOMO–LUMO gap when compared with molecular systems. The variation of the band gap with particle size strongly affects the absorption and emission properties of the particles and as expected the photoluminescence energy increases with decreasing particle size. The color observed for small particles also depends on their sizes, e.g. HgSe particles with a size of about 500 Å are black, while particles with a size of 30 Å are transparent, with an absorption threshold of 380 nm.

Another difference between bulk semiconductors and nanocrystalline particles is the extent of band bending when in contact with electrolytes. For a macroscopic flat electrode, interfacial charge transfer can be envisaged when the electrolyte contains an electroactive species. The latter may act as either an electron donor or an electron acceptor and its presence will promote the transfer of mobile charge carriers across the solid–liquid interface. The presence of such an interfacial redox reaction results in the formation of a space charge layer at the interface and within these layers the valence and conduction bands are bent. A typical example of this is shown in Figure 6.4. Without the presence of a redox-active species, flat band edges are observed. In the presence of an electroactive species, various types of behavior can be observed, depending on the nature of this redox couple. The case shown is that where positive charges are accumulated at the surface. As a result of this process,

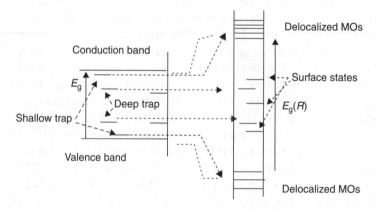

Figure 6.3 Correlation between the electronic structures of bulk (left) and nanocrystalline (right) semiconductor particles. Reprinted with permission from A. Hagfeldt and M. Grätzel, *Chem. Rev.*, **95**, 49 (1995). Copyright (1995) American Chemical Society

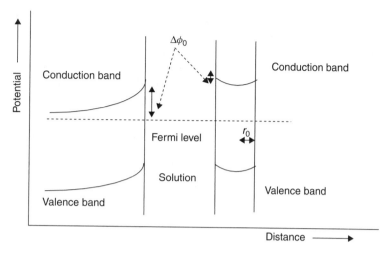

Figure 6.4 Correlation between space layer formation in bulk and nanocrystalline semiconductor particles

an accumulation layer is formed and band bending is observed. Where charges are depleted (electron injection into the solution) a depletion layer is established and the band bending is upwards. This band-bending process considerably changes the interaction between the solution and the semiconductor. This is illustrated by the differences observed between the band edge and the Fermi level.

For nanoparticles with a size relevant to the photoinduced processes discussed later in this chapter, the situation is very different. For small particles, the potential drop, $\Delta\phi_0$, between the surface and the center of particles can be approximated by the following equation:

$$\Delta\phi_0 = kT/6e(r_0/L_D)^2 \tag{6.4}$$

where r_0 is the radius of the particle and L_D is the Debye length, which is a function of the number of ionized dopant molecules per area (cm²). For large particles, the behavior is the same as that observed for macroscopic electrodes. Equation (6.4) shows that since the number of dopants in TiO₂ particles is expected to be rather low, the electric field in nanocrystalline particles will be small. As a result, no substantial band bending is expected and this is indeed observed. With a dielectric constant of 130 and an ionized donor concentration of 10^{17} cm^{-3}, the Debye length is 30 nm and the voltage difference between the center and the surface of a 16 nm sized TiO₂ particle is 0.3 mV. However, as illustrated in Figure 6.4, since no space-charge layer is created a shift in the band edge energies of nano-scale semiconductor particles is observed. These shifts are such that the Fermi level will be situated halfway between the valence and conduction bands.

6.2.3 Preparation and Structural Features of Nanocrystalline TiO₂ Surfaces

Nanocrystalline TiO₂ surfaces are prepared by coating conducting glass with a paste containing colloidal semiconductor particles, followed by a sintering process. For the solar cell type applications of nanocrystalline surfaces under discussion here, there are two central structural features that need to be controlled. These are the presence of an efficient contact between the nanoparticles and a large surface area. The first criterion is related to charge percolation through the semiconductor, while the latter facilitates a high light-to-electricity or light-to-charge separation ratio. The development of preparation methods has concentrated on optimizing these two factors. Importantly, the surface area can be controlled to a large extent by the sintering process. In the ideal material, the pores are interconnected and in contact with the bulk electrolyte. Upon modification of such a surface with a sensitizer, the amount of light absorbed is then optimized.

In a typical preparation method [5], acetic acid is added to titanium isopropoxide and the modified TiO_2 precursor is added in a single batch to a large volume of deionized water at room temperature under vigorous stirring. A white precipitate forms immediately and the reaction mixture is then left stirring for an additional hour to ensure completion of the hydrolysis reaction. At this stage nitric acid is added to bring the pH of the suspension to 1. The reaction mixture is then sonicated and subsequently stirred for 2 h at 80 °C under reflux in order to destroy any agglomerates formed in the reaction process. The reaction mixture, which now contains the redistributed primary particles, is subsequently heated for 12 h at 230 °C, which results in the formation of particles up to 20 nm in size. During this autoclaving process, sedimentation takes place and sonication is needed to redisperse the particles. The colloidal suspension obtained is concentrated by rotary evaporation to a final TiO_2 concentration of 11 wt%. To prevent cracking, poly(ethylene glycol) is added in a proportion of 50 % TiO_2. The colloidal pastes are typically deposited by using a scalpel to create thin layers on glass slides coated with indium–tin oxide (ITO). The layer thus formed is dried in air at room temperature for 10 min, after which 15 min heating at 50 °C takes place. Finally, the film is heated to 450 °C at a rate of 20–50 ° min^{-1} and left at that temperature for a further 30 min. The material obtained in this manner has a specific surface area of 91 $m^2 g^{-1}$, which indicates a particle size of 17 nm.

In the above preparation, an autoclaving temperature of 230 °C was used. If, however, higher temperatures are used then more opaque films are obtained which contain particles with sizes of 100 nm and more. It is important to realize that as with all heterogeneous systems the structure of the substrate layer is of great importance for the properties of the ultimate interfacial supramolecular assembly. For solar cell applications, very open sponge-like layers with a thickness of about 10 µm are normally preferred. The presence of some larger particles are found to be beneficial for the building up of layers with the optimum thickness, and as a result autoclaving temperatures of between 240–250 °C have been used. However, for electrochromic applications, a thickness of about 4 µm is more appropriate and sintering temperatures of 200 °C are used. An important issue is that in fundamental studies of the photoinduced charge-separated processes at TiO_2 surfaces modified with molecular dyads or triads, the methods used to prepare semiconductor substrates have been those normally applied for solar cell applications. Although these optimize the absorbance features of the modified surfaces, they also promote the interaction of all components of the molecular assemblies with the semiconductor surface. This gives rise to the presence of various charge-injection pathways and precludes the design of interfacial supramolecular structures with uniquely defined electron-injection pathways. For such studies, the use of films which are less open and contain large particles seems more appropriate. However, a reduction of the surface area would automatically lead to a reduced absorbance. Therefore, it is important that considerable thought is given to the surface structure most appropriate for a particular type of study.

Surface characterization techniques have been used extensively to investigate the structural features of the nanocrystalline surfaces obtained. Some examples are shown in Figures 6.5 and 6.6. In Figure 6.5, nitrogen adsorption measurements indicate, upon autoclaving at 230 °C, that the average pore size is about 20 nm, in agreement with the particle size calculated. The particles obtained are composed of anatase (a naturally occurring crystalline form of TiO_2). Figure 6.6 shows a scanning electron micrograph of a nanocrystalline TiO_2 layer. According to transmission electron microscopy of such films, the (101) orientation is most exposed, followed by the (100) and (001) faces.

The films typically have a porosity of about 50% and allow for a contacting solution to reach to the underlying conducting layer. This permeability allows for efficient penetration of solution-based species all through the layer. The effective surface area can be a factor of 1000 higher than that of a flat surface, and the layer thickness is typically up to 10 µm. This large surface area allows for strong light absorption even though only a monolayer of the dye has been immobilized on the metal oxide surface. Importantly, it has been shown that the injection of charge from the molecular component to the semiconducting surface is most effective with monolayers. For thicker layers, the outer components interact less strongly (or not at all) with the solid substrate and charge injection is limited. From the applications, solar cell's or electrochromic device's point of view, availability of the large surface area is therefore important. A large surface area is also important from the experimental point of view. High absorbances allow the use of spectroscopic

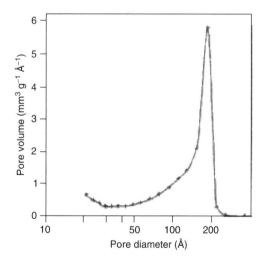

Figure 6.5 Pore size distribution in a nanocrystalline TiO_2 film as determined by nitrogen adsorption. Reprinted from P. Bonhote, E. Gogniat, S. Tingry, C. Barbe, N. Vlachopoulos, F. Lenzmann, P. Comte and M. Grätzel, *J. Phys. Chem., B*, **102**, 1498 (1998). Copyright (1998) American Chemical Society

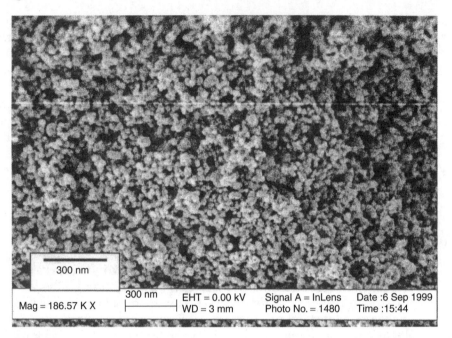

300 nm

| 300 nm | EHT = 0.00 kV | Signal A = InLens | Date :6 Sep 1999 |
| Mag = 186.57 K X | ⊢————⊣ | WD = 3 mm | Photo No. = 1480 | Time :15:44 |

Figure 6.6   Scanning electron micrograph of a nanocrystalline TiO_2 film. Reproduced from H. Lindstrom, E. Magnusson, A. Holmberg, S. Sodergren, S.-E. Lindquist and A. Hagfeldt, *Solar Energy Mat. Solar Cells*, (2002), **73**, 91–101. (Elsevier)

techniques for the study of the photophysical behavior of these modified layers. With flat surfaces, the absorbances obtained can be too low to allow for detailed mechanistic studies.

6.3   Physical and Chemical Properties of Molecular Components

Nanocrystalline semiconductor surfaces can be easily modified with monolayers by dipping them into a solution containing the surface active molecular component. In this manner, a monolayer of the molecular component is obtained. Adsorption is relatively slow but is normally completed in 12 h. In this section, the properties of assemblies incorporating molecular components capable of inducing photoinduced electron transfer processes will be discussed. Figure 6.7 shows some typical molecules which have been used as light-absorbing components. The compounds normally contain carboxy, phosphonate or ester groupings, which serve to attach the compound covalently to the oxide surface and to establish a good electronic coupling between the adsorbed species and the TiO_2 surface. Quantum mechanical calculations and infrared spectroscopy have provided direct evidence for this interaction and indicate that carboxy compounds are attached to the oxide surface, by two carboxylate bridges, to either hydroxyl groups or Ti ions. Figure 6.8 shows a schematic illustration for the case of biisonicotinic acid, which indicates the interaction between the carboxy groups and the TiO_2 surface. Another prerequisite is that to facilitate efficient injection into the solid substrate, the energy levels of the

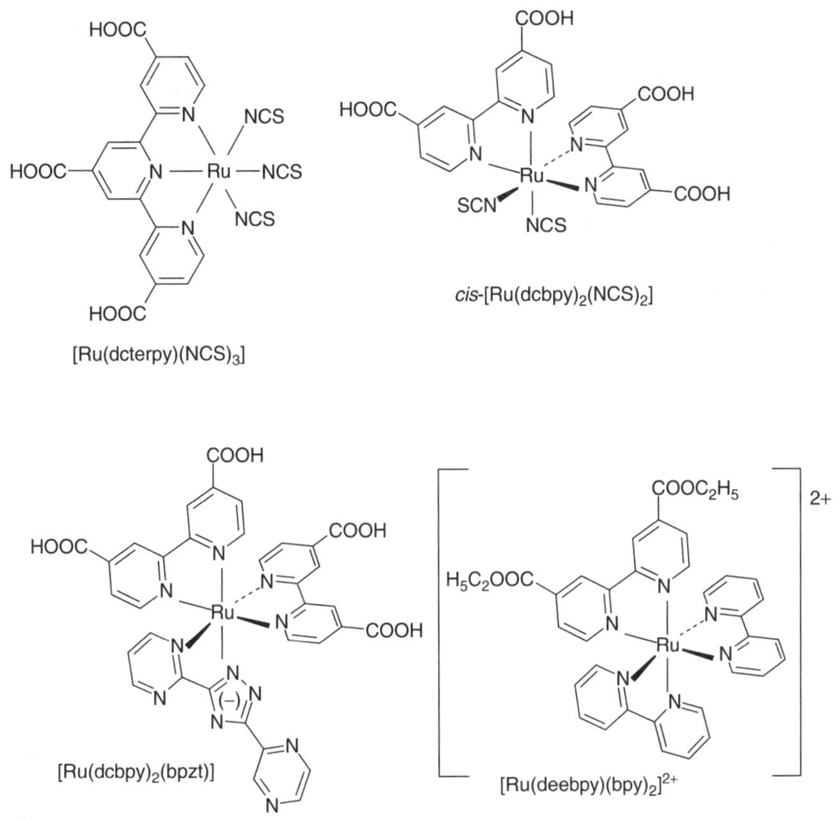

[Ru(dcterpy)(NCS)$_3$]

cis-[Ru(dcbpy)$_2$(NCS)$_2$]

[Ru(dcbpy)$_2$(bpzt)]

[Ru(deebpy)(bpy)$_2$]$^{2+}$

Figure 6.7 Molecular structures of some typical materials used as light-absorbing components in the modification of metal oxide surfaces

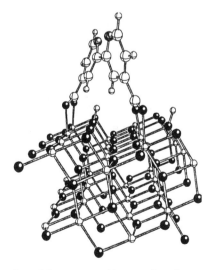

Figure 6.8 Schematic illustration of the proposed interactions between carboxy groups and a TiO$_2$ surface, as suggested by quantum mechanical calculations and infrared spectroscopy studies. Reproduced from H. Remsmo, K. Westermark, S. Sodergren, O. Kohle, P. Persson, S. Lunnell and H. Siegbahn, *J. Chem. Phys.*, **111**, 2744–2750. (American Institute of Physics, Woodburg, USA). (1999)

semiconductor and the molecular species need to be carefully matched. To allow injection, the excited-state energy of the dye needs to be higher than the lower edge of the conduction band of the semiconductor. This prerequisite will be considered in more detail later. After the injection has taken place, the electrons travel across the film towards the conducting support at the back of the cell which serves as a current collector. An energy of about 0.1 V is required to drive this process. These steps put certain conditions on the electronic properties of suitable compounds. In the next section, this broad picture will be discussed in more detail on the basis of some specific examples.

Ruthenium polypyridyl complexes have been studied extensively on semiconductor surfaces. Most compounds contain substituted 2,2'-bipyridyl (bpy) ligands, containing carboxy, ester or phosphonate functional groups. Their general behavior when attached to TiO_2 is similar to that found in solution, but there are specific differences depending on the energy match between the molecules and the surface. The most important reason for the popularity of these compounds as photosensitizers for TiO_2 is because of the energies of both the ground and excited states and in the nature of their photophysical properties. In addition, this type of compound has proved to be very popular because of the availability of a well-defined synthetic chemistry, thus allowing for 'fine tuning' of their electronic and redox properties.

In order to understand the behavior of the surface-bound compounds, the photophysical properties of ruthenium polypyridyl compounds in solution will be considered first. Figure 6.9 shows in a schematic diagram the electronic properties observed for typical ruthenium polypyridyl compounds. Upon excitation of the molecule, an electron is promoted from a metal-based ground state, which is d–d in character, to a π^*-orbital of the 2,2'-bipyridyl ligand. This is considered to be a metal-to-ligand charge-transfer transition (MLCT) with singlet character. In solution, fast and efficient intersystem crossing occurs from this singlet state to a triplet ^{3}MLCT state. From this state, several deactivation processes can occur, which can either be radiative or non-radiative pathways. The first will lead to emission, while in the latter the excess energy is dissipated by non-radiative deactivation with solvent or via population of the metal-centered triplet (^{3}MC) state. Population of this state has important implications, since it can lead to photochemical decomposition of the complex by photoinduced ligand exchange. In view of the potential application of this class of compounds in solar energy conversion systems, much research has

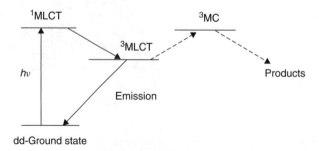

Figure 6.9 Schematic diagram of the electronic properties of typical ruthenium polypyridyl complexes

been carried out in the last twenty years on improving the photostability of these compounds by increasing the energy of this deactivating ^{3}MC state. It is somewhat ironic that with the development of the present generation of solar cells based on nanocrystalline TiO$_2$ films sensitized by ruthenium polypyridyl complexes, this photolability no longer constitutes a major problem (see below).

As an example of the photophysical behavior of a surface-bound ruthenium compound, the polypyridyl complex [Ru(dcbpy)$_2$(NCS)$_2$] (see Figure 6.7), where dcbpy is 4,4'-dicarboxy-2,2'-bipyridyl, will be discussed. The carboxy substituents are introduced to provide a strong interaction between the molecular components and the metal oxide surface. The presence of up to four carboxy groupings makes an accurate assessment of the molecular composition of the dyes difficult. For example, the presence of several acid–base equilibria makes it difficult to determine the protonation state of these ligands, and therefore the charge of the metal complex. For simplification, these issues have been ignored in the molecular composition of the compounds discussed, with the molecular formulae given assuming the presence of a neutral metal complex unless otherwise stated.

The thiocyanate-based ruthenium polypyridyl complex has been studied in great detail because of its high efficiency as a sensitizer in regenerative photovoltaic cells. In Figure 6.10, some photoelectron spectroscopy data are given, which shows that for this compound a good match is obtained between the energy of its excited state and the level of the conduction band of the semiconductor. It is evident from this figure that for the unmodified semiconductor, direct excitation into the conduction band is only possible in the UV part of the spectrum. However, the [Ru(dcbpy)$_2$(NCS)$_2$] complex has a visible absorption with a maximum at 530 nm and the electronic properties of the compound ensure that upon irradiation at this wavelength the excited state reached is 0.25 eV above the conduction band edge. The relative energies can be estimated from the differences between the valence band edge position, the HOMO level of the dye of 1.3 eV, the dye's absorption maximum of 2.3 eV and the band gap of 3.2 eV of anatase TiO$_2$.

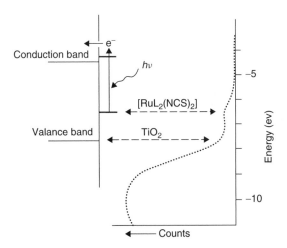

Figure 6.10 Energy diagram outlining the valence band energies for TiO$_2$ and the energy levels for the dye [Ru(dcbpy)$_2$(NCS)$_2$]. Reprinted with permission from A. Hagfeldt and M. Grätzel, *Acc. Chem. Res.*, **33**, 269 (2000). Copyright (2000) American Chemical Society

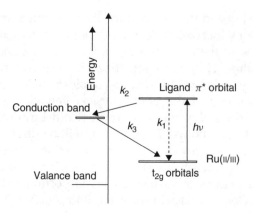

Figure 6.11 Illustration of the photophysical processes expected for a TiO$_2$-bound ruthenium polypyridyl dye

UV–visible absorption spectra of surfaces modified with ruthenium polypyridyl complexes such as [Ru(dcbpy)$_2$(NCS)$_2$] indicate that the electronic properties of the compound do not undergo any fundamental change when immobilized on a semiconductor surface. However, for the excited state a different deactivation pathway is created, namely interaction with the solid substrate and as a result kinetic factors may ensure that different excited-state processes become dominant. Figure 6.11 outlines the fundamental photophysical processes expected for a ruthenium compound immobilized as a monolayer on a TiO$_2$ surface. Upon excitation of the immobilized compound, an electron is promoted from the metal based t_{2g} ground state to a ligand-based π^*-orbital. This excitation process formally creates a negatively charged polypyridyl ligand and a hole or Ru(III) site on the metal. From this ligand-based excited state, the electron can either be deactivated via the emission process observed in solution (k_1), by non-radiative interaction with solvents, by population of the ^{3}MC level (see Figure 6.9) or by injection into the semiconductor surface with a rate k_2. The direct binding of the dye to the TiO$_2$ surface ensures a strong electronic coupling, consequently electron injection into the semiconductor surface is very fast. It is generally assumed that the presence of a covalent bond between the adsorbed species and TiO$_2$ greatly enhances the interaction between the ligand-based excited state and the Ti-based 3d orbitals. Detailed investigations of the injection process indicate a biexponential injection process with rise times of 50 ± 25 fs and 1.7 ± 0.5 ps, in an 84 to 16 % ratio (see below). The slower component depends strongly on the sample condition and its origin requires further elucidation. An intriguing possibility is that the two injection pathways are related to the nature of the injection orbital on the dye. One would expect from the photostability of the adsorbed dyes that injection takes place from the ^{1}MLCT state; intersystem crossing to the emitting ^{3}MLCT level is, however, in the picosecond range and could conceivably compete with injection. The latter, from both levels, would then give rise to a dual exponential process with different rise times. The fundamental observation that can be made is that for kinetic reasons the excited state is exclusively deactivated by injection into the semiconductor surface and that no photodecomposition and no emission are observed. This is an important observation, since [Ru(dcbpy)$_2$(NCS)$_2$] is extremely photolabile in solution.

The photostability observed upon immobilization is clearly important from the application point of view but also for fundamental reasons. The photostability of the assembly shows that upon interaction between the molecular component and the semiconductor surface, fundamental changes in the photophysical behavior can be obtained provided that the energy levels of modifier and surface are tuned. Importantly, these fundamental changes are not related to changes in the electronic properties of the molecular component, but are caused by the interaction of the excited molecular component with available surface states.

6.3.1 Charge Separation at Nanocrystalline TiO$_2$ Surfaces

TiO$_2$ surfaces have been applied to a great extent as photocatalysts. For such processes to be effective, irrespective of whether the photoexcitation is direct or sensitized, the light-induced charge separation needs to be fast. Light absorption leads to the generation of electrons and holes and the destination of these charge carriers has to be considered. They may recombine, diffuse apart and undergo chemical reactions, or produce a photocurrent. For n-type semiconductors such as TiO$_2$, the electrostatic field is such that holes migrate to the surface while electrons move to the back contact. This is an important observation, since it means that for sensitizers which photoinject into the conduction band, the hole will not diffuse into the semiconductor surface. In the case of nanoparticles, where the space-charge layer, and therefore the electrostatic field, are very small, the charge separation occurs via diffusion. Importantly, in small particles, diffusion to the surface is faster than recombination and in principle high yields for photoinduced charge separation can be obtained. The effectiveness of this process is, however, also determined by the interfacial charge-transfer process involving the semiconductor and a solution-based or covalently attached redox couple. For the latter, the rate for this process is controlled simply by the time required for tunneling across the interface and the thickness of the reaction layer.

One of the most surprising and interesting results obtained from these studies has been the large difference between this photoinduced charge injection and the reverse process. Charge injection is normally in the pico- or femtosecond range but the subsequent recombination occurs in the microsecond time scale. Therefore, for these systems, light-induced charge separation is very efficient. Several explanations of this behavior have been put forward, some of which are related to the properties of the semiconductor surface and some to the nature of the interaction between the surface and the attached molecule. It has been pointed out that charge recombination has a large driving force and that this process is in the Marcus inverted region where the rate of the reaction drops with increasing driving force. In addition, entropy factors have been considered. In a nano-sized TiO$_2$ particle, an electron is delocalized over several thousand conduction band states, and if there is only one sensitizer available on that particle than the recombination of the electron with the oxidized molecule is associated with a significant entropy loss. Hasselmann and Meyer [6] have carried out a study of the recombination process for a series of ruthenium, osmium and rhenium photosensitizers. These authors observed that although there was a difference of about 960 mV in the apparent driving force of recombination for the different dyes, the recombination rates were the same. It was suggested that the charge combination rate is limited by diffusional encounters in the injected electron with the oxidized sensitizer. The fact that the recombination rate is independent of the driving force seems to suggest that invoking the Marcus inverted region to explain the slow recombination is inappropriate. Other arguments involve the consideration of the molecular orbitals involved, as outlined in the next section.

The fundamental and practical attraction of the dye-modified TiO_2 surfaces lies in the next step, i.e. the back reaction (k_3 in Figure 6.11). While fast injection is very important since it will inhibit side processes such as photodecomposition of the compound, for any application envisaged the back-reaction needs to be slow. In other words, charge separation needs to be both fast and long-lived. While nature has developed systems that encourage long-lived charge separation, artificial systems have been far less effective. Detailed studies have shown that the back reaction k_3 can be up to a million times slower that the forward injection process. Several arguments to explain this slow recombination based on the properties of the solid substrate and on entropy considerations have already been put forward in the last section. Here, the nature of the electronic orbitals of the molecular component that are involved in the forward and recombination reactions will be considered. In Figure 6.12, a schematic diagram outlining the involvement of the different components of the assembly in both processes is shown. It is important to point out that recombination can only occur at the surface of the semiconductor, since only there can injected electrons recombine with the oxidized sensitizer. Figure 6.12 shows that while the forward reaction involves a ligand-based π^* orbital which is strongly coupled to the surface, the back-reaction is to metal-based orbitals which have a much less favorable overlap with the conduction band of the semiconductor. This lack of coupling and its remoteness from the surface is thought to be at least partly responsible for the long-lived charge separation. There is at present, however, only a very limited understanding of the recombination process. The recombination is non-exponential and it seems likely that this is related to issues such as the particle size and the number of adsorbed dye molecules per particle. The Butler–Volmer equation and an Auger capture model have been used to model the recombination

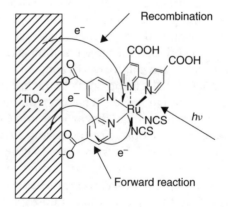

Figure 6.12 Orbital diagram for the surface-bound ruthenium complex, showing the electron-injection and recombination processes in terms of the components of the assembly involved

processes for SnO$_2$ surfaces modified with adsorbed [Ru(bpy)$_3$]$^{2+}$-type dyes [7]. With these two approaches, the non-linear dependence of the recombination rate on these surfaces have been modeled. However, much more work is needed in this area, in particular since the nature of the semiconductor material and its structural features are controlling elements in the recombination process.

6.4 Photovoltaic Cells Based on Dye-Sensitized TiO$_2$

One area which has been very instrumental in promoting the study of the interaction between molecular components and semiconductor surfaces, has been the development of the regenerative-dye-sensitized solar cell by Grätzel and co-workers [2]. Over the last 50 years there has been a constant improvement in the development of solar cells based on silicon and recently efficiencies of up to 24 % have been obtained. The cells have a long lifetime but a disadvantage is that they are rather expensive to produce. In addition, they are only able to absorb light in the high-energy region. In an attempt to solve these limitations, a novel solar cell based on adsorbed molecular dyes has been developed. In these cells, the excitation of dyes capable of adsorbing visible, sub-band-gap light inject electrons into the conduction bands of nanocrystalline TiO$_2$, as discussed above. In order to achieve a closed circuit and to obtain a cell that is constantly regenerated, the, now oxidized, dye is re-reduced by a solution-based relay, while the latter is re-reduced at a counter electrode. For the production of an efficient solar cell, the dye has to fulfil a number of criteria. First, the compound has to be an efficient light absorber in the visible region. Secondly, to allow injection into the conduction band of the semiconductor, the excited-state redox potential has to be carefully controlled. Thirdly, to allow the re-reduction of the dye after injection, the ground-state redox potential needs to be more positive than that of the solution-based redox couple and ideally as positive as possible. Fourthly, the reorganization energies for excited-state and ground-state redox processes needs to be as small as possible in order to minimize free energy losses during the electron transfer processes. Finally, the dye must be firmly bound to the oxide surface.

The [Ru(dcbpy)$_2$(NCS)$_2$] complex discussed above has become the benchmark for the study of photovoltaic cells based on modified TiO$_2$. In a typical Grätzel-type solar cell [2], this compound is immobilized on a mesoporous titanium oxide layer composed of nanocrystalline anatase TiO$_2$ particles. Figure 6.13 shows a schematic of the relevant energies, components and electron transfer processes of regenerative solar cells in general terms. This figure outlines the adsorption and injection process (k_2) and the electron transport through the oxide layer (k_5) already discussed above, but another important component has been added, namely a solution-based reagent capable of reducing the oxidized dye. In general, the I$^-$/I$_3^-$ redox couple dissolved in acetonitrile is used for this process. By the introduction of this redox couple, the original oxidation state of the dye is restored by electron donation (k_4) from I$^-$ to the Ru(III) metal center. In principle, this reduction process prevents the recombination of the conduction band electrons with the oxidized dye (k_3). The potentially energy

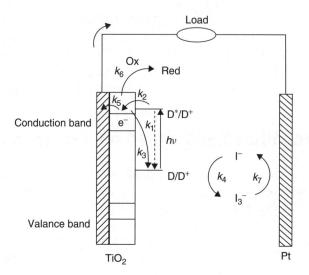

Figure 6.13 Schematic of a solar cell based on TiO_2 sensitized with an electron donor (D) and the electron relay, iodine/iodide. Reproduced by permission of Wiley-VCH from A. C. Lees, B. Evrard, T. E. Keyes, J. G. Vos, C. J. Kleverlaan, M. Alebbi and C. A. Bignozzi, *Eur. J. Inorg. Chem.*, 2309 (1999)

wasting reaction between the injected electrons and iodine produced in the solution is also considered (k_6). The triiodide produced is in turn reduced by the counter electrode (k_7). The voltage generated upon irradiation depends on the difference in energy between the Fermi level of the electrons in the metal oxide and the redox potential of the oxidant in the electrolyte.

Two experimental factors are important in the design of practical cells, i.e. the photovoltaic efficiency and the cell voltage. The efficiency is normally given as the incident photon-to-current conversion efficiency (IPCE) as a function of the wavelength λ. This efficiency is determined by three different components. The first of these is the light harvesting efficiency (LHE), which represents the fraction of radiant power that is absorbed by the modified surface. This value is expected to be wavelength-dependent. The second factor is the quantum yield for charge injection, ϕ, while the third is the efficiency of electron collection in the external circuit, η. IPCE(λ) is then given by the following equation:

$$IPCE(\lambda) = LHE(\lambda)\phi\eta \tag{6.5}$$

The latter two factors in this equation depend on kinetic factors, while LHE(λ) depends on the nature of the light-absorbing sensitizer and the active (surface) area of the nanocrystalline surface. Measurements are normally carried out by using monochromatic light and the IPCE values are then calculated using the following equation:

$$IPCE(\lambda) = \frac{1.24 \times 10^3 (eV\,nm) \times photocurrent\ density(\mu A\,cm^{-2})}{wavelength\ (nm) \times photon\ flux(\mu W\,cm^{-2})} \tag{6.6}$$

This equation shows that the efficiency of a solar cell of this type is controlled by many factors and not only by light absorption by dye.

The processes outlined in Figure 6.13 above can be represented by the following reaction scheme [8]:

$$S \xrightarrow{h\nu, \phi} S^+ + e^-$$

$$S^+ + e^- \xrightarrow{k_3} S$$

$$S^+ + I^- \xrightarrow{k_4} S + I^\bullet$$

$$e^- \xrightarrow{k_5} e^-_{ex}$$

$$e^- + I^\bullet \xrightarrow{k_6} I^-$$

$$e^-_{ex} + I^\bullet \xrightarrow{k_7} I^-$$

Under steady-state conditions, that is, under constant irradiation, the incident photon-to-current efficiency (IPCE) as a function of the wavelength λ can then be given in terms of these reactions as follows:

$$\text{IPCE}\lambda = (\text{LHE}\lambda)(\phi)\left(\frac{k_4[I^-]}{k_3[e^-] + k_4[I^-]}\right)\left(\frac{k_7[e^-_{ex}]}{k_6[e^-] + k_7[e^-_{ex}]}\right) \tag{6.7}$$

Making the reasonable assumption that the regeneration of the oxidized redox couple at the counter electrode is faster than its recombination with the TiO_2 surface ($k_6[e^-] \ll k_7[e^-]$), the efficiency of electron collection in the external circuit, η, the last term in the equation given above, can be simplified as follows:

$$\eta = \frac{k_4[I^-]}{k_3[e^-] + k_4[I^-]} \quad \text{or} \quad \eta = 1 - \frac{k_3[e^-]}{k_3[e^-] + k_4[I^-]} \tag{6.8}$$

This analysis clearly shows that the regeneration of the oxidized dye by the oxidant in the solution is of central importance to the efficiency of the cell, both in terms of collection yields and long-term stability. For [Ru(dcbpy)₂(NCS)₂], regeneration takes place in about 19 ns. Importantly, this re-oxidation is several orders of magnitude faster than the recombination of the oxidized dye with the semiconductor, which is on the ms time-scale. For other compounds, slower values for k_4 have been observed but these are still in the ns range. A slower reaction of the Ru(III) dye with the redox center in solution will inevitably result in a decrease of the efficiency of the cell and may also lead to decomposition of the oxidized dye. Therefore, it is important that for [Ru(dcbpy)₂(NCS)₂] the regeneration of the Ru(II) species is eight orders of magnitude faster than decomposition of the dye in the oxidized state. As a result, 'turn-over numbers' of 10^8 have been obtained under continuous operation. This stability is important for practical applications since, unlike in natural photosynthesis where porphyrins are constantly renewed, such a process is not feasible in artificial systems.

Other parameters that need to be considered to achieve efficient dye-sensitized solar cells include the open circuit photovoltage, V_{oc}, which is determined theoretically by the potential difference between the Fermi level of the semiconductor and the Nernst potential of the redox couple in solution, i.e. I^-/I_3^- in the case of the cell under consideration here. However, there are kinetic reasons why the theoretical value is not obtained in practice. One issue to be considered, for example, is the energy-wasting reaction of a recombination of the injected electrons with the I_3^- produced upon reduction of the oxidized dye.

A typical example of the IPCE obtained for $[Ru(dcbpy)_2(NCS)_2]$ as a function of wavelength is given in Figure 6.14, while the photocurrent–voltage characteristics for this compound are shown in Figure 6.15. The first of these figures shows the high efficiency for the photon-to-current conversion process. The spectral range covered by $[Ru(dcbpy)_2(NCS)_2]$ is also quite considerable. As also shown in Figure 6.14, this spectral performance in the infrared can be improved considerably by using the dye $[Ru(tcterpy)(NCS)_3]$.

The overall efficiency, η_{global}, of the photovoltaic cell depends on the integral photocurrent density, i_{ph}, which represents the overlap between the solar light envelope and the monochromatic current yield, the open-circuit voltage, V_{oc}, the fill factor of the cell (ff), and the light intensity, I_s, as shown in the following:

$$\eta_{global} = i_{ph}V_{oc}(ff)/I_s \qquad (6.9)$$

At present, the overall efficiency of the devices produced is of the order of 10 % for $[Ru(dcterpy)(NCS)_3]$. Further improved performance depends on increasing the light absorption in the near-infrared region and an improvement of the short-circuit photocurrent. Significant gains could also be made by increasing the open-circuit voltage. However, this would involve the use of a different redox couple than I^-/I_3^- and the identification of a superior reductant seems unlikely at the present time.

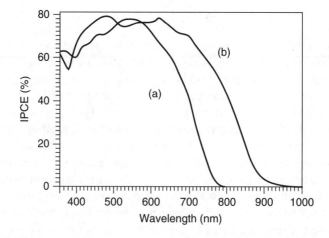

Figure 6.14 Photocurrent as a function of the wavelength for nanocrystalline TiO_2 modified with (a) $[Ru(dcbpy)_2(NCS)_2]$, and (b) $[Ru(tcterpy)(NCS)_3]$. Reprinted with permission from A. Hagfeldt and M. Grätzel, *Acc. Chem. Res.*, **33**, 269 (2000). Copyright (2000) American Chemical Society

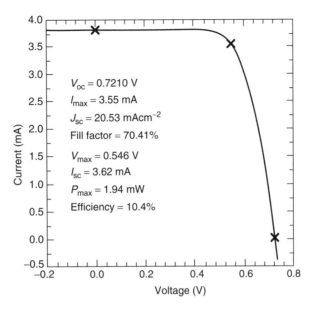

Inside the figure:

$V_{oc} = 0.7210$ V

$I_{max} = 3.55$ mA

$J_{sc} = 20.53$ mAcm^{-2}

Fill factor $= 70.41\%$

$V_{max} = 0.546$ V

$I_{sc} = 3.62$ mA

$P_{max} = 1.94$ mW

Efficiency $= 10.4\%$

Axis labels: Current (mA) (vertical), Voltage (V) (horizontal)

Figure 6.15 Photocurrent–voltage characteristics for a nanocrystalline TiO$_2$ surface modified with [Ru(tcterpy)(NCS)$_3$]. Reprinted with permission from A. Hagfeldt and M. Grätzel, *Acc. Chem. Res.*, **33**, 269 (2000). Copyright (2000) American Chemical Society

6.5 Photoinduced Charge Injection

As already discussed above, one of the principal reasons for the introduction of semiconductor surfaces in supramolecular assemblies is to achieve long-lived charge separation based on a fast electron injection into the substrate, coupled to a slow back-reaction. In this section, the charge-injection process will be considered in more detail by using some examples.

Photoinduced electron injection is by no means a new development. This process has already been applied in areas such as silver halide photography. In this discussion, only sensitized TiO$_2$ surfaces will be considered. Many experiments have shown that the charge injection into the semiconductor surface is very fast. In order to study these processes, fast spectroscopic techniques are preferred. Whether or not charge injection takes place can be studied conveniently on the nanosecond time-scale by using transient absorption spectroscopy. However, to address the injection process directly, experiments are carried out on the femtosecond time-scale, while recombination and charge separation require the nanosecond to microsecond range.

The fundamentals of electron transfer from an adsorbed molecule to a solid substrate has been considered by Willig and co-workers [9]. In this study, the photoinduced electron injection from the excited state of [Ru(dcbpy)$_2$(NCS)$_2$] into nanocrystalline TiO$_2$ was investigated both in high vacuum and in methanol containing 0.3 M LiCl. The central issue addressed in this investigation was to determine the time-scale on which the photoinduced electron injection from the molecular component to the solid substrate occurs. Transient absorption spectroscopy was used to follow this charge separation. The measurement was based on

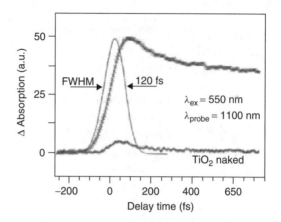

Figure 6.16 Transient absorption signal detected at 1100 nm for [Ru(dcbpy)$_2$(NCS)$_2$]-coated TiO$_2$ in high vacuum. For comparison, the signal obtained for 'naked' TiO$_2$, obtained under the same conditions, is also shown. Reprinted with permission from T. Hannappel, B. Burfeindt, W. Storck and F. Willig, *J. Phys. Chem., B*, **101**, 6799 (1997). Copyright (1997) American Chemical Society

the knowledge that upon injection of electrons into the semiconductor an absorption band is observed with a maximum at 1200 nm. The rise time of this absorption can be taken as an indication for the rate of electron injection from the dye into TiO$_2$. The results obtained in this experiment are shown in Figure 6.16. Upon excitation of the dye at 550 nm, a rise of the absorption feature, indicating injection into empty electronic states at the surface of the semiconductor, is observed in the femtosecond range. A detailed analysis of this rise shows that injection is faster than 25 fs. No evidence is obtained for the formation of an excited-state [Ru(dcbpy)$_2$(NCS)$_2$] species, again indicating fast injection from the molecular component to the solid substrate. It was furthermore shown that the electron injection is independent of temperature over the 20–295 K range. This behavior is explained by the presence in the conduction band of many electronic acceptor levels of different energies which can be populated from the excited state of the dye.

The very short time-scale observed for electron transfer suggests that it occurs by a mechanism quite different from that normally observed for larger distances. In the latter case, vibrational deactivation within the excited state of the dye is faster than electron transfer. For the modified TiO$_2$ surfaces, the electron transfer reaction does not involve vibrational relaxation and is, therefore, different from the conventional Marcus–Levich–Jortner–Gerischer type of electron transfer mechanism devised for weakly coupled electronic states. It is also not correct to consider it as an optical charge-transfer process from a donor to an acceptor level, but is better described as an ultrafast electron-transfer reaction with a finite reaction time controlled by electron tunneling. Therefore, the injection reaction is controlled by the electronic tunneling barrier, which is controlled by the nature of the linker, by the Franck–Condon overlap for the interacting states of molecular component and semiconductor, and by the escape time for the initially prepared wave packet describing the 'hot' electron from the reaction distance of the oxidized dye. The covalent bond formed between the molecular component and the oxide surface greatly facilitates this fast injection rate.

Electron transfer takes place prior to any significant redistribution of vibrational energy in the dye's excited state. This mechanism means that injection takes place from the ^{1}MLCT state which is populated by the initial absorption process and that intersystem crossing to the lower-lying ^{3}MLCT state normally observed in solution is not competitive for the adsorbed species (see Figure 6.12 above).

6.5.1 External Factors which Affect Photoinduced Charge Injection

In the following, factors external to the actual interfacial supramolecular assembly, which are capable of modifying the photoinduced electron injection process, are considered. This discussion will concentrate on how the rate of charge injection can be manipulated by changing the composition of the electrolyte and by changing the external potential applied to the semiconductor film.

6.5.2 Composition of Electrolyte

In typical investigations on the behavior of modified TiO_2 surfaces upon irradiation, measurements are carried out in acetonitrile containing 0.1 M $LiClO_4$. Under these conditions, very fast electron injection into the semiconductor surface is observed. However, it has been noted that this injection process depends strongly on the lithium concentration of the contacting acetonitrile solution [10,11]. In the absence of lithium, *no* injection is observed. This is an important observation since this opens up the possibility of modulating the photophysical behavior of the interfacial supramolecular assembly by external manipulation of the conditions. In this section, this observation is discussed in more detail, and, in addition, the possibility to use the surface potential of the semiconductor surface as a driving force will be considered.

A number of different approaches can be taken to investigate the charge-injection process. The first one, outlined in the last section, is based on the absorption rise observed at about 1200 nm which is associated with the presence of electrons in TiO_2. A second method is based on the measurement of the IPCE values for the assemblies in the presence of iodide, while the third approach is based on the intrinsic spectroscopic features of the sensitizer. In this present section, the focus is on the latter two approaches since the absolute rate for charge injection is not of direct interest but simply whether or not injection is taking place. To estimate the injection and charge-separation process, the transient absorption spectra of the sensitizer in solution are compared with those obtained in the interfacial supramolecular assembly. A typical example of this approach is shown in Figure 6.17 for the compound [Ru(dcbpy)$_2$(bpzt)] [8], (see Figure 6.7 above for the structure).

Figure 6.17(a) shows the time-resolved transient absorption spectra obtained for [Ru(dcbpy)$_2$(bpzt)] in aqueous solution at pH 7. The important features are the bleaching of an absorption feature with a maximum at 450 nm and the growing in of a new band at 350 nm. The bleaching process can be attributed to the disappearance of the metal-to-ligand charge-transfer (MLCT) transitions in the excited state, while the feature at 350 nm is indicative of intraligand transitions of the dcbpy radical formed upon excitation of the complex. This type of behavior which is typical

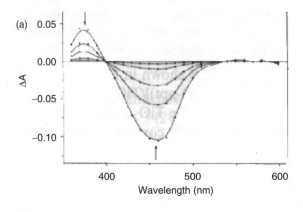

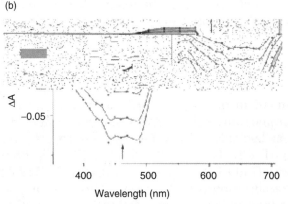

Figure 6.17   Time-resolved transient absorption spectra of [Ru(dcbpy)$_2$(bpzt)]: (a) in aqueous solution (pH 7), with spectra taken at 0, 25, 50, 100 and 150 ns, respectively; (b) immobilized at nanocrystalline TiO$_2$, with spectra taken at 0, 50, 250, 1000 and 2000 ns, respectively. Reproduced by permission of Wiley-VCH from A. C. Lees, B. Evrard, T. E. Keyes, J. G. Vos, C. J. Kleverlaan, M. Alebbi and C. A. Bignozzi, *Eur. J. Inorg. Chem.*, 2309 (1999)

of ruthenium polypyridyl complexes with polypyridyl-localized excited states. The lifetime of the excited state for this compound under these conditions is 45 ns.

When the complex is adsorbed on TiO$_2$, a quite different behavior is observed, as shown in Figure 6.17(b). In acetonitrile containing 0.1 M LiClO$_4$, the bleaching at 450 nm is observed, indicating that after excitation the molecule is no longer in the ground state, while the absorbance increase at 350 nm, indicative of the formation of the dcbpy radical, is not observed. This observation indicates that the electron promoted upon photoexcitation is no longer located on the dcbpy ligand as it is in solution, but has been injected into the semiconductor surface. In addition, no emission is observed from the surface-bound species. This behavior is characteristic for a wide range of ruthenium polypyridyl complexes and agrees with very fast electron injection into the semiconductor.

This method of analysis will now be applied to investigate charge injection under various conditions. First of all, the observation made by Meyer and co-workers [10] that in pure acetonitrile no injection is taking place will be considered. They reported that upon excitation of a TiO$_2$ surface modified with [Ru(deebpy)(bpy)$_2$] (for the structure, see Figure 6.7 above) in pure acetonitrile, a transient absorption

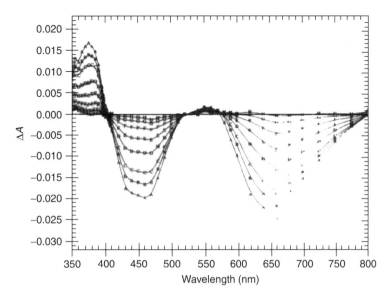

Figure 6.18 Transient absorption spectra, uncorrected for luminescence, observed after 532 nm excitation of a $[Ru(deebpy)(bpy)_2]^{2+}$-modified TiO_2 film in neat CH_3CN under argon. The apparent negative absorption change observed beyond 570 nm is due to emission. This part of the spectrum is included to illustrate the correspondence between absorption and luminescence kinetics. Reprinted with permission from C. A. Kelly, F. Farzad, D. W. Thompson and G. J. Meyer, *Langmuir*, **15**, 731 (1999). Copyright (1999) American Chemical Society

spectrum typical of solution-based ruthenium polypyridyl compounds is observed (see Figure 6.18). In addition, a strong emission band is observed at 650 nm. In particular, the presence of this emission signal is a clear indication that under these conditions injection into the semiconductor is not taking place. Supporting evidence for this observation comes from experiments carried out by Solbrand *et al.* [11] who investigated the magnitude of the photocurrent of TiO_2 sensitized by $[Ru(dcbpy)_2(NCS)_2]$ as a function of the Li^+ concentration. In these experiments, the photocurrent transients obtained for a derivatized TiO_2 surface were measured in 0.1 M KI in the presence of increasing amounts of Li^+. The results obtained in this study are shown in Figure 6.19. In this experimental arrangement, iodide is used to reduce the oxidized dye. The presence of this rather high iodide concentration should be more than sufficient to allow for a substantial photocurrent. However, as illustrated in curve (d) of Figure 6.19, no appreciable photocurrent is observed in pure 0.1 M KI. The photo-current is increased systematically upon the addition of lithium trifluoromethanesulfonate. Such an observation clearly indicates that charge injection only takes place in the presence of lithium ions. Significantly, irradiation of the same surface in pure lithium trifluoromethanesulfonate (without iodide) does not result in a photocurrent either. Thus, clearly the largest photocurrent is obtained by a combination of both iodide and lithium ions in the contacting electrolyte.

Therefore, both types of experiments indicate that the photoinduced charge injection into the semiconductor surface can be controlled through the lithium concentration. Other cations, such as calcium, barium and magnesium, show similar increases in the injection yield, while sodium and potassium are less effective. This

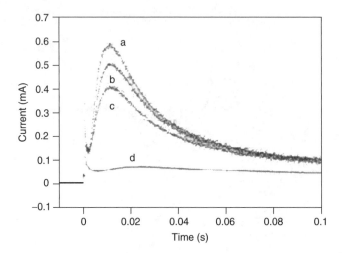

Figure 6.19   Photocurrent as a function of time for a 10.0 mm thick nanocrystalline TiO_2 electrode modified with [Ru(dcbpy)$_2$(NCS)$_2$], where the electrolyte is propylene carbonate containing 0.1 M KI with different concentrations of lithium trifluoromethanesulfonate: (a) 0.1 M; (b) 0.05 M; (c) 0.025 M; (d) 0 M. Reprinted with permission from A. Solbrand, A. Henningson, S. Sodergren, H. Lindstrom, A. Hagfeldt and S.-E. Lindquist, *J. Phys. Chem., B*, **103**, 1078 (1999). Copyright (1999) American Chemical Society

surprising behavior is, at this stage, not fully understood but it appears that the presence of Li$^+$ affects the energy of the acceptor levels in the semiconductor. It is proposed that in the presence of lithium the energy of these levels become more positive and that this improves the injection efficiency. Considering the known intercalating properties of lithium ions into TiO_2, such an explanation is not unreasonable. The presence of these ions may act as charge-compensating ions for the injecting electrons.

The observation that the injection efficiency can be modulated by changing the composition of the contacting solution opens up the possibility of controlling the competition between *electron injection into* the semiconductor surfaces and *intermolecular processes at* the semiconductor surfaces [12]. For example, in the absence of lithium, excitation of a TiO_2 surface modified with a 1:1 mixture of [Ru(dcbpy)(bpy)$_2$] (for the structure see Figure 6.7 above) and [Os(dcbpy)(bpy)$_2$] leads to a mixed ruthenium/osmium emission. Important also is the observation that the osmium-based emission is much stronger. This means that efficient energy transfer at rates $> 10^8$ s^{-1} is taking place at the semiconductor surface, or otherwise the ruthenium emission signal would be expected to be stronger. Therefore, clearly in the absence of lithium, intermolecular energy transfer across the semiconductor surface from the ruthenium to the osmium centers is faster then charge injection into the semiconductor surface. Upon addition of Li$^+$, interfacial electron transfer from both centers to the oxide surface is observed instead. These experiments clearly show that the interaction between the semiconductor surface and the molecular components can be manipulated by the nature of the contacting solution. In neat acetonitrile, the photophysics of the compounds observed in solution is maintained and electron injection into the TiO_2 surface does not take place. This is a very important result because it shows that the competition between charge injection into the solid substrate and radiative deactivation of the sensitizer can be manipulated by the composition of the contacting solution phase.

6.5.3 The Effect of Redox Potential

The picture developed so far for modified TiO$_2$ surfaces is that upon excitation of the dye, injection of the excited electron into the conduction band takes place. It is expected that this interfacial process will be influenced by the electrochemical potential of the surface. In this section, the effect of applying an external bias on the electron-injection process will be discussed [13]. The system to be considered will be a typical TiO$_2$ surface modified with [Ru(dcbpy)$_3$]. In solution, this compound emits strongly but upon immobilization injection into the semiconductor surface is possible instead of emission. There will be competition between these two processes as can be seen from Figure 6.11 above. In order to investigate the dynamics of these two processes, the emission intensity and the quantum yield for photosensitized electron injection were measured as a function of the externally applied bias. The results obtained are shown in Figure 6.20. In this experiment, the modified surface was excited at 500 nm, coinciding with the absorption maximum of the dye, which has a λ_{max} of 450 nm. Figure 6.20 shows that the most efficient electron injection is obtained when a positive bias is applied to the electrode. At a positive bias, the injection quantum yield is equal to 1 and no emission is observed. However, when a negative potential is applied the injection quantum yield is reduced and instead an emission signal is observed with a maximum at about 640 nm, which can be attributed to the immobilized dye. This shows that applying a potential to the electrode surface influences the relative importance of the recombination reaction and electron injection. By applying a voltage to an electrode, the Fermi level of the semiconducting layer will be affected. The results obtained are consistent with

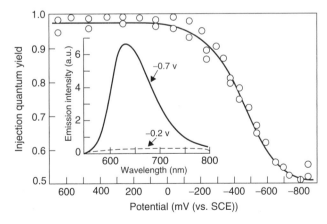

Figure 6.20 The effect of the electric potential on the quantum yield of photoinduced electron injection from immobilized [Ru(dcbpy)$_3$] into the conduction band of a TiO$_2$ substrate. The inset shows the luminescence intensity of the same modified surface at two different electrode potentials. An excitation wavelength of 500 nm was used, with an electrolyte of 0.2 M LiClO$_4$ at pH 3. Reprinted with permission from B. O'Regan, J. Moser, M. Anderson and M. Grätzel, *J. Phys. Chem.*, **94**, 8720 (1990). Copyright (1990) American Chemical Society

the filling up of electron traps when a negative potential, close to that of the flat-band potential (at pH 3, the band-gap for nanocrystalline TiO_2 (anatase) is -0.52 V (vs. SCE) is applied. When these traps are filled, charge separation is not possible since there are no acceptor states and recombination, resulting in emission, is observed. At more positive potentials, the conduction band-edge empties and the recombination reaction leading to emission is now no longer competitive with electron injection.

Another important observation is that changing the electrode potential does not create a significant electric field across the semiconductor film and electron transport through the film is occurring by diffusion rather than migration. This is explained by the absence of the formation of a significant depletion layer, as expected for nanoparticles of the size being used. As outlined in Section 6.2.2, the depletion layer of nano-sized particles is controlled by the size of the particle. For particles with a size of 10–30 nm, the difference in potential between the center and the surface of these particles can be estimated to be only 0.3 mV under conditions of maximum depletion and this is not big enough to create a significant electric field gradient.

6.6 Interfacial Supramolecular Assemblies

The competition between molecular-based and molecule–substrate interactions is one of the features that make supramolecular assemblies based on the combination of molecular components and solid substrates so exciting and also potentially useful from the applications point of view. The control issue is whether can one achieve long-lived charge separation between molecular components when immobilized on a surface, and from the fundamental perspective, can the interactions between the surface and molecular components be manipulated? In this section, the immobilization of molecular components consisting of at least two electroactive and/or photoactive units will be discussed. The intramolecular interactions within these 'dyads' in solution, as well as their behavior as interfacial supramolecular 'triads' when immobilized on nanocrystalline TiO_2, will be compared.

6.6.1 Ruthenium Phenothiazine Assembly

In the design of such supramolecular dyads, a number of prerequisites should to be considered. In order to obtain a true supramolecular assembly there needs to be substantial interaction between the different components of the assembly. There should, however, not be any substantial changes in the physical properties of these components, but their combination should lead to some new and novel characteristics. The combination of components should have properties over and above those of the separate components, without destroying their individual characters. Molecular dyads may, for example, contain a photosensitizer and an electron or energy donor or acceptor. An example of such a combination of a sensitizer and an electron donor is the Ru–PTZ dyad [14] shown in Figure 6.21. In this assembly, the ruthenium center is the sensitizer, S, and the phenothiazine

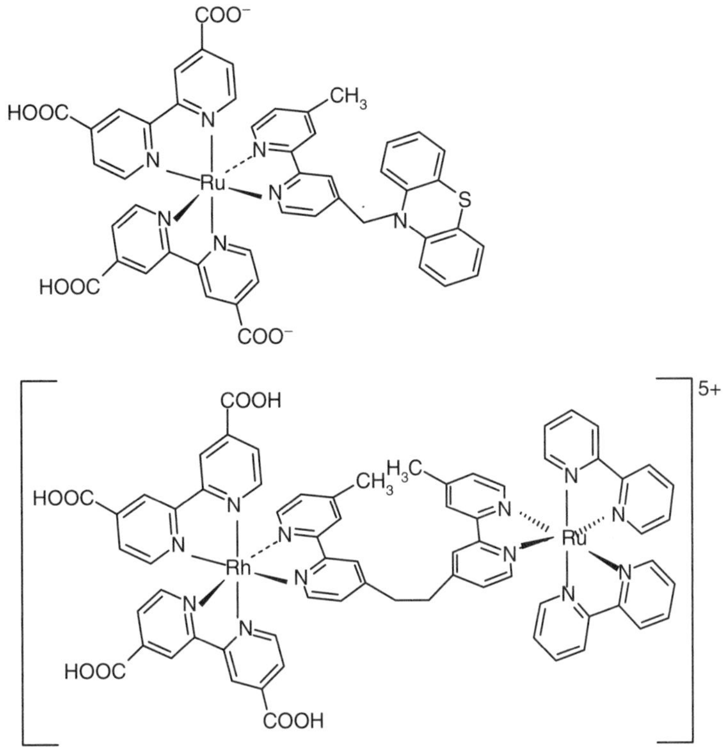

Figure 6.21 Molecular structures of the Ru–PTZ and Rh–Ru dyads

groupings acts as the electron donor, D. The photophysics of this dyad in solution is characterized by the following processes:

$$Ru(II)(dcbpy)-PTZ + h\nu \longrightarrow Ru(III)(dcbpy^-)^*-PTZ \tag{6.10}$$

$$Ru(III)(dcbpy^-)^*-PTZ \longrightarrow Ru(II)(dcbpy^-)-PTZ^+ \tag{6.11}$$

$$Ru(II)(dcbpy^-)-PTZ^+ \longrightarrow Ru(II)(dcbpy)-PTZ \tag{6.12}$$

In other words, upon excitation of the ruthenium center (Equation (6.10)), electron transfer from the donor PTZ grouping to the formally Ru(III) metal center takes place with a rate constant of about 2.5×10^8 s^{-1} (Equation (6.11)). The corresponding back-reaction (Equation (6.12)) is faster and cannot be resolved. Thus, in solution the PTZ grouping acts as an efficient reductive quenching agent but the charge separation is only shortly lived. Upon immobilization of this dyad on a semiconductor surface, new electron transfer pathways become available since the solid substrate becomes an active partner in the interfacial supramolecular triad. The processes that now need to be considered are shown in Figure 6.22. The aim of the immobilization of this assembly is to obtain a charge-separated state of the type TiO$_2$$^-$–Ru(II)–PTZ$^+$. Such a state can be reached by two different pathways, i.e. (a) charge injection which is followed by electron donation from the PTZ group to the ruthenium donor, and (b) electron donation from the reductive quencher which precedes injection. Laser flash photolysis studies have shown that excitation with visible light results in

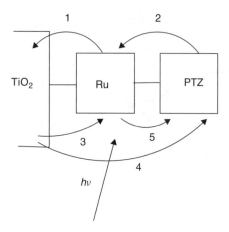

Figure 6.22 Schematic illustration of a sensitizer–donor assembly such as the Ru–PTZ system immobilized on TiO_2

the formation of Ru(III)(dcbpy$^-$)–PTZ and that fast injection from this ruthenium polypyridyl excited state into TiO_2 occurs as the next step. Electron transfer from the PTZ unit to the ruthenium center then sets up the charge-separated state. The important result is that the recombination reaction ('4') takes place at a rate of 3.6×10^3 s^{-1}. This recombination is three orders of magnitude *slower* than that observed for the model compound [Ru(dcbpy)$_2$(dmbpy)](dmbpy = 4,4'-dimethyl-2,2'-bipyridyl). For this compound, which does not contain a phenothiazine moiety, a value of 3.6×10^6 s^{-1} was obtained for process '3'. The charge separation obtained is also many orders of magnitude larger than that observed for the unsupported dyad in solution.

6.6.2 Rhodium–Ruthenium Assembly

In the example discussed above, the heterotriad consists of a photosensitizer and an electron donor. In the following example, a ruthenium polypyridyl sensitizer is combined with an electron acceptor, in this case a rhodium(III) polypyridyl center [15]. The structure of this dyad is shown in Figure 6.21 above. The absorption characteristics of the dyad are such that only the ruthenium moiety absorbs in the visible part of the spectrum. Irradiation of a solution containing this ruthenium complex with visible light results in selective excitation of the Ru(II) center and in an emission with a λ_{max} of 620 nm. This emission occurs from the ruthenium-polypyridyl-based triplet MLCT level, the lifetime of which is about 30 ns. This lifetime is very short when compared with the value of 700 ns obtained for the model compound [Ru(dcbpy)$_2$dmbpy)], which does not contain a rhodium center. Detailed solution studies have shown that this rather short lifetime can be explained by fast oxidative quenching by the Rh center as shown in the following equation:

$$Rh(III)-{}^*Ru(II) \longrightarrow Rh(II)-Ru(III) \tag{6.13}$$

This process is followed by a back electron transfer which is so fast that the charge-separated state cannot be observed. Compared to the Ru–PTZ example discussed

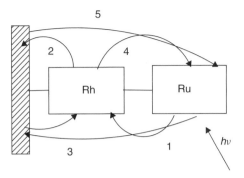

Figure 6.23 Schematic illustration of an interfacial supramolecular assembly incorporating a Ru–Ru diad, showing the potential electron transfer pathways in the interfacial supramolecular triad TiO_2–Ru–Ru

above, the arrangement chosen for this assembly is different in that the sensitizer is not directly bound to the surface. Therefore, an important question is whether electron donation from the sensitizer to the semiconductor surface can occur, and if it does, by what mechanism? The possible pathways for the electron transfer processes are shown in Figure 6.23.

When the interfacial supramolecular triad is irradiated in the presence of I^- under solar cells conditions, appreciable photocurrents are obtained. The profile of the photoaction spectrum shows clearly that photoinjection into TiO_2 takes place upon excitation of the ruthenium center. However, the IPCE values obtained are lower than those observed for the model compound, thus suggesting that injection is less efficient in the heterotriad. Of major interest is the mechanism for charge injection. Two different pathways can be envisaged. First, the charge injection may be a two-step process and takes place via the rhodium center as shown in the following equations:

$$TiO_2-Rh(III)-^*Ru(II) \longrightarrow TiO_2-Rh(II)-Ru(III) \tag{6.14}$$

$$TiO_2-Rh(II)-Ru(III) \longrightarrow TiO_2(e)-Rh(III)-Ru(III) \tag{6.15}$$

Secondly, direct injection from the ruthenium center to the TiO_2 surface needs to be considered, as follows:

$$TiO_2-Rh(III)-^*Ru(II) \longrightarrow TiO_2(e)-Rh(III)-Ru(III) \tag{6.16}$$

In this mechanism, the rhodium moiety is bypassed and injection will not be facilitated by the carboxy linker. It is not straightforward to differentiate between these two mechanisms. It can be argued that if the injection takes place via the rhodium center then the excited-state properties of the immobilized dyad should be similar to those of the dyad in solution. A comparison of these properties can be made on the basis of their emission decays and since the emission lifetimes for the dyad and the triad are very similar it appears that injection is taking place via the two-step pathway outlined in Equations (6.14) and (6.15). However, a detailed analysis of transient absorption data shows that the situation is more complicated. The transient absorption spectra under consideration are given in Figure 6.24. The

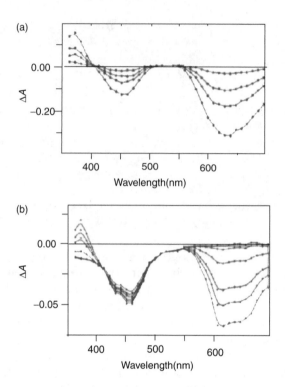

Figure 6.24   Time-resolved transient absorption spectra in 0.1 M LiClO$_4$/acetonitrile, at a λ_{exc} of 532 nm: (a) Ru(III)–Ru(II) in solution, with transients taken at 0, 20, 40 and 70 ns; (b) the TiO$_2$–Rh(III)–Ru(II) triad, with transients taken at 0, 10, 20, 50, 100, 500 and 2000 ns. Reprinted with permission from C. J. Kleverlaan, M. T. Indelli, C. A. Bignozzi, L. Pavanin, F. Scandola, G. M. Hasselmann and G. J. Meyer, *J. Am. Chem. Soc.*, **122**, 2840 (2000). Copyright (2000) American Chemical Society

positive absorbance at 370 nm and the bleaching at 450 nm are as already discussed earlier indicative of the formation of the dcbpy radical anion. The apparent bleaching at 650 nm can be interpreted as the emission. Importantly, for the model compound [Ru(dcbpy)$_2$(dmbpy)], no emission is observed upon immobilization on the surface and no evidence fore the formation of the dcbpy-based radical anion is observed. The spectral features observed for this mononuclear compound are similar to those shown above for [Ru(dcbpy)$_2$(bpzt)] in Figure 6.17. By comparison of the two sets of spectra presented in Figure 6.24, it is clear that although they are similar they are by no means identical. The ratio between the positive absorption and the bleaching at 450 nm is smaller. In addition, at 395 nm, where no spectral changes are expected, a definite bleaching is observed. This clearly indicates that another species is formed upon excitation of the ruthenium center. This can be explained by assuming that in addition to the ruthenium-based excited states, oxidized Ru(III) centers are also formed upon excitation. These states are created by direct injection from the ruthenium moiety into TiO$_2$, as shown in Equation (6.16). It is important to note that while the direct injection process is clearly very fast, since it takes place on the same time-scale as the laser flash, the lifetime of the excited state in the triad is the same as that found for the dyad in solution. This suggests that the two observable exited state decay and direct injection processes are not coupled and must involve structurally different triads. Considering the flexible nature of the linker (See Figure 6.21 above), the presence of different orientations on the surface is not unexpected. Careful analysis of the data show that about 35 % of the triads inject directly into the semiconductor surface. This is an important observation since there is no direct link between the ruthenium component and the TiO$_2$ surface.

For the assemblies involved in the two-step process, the excited state decays by intramolecular electron transfer, with a time constant of 30 ns. The initial intramolecular electron transfer step (Equation (6.14)) takes place with unit efficiency. The lifetime of the charge-separated state which is formed depends on competition between charge recombination and further charge separation. In this case, charge separation into TiO$_2$ has an efficiency of 40 %, while the remainder of the state recombine to the ground state. This relatively low injection value may well be responsible for the rather low photocurrent observed for the heterotriad. Figure 6.25 shows a general summary of the processes taking place upon excitation of the heterotriad.

Apart from the injection process, the nature of the back-reaction is also of importance. Unfortunately, however, the complexity of this process and its strong dependence on the laser power prevents a detailed analysis of the data. However, by comparing the rates of the back-reaction of the non-rhodium-containing heterodyad (Equation (6.17)) with that of the heterotriad (Equation (6.18)), a substantial increase in the duration of the charge-separation is obtained:

$$TiO_2(e)-Ru(III) \longrightarrow TiO_2-Ru(II) \tag{6.17}$$

$$TiO_2(e)-Rh(III)-Ru(III) \longrightarrow TiO_2-Rh(III)-Ru(II) \tag{6.18}$$

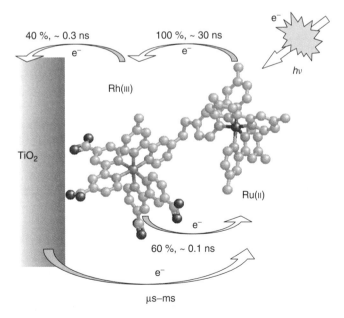

Figure 6.25 Summary of the photoinduced electron transfer processes at a TiO$_2$ surface modified with a Ru(III)–Ru(II) dyad. Figure kindly provided by Professor F. Scandola

Increasing the distance between the donor and acceptor sites and, in particular, the introduction of a solid substrate charge clearly stabilizes the charge-separated state.

6.6.3 Ruthenium Osmium Assembly

In the Rh–Ru example discussed above, the presence of a flexible linker leads to two independent processes. In the next case study, a very rigid ruthenium–osmium dinuclear complex [16] is considered (see Figure 6.26). Before dealing with the photophysical aspects of the assembly, the synthetic approach taken in the preparation of this and other dinuclear structures will be discussed briefly. The Ru–Os dinuclear compound was prepared in a two-step process, as shown in Figure 6.26. It is well documented that in the mononuclear compounds obtained from Hbpt, this ligand is coordinated to one of the pyridine rings and to the N2 position of the central triazole ring. A small amount of N4-coordinated material, together with some dinuclear material, are removed by chromatographic techniques. Thus, since the ruthenium carboxy grouping is coordinated first, the Ru–Os complex is bound to the titanium oxide surface by a carboxy-containing metal center coordinated to a pyridine ring and the N2 nitrogen of the 1,2,4-triazole ring. Consequently, the Os(bpy)$_2$ moiety is bound to the second pyridine ring and to the N4 atom of the triazole ring. In a typical reaction, Na$_3$[Ru(dcb)$_2$(bpt)]·3H$_2$O (4×10^{-4} M) was reacted under reflux with [Ru(bpy)$_2$(Cl$_2$)]·2H$_2$O (4×10^{-4} M) in ethanol/water (2:1) for 8 h. Subsequently, the product was precipitated by lowering the pH to 2.7 with HCl. Purification was achieved by column chromatography with Sephadex LH20 resin. Isolation of the product after chromatography was achieved by adjusting the pH with HCl as before, but now in the presence of NH$_4$PF$_6$. The compounds obtained show well-defined ^{1}H NMR spectra and the formation of the dinuclear species can be confirmed by mass spectrometry. The elemental composition of such compounds is not necessarily straightforward. Because of the presence of various acidic protons, the exact composition of the compounds in the solid state is difficult to determine.

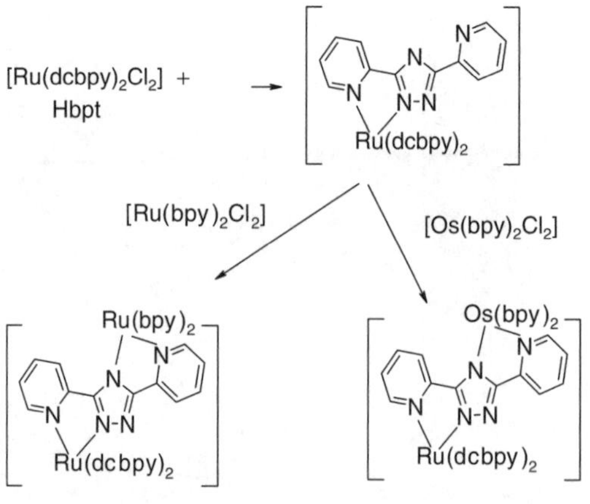

Figure 6.26 Molecular structure and the synthetic pathway for the Ru–Os diad. Reprinted with permission from A. C. Lees, C. J. Kleverlaan, C. A. Bignozzi and J. G. Vos, *Inorg. Chem.*, **40**, 5343 (2001). Copyright (2001) American Chemical Society

In this case, the best results were obtained by precipitating the dinuclear metal complexes as the PF_6 salts. Elemental analysis suggests, as already indicated above, that the products are isolated as sodium salts such as $Na_3[Ru(dcb)_2(bpt)]\cdot 3H_2O$ for the mononuclear precursor and $Na_3[Ru(dcb)_2(bpt)Os(bpy)_2](PF_6)_2\cdot 3H_2O$ for the Ru–Os dyad. The number of sodium atoms incorporated is, however, not always the same but appears to depend on the manner in which the materials are precipitated. Nevertheless, the uncertainty in the composition of the compounds in the solid state does not affect the measurements, since in all cases the protonation state of the compounds in solution is controlled by the pH of the solutions.

In aqueous solution at pH 7, the Ru–Os compound shows a ruthenium-based MLCT transition in the 450 nm region and an additional osmium-based absorption feature at 650 nm. This latter absorption can be assigned to a transition from the osmium center to the ^{3}MLCT state. No major shift in the absorption features of the compound is observed upon binding to the TiO_2 surface. For the Ru–Os diad, a single emission signal is observed in aqueous solution (pH 7) with a λ_{max} of 770 nm. The energy maximum observed, together with its independence of pH, confirms this emission as being osmium-based. Since the ruthenium moiety contains acidic carboxy groupings, emission from this center will depend on the pH of the solution. This assignment is in agreement with the behavior observed for many other ruthenium–osmium dinuclear compounds and indicates energy transfer from the ruthenium to the osmium center after excitation of the former. However, the interfacial supramolecular TiO_2–RuOs assembly does not show any emission. Transient absorption spectra obtained for the Ru–Os assembly in solution, and on TiO_2, after excitation at 532 nm, are shown in Figure 6.27. In solution (Figure 6.27(a)), bleaching in the 500 nm region and at 600 nm are observed, while on TiO_2 (Figure 6.27(b)), bleaching signals are observed at 380, 500 and 600 nm.

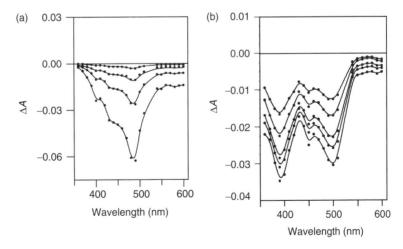

Figure 6.27 Transient absorption spectra of the Ru–Os assembly: (a) in solution (pH 7), at τ_d values of 0, 15, 30 and 50 ns, respectively ($\tau = 20$ ns), with $\lambda_{exc} = 532$ nm and $P = 6.5$ mJ per pulse; (b) immobilized at TiO_2, at τ_d values of 10, 250, 500, 2000 and 5000 ns, respectively, with $\lambda_{exc} = 532$ nm and $P = 0.5$ mJ cm^{-2}. Reprinted with permission from A. C. Lees, C. J. Kleverlaan, C. A. Bignozzi and J. G. Vos, *Inorg. Chem.*, **40**, 5343–5349 (2001). Copyright (2001) American Chemical Society

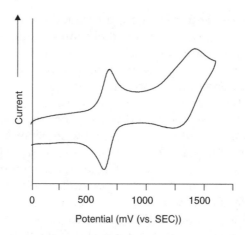

Figure 6.28   Cyclic voltammogram (voltage vs. SCE) of the TiO_2–Ru–Os system, with a surface coverage, Γ, of ~1.0×10^{14} molecules cm^{-2}, measured in 0.3 M $LiClO_4$/acetonitrile, at a scan rate of $v = 50$ mV s^{-1}. Reprinted with permission from A. C. Lees, C. J. Kleverlaan, C. A. Bignozzi and J. G. Vos, *Inorg. Chem.*, **40**, 5343 (2001). Copyright (2001) American Chemical Society

Electrochemical studies show that no major changes in the electrochemistry are observed when transferring the compounds from the solution phase to the TiO_2 surface. Since Os–polypyridyl complexes tend to have less positive M(II/III) redox processes than their ruthenium analogues, the oxidation at 0.66 V can be assigned to the Os center, while the process at 1.37 V is assigned to the Ru(II/III) redox process (see Figure 6.28). For the Ru–Os system, the intrinsic difference between the redox potentials of the two centers (ignoring metal–metal interactions) is estimated at about 400 mV.

Electron injection

A comparison of the absorption spectra obtained for the heterotriad with those of the solution-based dyad indicates that there are no major changes in the absorption features upon attachment to the semiconductor surface. However, a comparison of the two sets of spectra shown above in Figure 6.27 clearly indicates that the excited-state behavior of the interfacial supramolecular triad is substantially different from that of the molecular dyad in solution. At TiO_2, strong bleaching is observed at 380 and 500 nm for the Ru–Os assembly. This is assigned to bleaching of the Os(II) MCLT bands. Such a result indicates that injection into the oxide surface is very efficient, thus resulting in reduced TiO_2 and an oxidized dye. The absence of emission from the Ru–Os diad is initially surprising. The design of the electronic levels for the dyad is such that the energy transfer pathway leads away from the TiO_2 surface to the osmium center. The absence of an emission from the osmium center indicates that injection from the osmium moiety to the oxide surface is faster than energy transfer from the ruthenium to the osmium center or emission.

This behavior can be described by the following reaction sequence. Since both metal centers have a significant absorption at 532 nm, excitation at this wavelength will yield two different excited species, as follows:

$$\text{TiO}_2-\text{Ru(II)Os(II)} \longrightarrow \text{TiO}_2-{}^*Ru(II)\text{Os(II)} + \text{TiO}_2-\text{Ru(II)}{}^*Os(II) \qquad (6.19)$$

At this stage, the behavior observed for the mixed metal triad becomes significantly different from that observed for the solution-based dyad. For the molecular dyad in solution, a well-defined osmium-based emission is observed, while for the interfacial supramolecular triad no emission is detected. The transient absorption data indicate that injection by both centers of the dyad into the oxide surface is fast (<10 ns) (see Equations (6.20) and (6.21) below). An alternative process for the reaction shown in Equation (6.20) is energy transfer to the osmium moiety rather than into the semiconductor, as observed for the Ru–Os diad in solution, so resulting in $\text{TiO}_2-\text{Ru(II)}{}^*Os(II)$. Since no emission is observed for the assembly, it appears that this process is unable to compete with charge injection. More importantly, the formation of the charge-separated state, $\text{TiO}_2-\text{Ru(II)Os(III)}$, in the reaction shown in Equation (6.21) is not trivial. Since neither energy nor electron transfer from ${}^*Os(II)$ to the ruthenium center is energetically favorable, the absence of any osmium-based emission indicates that remote injection into the semiconductor from the osmium center is taking place:

$$\text{TiO}_2-{}^*Ru(II)\text{Os(II)} \longrightarrow \text{TiO}_2(e)-\text{Ru(III)Os(II)} \qquad (6.20)$$

$$\text{TiO}_2-\text{Ru(II)}{}^*Os(II) \longrightarrow \text{TiO}_2(e)-\text{Ru(II)Os(III)} \qquad (6.21)$$

The redox chemistry suggests that for both of these processes the final species is expected to be $\text{TiO}_2(e)-\text{Ru(II)Os(III)}$. This can be formed from the Ru(III) intermediate produced in the reaction shown in Equation (6.20), via the following reaction:

$$\text{TiO}_2(e)-\text{Ru(III)Os(II)} \longrightarrow \text{TiO}_2(e)-\text{Ru(II)Os(III)} \qquad (6.22)$$

What is particularly attractive in the Ru–Os diad system is that irrespective of the exact nature of the photoinduced electron injection, the electrochemical properties of the dyad ensures that the species formed will be $\text{TiO}_2(e)-\text{Ru(II)Os(III)}$. As a result, all irradiated sites will react in a similar manner and the photoinduced electron injection can be illustrated as shown in Figure 6.29. More importantly, all irradiated molecular components yield the same photoproduct.

Charge separation

One of the aims of studies involving solid components is to reduce the rate of back-reaction. The charge-recombination reaction expected in the absence of a solution-based redox couple is shown as follows:

$$\text{TiO}_2(e)-\text{Ru(II)Os(III)} \longrightarrow \text{TiO}_2-\text{Ru(II)Os(II)} \qquad (6.23)$$

The recombination behavior of this immobilized dyad is illustrated in Figure 6.30, which shows the time-resolved transient absorption spectra both without and in the presence of I^-/I_3. This figure shows clearly that recombination in the absence of the redox relay is slow and takes place on the microsecond time-scale.

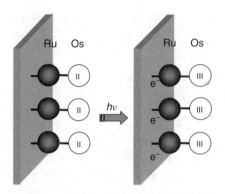

Figure 6.29 Illustration of the photoinduced phase-separation for a TiO_2–Ru–Os interfacial supramolecular triad. Reprinted with permission from A. C. Lees, C. J. Kleverlaan, C. A. Bignozzi and J. G. Vos, *Inorg. Chem.*, **40**, 5343–5349 (2001). Copyright (2001) American Chemical Society

Considering that the injection is faster than 10 ns, i.e. faster than the excited-state lifetime, a considerable degree of charge separation is obtained in these interfacial supramolecular triads. As shown for other systems, the incorporation of a solid substrate is very beneficial in this respect. Particularly surprising at first instance is the slow recombination reaction in the presence of the redox relay. In the presence of the I^-/I_3^- redox couple, the following reaction is expected:

$$TiO_2(e)-Ru(II)Os(III) + I^- \longrightarrow TiO_2(e)-Ru(II)Os(II) + I_3^- \qquad (6.24)$$

However, Figure 6.30(b) shows that although some recombination is taking place, the above reaction is slow and far from complete on the microsecond time-scale.

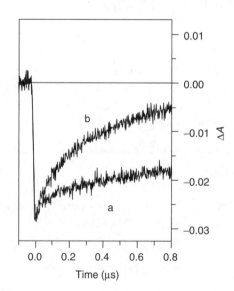

Figure 6.30 Microsecond decay kinetics, measured at 480 nm, of the TiO_2–Ru–Os system in the presence of (a) MeCN/0.3 M $LiClO_4$, and (b) MeCN/0.3 M LiI ($\lambda_{exc} = 532$ nm; $P = 0.3$ mJ cm^{-2}). Reprinted with permission from A. C. Lees, C. J. Kleverlaan, C. A. Bignozzi and J. G. Vos, *Inorg. Chem.*, **40**, 5343–5349 (2001). Copyright (2001) American Chemical Society

This is considerably different from the recombination reaction with, for example, typical ruthenium dyes. This slow re-reduction of the dyad is explained by the low redox potential of the osmium center, the value of 0.66 V (vs. SCE) observed, points to a small driving force for the redox process. This observation is important for the design of dyes for solar cell applications. Osmium compounds have very attractive absorption features, which cover a large part of the solar spectrum. However, their much less positive metal-based oxidation potentials will result in a less effective re-reduction of the dyes based on that metal and this will seriously affect the efficiency of solar cells. In addition, for many ruthenium-based dyes, the presence of low energy absorptions, desirable for spectral coverage, is often connected with low metal-based redox potentials. This intrinsically hinders the search for dyes which have a more complete coverage of the solar spectrum. Since electronic and electrochemical properties are very much related, a lowering of the LUMO–HOMO distance also leads to a less positive oxidation potential.

In a more general sense, these observations show that upon immobilization of photoactive compounds onto a solid substrate a substantial difference is detected between the photophysical processes observed for the heterotriad and the dyad in solution. More importantly, direct injection from those moieties not directly bound to the oxide surface can be efficient – this is always fully realized and such an observation is important for the further development of real devices. As a result of this through-space interaction, no osmium-based emission is observed and injection from both the ruthenium and the osmium centers is faster than the laser pulse. An interesting observation is also that upon irradiation of the heterotriad TiO_2–Ru–Os, only one final product, i.e. $TiO_2(e)$–Ru(II)Os(III), is obtained. In view of the potential of these modified surfaces as potential molecular devices, this is an important feature. The presence of a rigid structure rather than a flexible one, as observed in the Ru–Rh case, clearly leads to a more uniform behavior.

6.7 Electrochemical Behavior of Nanocrystalline TiO₂ Surfaces

Because of their transparent nature, TiO_2 films are attractive candidates as substrates for developing electrochromic devices. However, since TiO_2 is a semiconductor material it has a electronic band-gap and it is to be expected that the nanostructured surfaces will only be conductive when the redox potential of an adsorbed species is higher that the conduction band-edge. For TiO_2, the flat band potential in aprotic solvents such as acetonitrile is about -2 V (vs. NHE) and although the band-edge can be moved to more positive potentials by the addition of Li^+ or by changing the solvent to water, the limited use that can be made of this substrate at moderate potential remains. It is therefore quite surprising that a rather well-defined electrochemistry is observed for many absorbed species, such as certain triarylamines (for an example, see Figure 6.31) with redox potentials well into the band-gap. An example of a cyclic voltammogram for a triarylamine-modified nanocrystalline TiO_2 surface is shown in Figure 6.32. As the latter shows, the redox potential of this compound is well within the band-gap region and therefore no electrochemical signal would be expected.

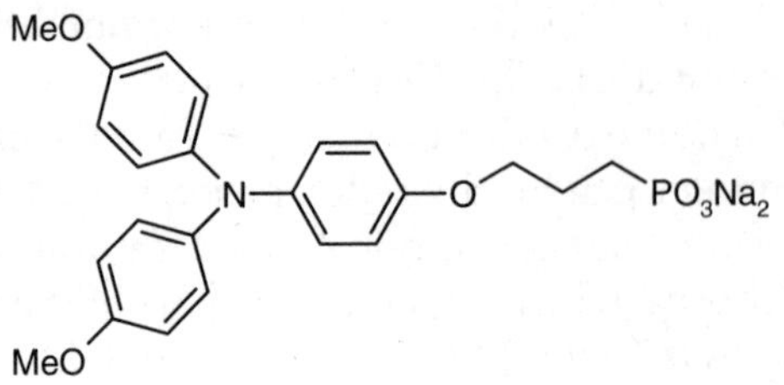

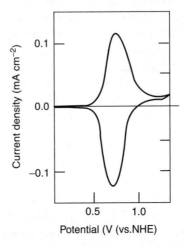

Figure 6.31 Molecular structure of a typical electroactive aromatic amine, with a redox potential well into the band-gap region

Figure 6.32 Cyclic voltammogram of a nanocrystalline TiO_2 electrode modified with a electroactive aromatic amine, measured at a scan rate of 2 mV s^{-1}. The electrolyte is acetonitrile containing 0.5 M lithium bis(trifluoromethylsulfonyl)imide. Reprinted with permission from P. Bonhote, E. Gogniat, S. Tingry, C. Barbé, N. Vlachopoulos, F. Lenzmann, P. Comte and M. Grätzel, *J. Phys. Chem., B*, **102**, 1498 (1998). Copyright (1998) American Chemical Society

A number of possible mechanisms could be responsible for this surprising behavior. First, electron transport could be carried by the semiconductor substrate. However, this is not very likely, since this would mean that electrons can be transported within the energy gap. In addition similar results are obtained on Al_2O_3, hence suggesting that the nanocrystalline substrate is not directly involved in the electrochemical process. This indicates that the charge is carried by the adsorbed triarylamine and is transferred to the underlying conducting ITO layer. The mechanism behind this process may either be via diffusion of the electroactive species to the back contact or via an electron-hopping process between the firmly bound redox species inside the monolayer.

Grätzel and co-workers [5] have studied the mechanism for this electron transport process. Their approach is based on the measurement of charge-transport rates for a range of modified surfaces using chronoabsorptiometry. In this technique, the absorbance of the modified surface is followed as a function of time upon oxidation (or reduction) of the redox center. A typical example of the absorption changes observed upon oxidation of the triarylamine shown in Figure 6.31 is presented in

Figure 6.33. From these data, the diffusion coefficient, D_{app}, can be obtained from the linear part of this curve by using the Cottrell equation, as follows:

$$\Delta A = 2\Delta A_f D_{app}^{1/2} t^{1/2}/d\pi^{1/2} \tag{6.25}$$

where ΔA is the absorbance change observed and ΔA_f is the total absorbance change observed upon oxidation or reduction or the redox-active species in the monolayer. From this equation, charge transport values of the order of 10^{-12} m^2 s^{-1} are obtained. This diffusion coefficient was shown to be strongly dependent on the loading of the redox species in the monolayer. This is shown in Figure 6.34, which presents the data obtained for mixed monolayers containing various ratios of the triarylamine (see Figure 6.31) a redox-inactive molecule. It can be seen that

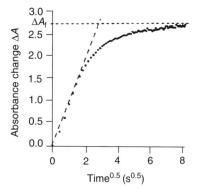

Figure 6.33 Transient absorbance plot for the modified electrode system described in Figure 6.32. Reprinted with permission from P. Bonhote, E. Gogniat, S. Tingry, C. Barbé, N. Vlachopoulos, F. Lenzmann, P. Comte and M. Grätzel, *J. Phys. Chem., B*, **102**, 1498 (1998). Copyright (1998) American Chemical Society

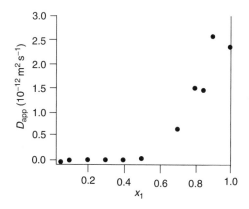

Figure 6.34 Variation of the diffusion coefficient, D_{app}, following a potential step from 0.2 to 1.0 V for the oxidation of a triarylamine in a mixed monolayer with a redox-inactive compound as a function of the mole fraction x_1. The electrolyte is 1-ethyl-3-methylimidazolium bis(trifluoromethylsulfonyl)imide. Reprinted with permission from P. Bonhote, E. Gogniat, S. Tingry, C. Barbé, N. Vlachopoulos, F. Lenzmann, P. Comte and M. Grätzel, *J. Phys. Chem., B*, **102**, 1498 (1998). Copyright (1998) American Chemical Society

a minimum surface coverage of about 50 % is needed to allow charge transport to occur. This is direct evidence for a charge percolation mechanism, where electron hopping between fixed molecules within the monolayer is rate-determining. Charge transport is reduced when the distance between the redox centers is increased and finally a cut-off distance is observed at which electron hopping is no longer possible. This observation indicates that an efficient lateral electron transfer process between redox sites, which have a high self-exchange rate, can explain the unexpected electrochemical behavior observed at these nanocrystalline semiconductor surfaces. In this process, the electron finally reaches the ITO back contact.

The fact that electrochemistry can be observed at nanocrystalline ITO/TiO$_2$ surfaces opens the way for their potential applications in electrochromic devices and as electrocatalysts, and also may lead to the development of novel sensing devices. In the following section, an approach to the development of an electrochromic device based on a modified nanocrystalline TiO$_2$ system will be discussed.

6.7.1 Electrochromic Devices

The inherent large surface areas of nanocrystalline films and the related high absorbances that can be obtained when using only monolayer coverages makes them particularly useful for device production. These include electrochromic windows, rear-view mirrors, and in high-contrast dynamic displays such as computer and television screens. However, the commercial application of many proposed systems has been limited by the relatively slow response times. Systems that have been proposed as potential devices include thin metal oxide films and solutions containing redox couples which change color upon oxidation or reduction. Thin films of polymer films containing such redox couples have also been studied. Devices based on these three approaches suffer from slow response times since the color changes are controlled by diffusion processes. The color changes that can be obtained for thin metal oxide films such as TiO$_2$, WO$_3$ and V$_2$O$_5$ are related to the injection of electrons, coupled with ion intercalation. In order to obtain substantial changes in the absorbances of the films this intercalation process must extend into the bulk of the material, a process which is slow, with switching times of typically tens of seconds. For devices containing solution-based redox couples, the switching is related to the diffusion of the species to the electrode surface. For devices containing thin polymer films incorporating chromophores, electron transport through the polymer layer becomes rate-limiting and again the response times of such devices is in the seconds time domain. In view of the fast photoinduced charge injection observed for the light-driven devices discussed above, the absorption of chromophores on nanostructured surfaces seems very attractive.

The results obtained for the solar cell discussed above suggest a strong interaction between the chromophore and the metal oxide layer, a large surface area, thus yielding large absorbances, and an efficient charge-separation upon injection. Several studies have indeed been carried out in an attempt to utilize the potential of nanocrystalline metal oxides as substrates for electrochromic devices. A particularly interesting approach has been reported by Fitzmaurice and co-workers [17]. These authors have constructed an electrochromic device based on the combination of two nanostructured metal oxide electrodes, where one of these surfaces acts as the negative electrode and the other as the cathode. The anode is based on nanocrystalline TiO$_2$, modified with a monolayer of a phosphate derivative of a viologen dye, while the positive electrode consists of antimony-doped tin dioxide (Sb–SnO$_2$). The latter material is coated with a single absorbed layer of a phosphoric acid derivative of phenothiazine. The structures of two of these electrochromic dyes are shown in Figure 6.35. The two electrodes are contacted by an electrolyte of 0.2 M LiClO$_4$ in γ-butyrolactone. Importantly, this electrolyte does not contain a redox relay. A schematic diagram of the cell used in these experiments is shown in Figure 6.36. In this assembly, the redox process is completely controlled by the absorbed chromophores. Viologens have been used in many studies as chromophores for electrochromic devices. In their neutral states, these molecule are colorless but upon one-electron oxidation strong blue colors are obtained. Phenothiazine is yellow in the neutral state but turns red upon oxidation. The device obtained in this way is yellow when no potential is applied. This yellow color is due to the phenothiazine, while the viologen is colorless in its normal state. Upon applying a potential of 1.2 V to the viologen-modified TiO$_2$, this electrode is biased negatively with respect to the phenothiazine-modified electrode and a blue–red color is observed. The spectral changes observed are shown in Figure 6.37. The color change is a result

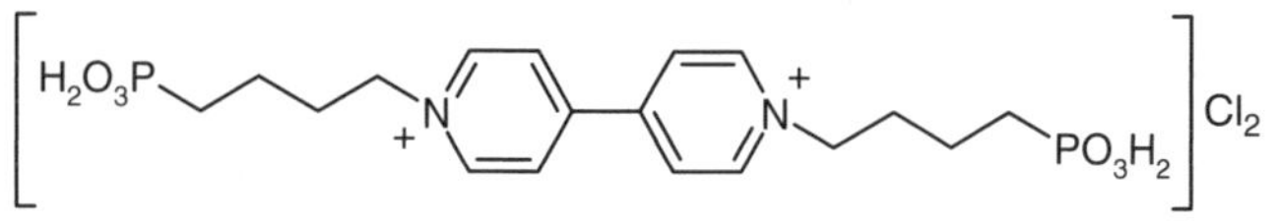

Bis(2-phosphonylethyl)-4,4'-bipyridinium dichloride

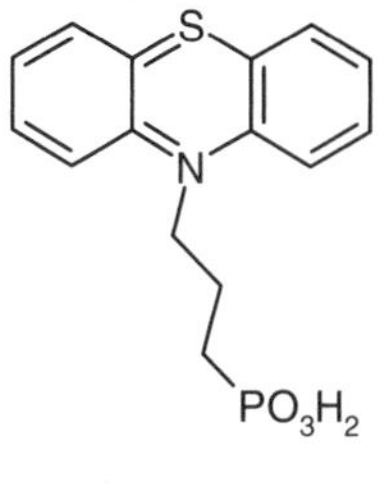

[(β-Phenothiazyl)propoxy]phosphonic acid

Figure 6.35 Molecular structures of two typical electrochromic dyes

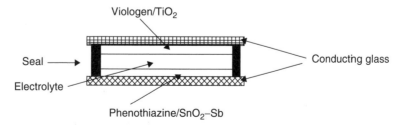

Figure 6.36 Schematic of the electrochromic cell device investigated by Fitzmaurice and co-workers [17]

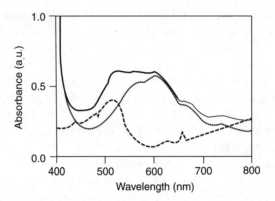

Figure 6.37 Optical absorption spectra of the electrochromic cell following the application of a negative bias of 1.2 V to the viologen-modified TiO$_2$ electrode: (——) optical spectrum of the cell; (······) contribution from reduced viologen molecules; (- - - -) contribution from oxidized phenothiazine molecules. Reprinted with permission from D. Cummins, G. Boschloo, M. Ryan, D. Corr, S. N. Rao and D. Fitzmaurice, *J. Phys. Chem.*, *B*, **104**, 11449 (2000). Copyright (2000) American Chemical Society

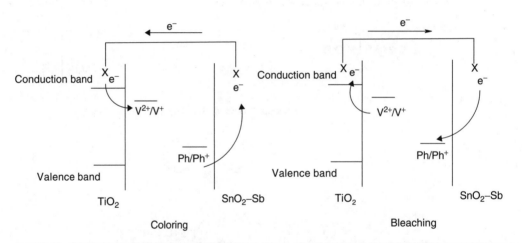

Figure 6.38 Electron flow diagram of the switching processes that take place for an electrochromic window based on viologen/TiO$_2$ and phenothiazine/SnO$_2$–Sb modified electrodes. Reprinted with permission from D. Cummins, G. Boschloo, M. Ryan, D. Corr, S. N. Rao and D. Fitzmaurice, *J. Phys. Chem.*, *B*, **104**, 11449 (2000). Copyright (2000) American Chemical Society

of two processes occurring simultaneously, namely the single-electron reduction of the viologen unit and the oxidation of the phenothiazine grouping. This is illustrated by the curves presented in Figure 6.37, which show the contribution of the individual electrodes to the overall electrochromic behavior of the device. An interesting observation is that the color obtained is very 'well-defined' since the number of oxidized and reduced molecules at the anode and cathode have to be the same. This is illustrated schematically in Figure 6.38, which shows the electron flow which takes place when an external bias is applied. Upon reversing the potential, the electron flow and the color change are reversed. Such a switching process can

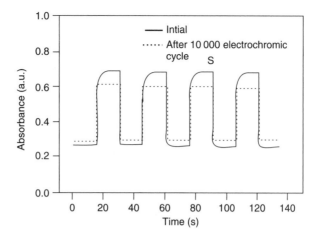

Figure 6.39 Changes in the optical absorption at 608 nm upon switching for an electrochromic device based on viologen-modified TiO_2 and phenothiazine-modified SnO_2–Sb (see text for further details). Reprinted with permission from D. Cummins, G. Boschloo, M. Ryan, D. Corr, S. N. Rao and D. Fitzmaurice, *J. Phys. Chem., B*, **104**, 11449 (2000). Copyright (2000) American Chemical Society

be repeated many times without serious degradation, and overall the stability of the device is very good. The most important observation that can be made with respect to this system is that the switching between the two states is very rapid. This is illustrated in Figure 6.39, which shows the absorption of the cell at 608 nm as a function of the applied potential. In this experiment, a voltage of 1.20 V is applied to the TiO_2 electrode, which biases this viologen-modified electrode negatively with respect to the phenothiazine-coated anode. This potential is applied for 15 s, followed by a voltage of 0.0 V for 15 s. The bleaching/coloring times observed in this experiment are of the order of hundreds of milliseconds, i.e. several orders of magnitude faster than those observed for conventional devices based on diffusing species. These findings clearly show that fast electrochromic systems can be obtained by immobilizing the electrochromic components at the electrode surfaces. The nanocrystalline nature of the electrode substrate allows for strong colors, while the fact that the electrochromic components are immobilized eliminates diffusion as being a rate-determining step. This, in turn, greatly increases the stitching speed of the device.

6.8 Alternative Semiconductor Substrates

The majority of the work carried out so far has concentrated on the use of nanocrystalline titanium oxide films. Much less attention has been paid to other potential substrates. Other semiconductor surfaces modified with sensitizers include ZnO, SnO_2, Fe_2O_3 and NiO. Like TiO_2, the first three of these are n-type semiconductors, while all four have very similar band-gaps and band-edge energies (see Figure 6.2 above). The behaviors of modified ZnO, SnO_2, and Fe_2O_3 surfaces are expected to be similar. [1] With some important differences, this is indeed observed. The photocurrents observed for sensitized SnO_2 surfaces are similar to those found for

TiO$_2$, although the IPCE values are considerably lower at about 1 %. Although the conditions at which this value was obtained had not been optimized, it seems likely that the relatively low injection rate observed for SnO$_2$ of $10^7 - 10^8$ s^{-1} plays an important role in this respect. In addition, for ZnO slow injection rates have been invoked to explain the relatively low photocurrents. It seems likely that the lower charge injection rate in this case is at least partly due to dye agglomeration resulting in more than monolayer coverage. The inefficiency of Fe$_2$O$_3$ has been explained by the recombination of charge carriers. It has been shown that this process can be reduced to some extent by the use of rod-like structure. The photocurrents obtained from films containing nano-sized rods rather than spherical particles proved to be up to a factor of 100 higher. [18] This observation has been explained in terms of improved electron transport along the rods because of the reduced number of grain boundaries. Such an observation is important since it shows that the morphology of the metal oxide film may affect the properties of the layer. However, more experiments are needed in this area.

It is to be expected that nickel oxide will behave quite differently. The band-gap given earlier in Figure 6.2 for this oxide is quite different to those observed for n-type semiconductors. At pH 7, the top of the valence band is found at about 0.54 V (vs. NHE), while the conduction band is expected at around −3 V. This material is expected to act as a p-type semiconductor and sensitization has to be approached quite differently since the high energy of the conduction band precludes the interaction with the excited state of a typical sensitizer with visible light. However, the position of the valence band allows for interaction with ground-state orbitals. This situation is illustrated schematically in Figure 6.40.

This diagram shows that NiO can be used as a photocathode rather than a photoanode, as is the case for semiconductors such as TiO$_2$, ZnO and SnO$_2$. For such a device, the HOMO of the dye should be located below the level of the top of the valence band, while the excited-state LUMO should be located above the redox potential of the A/A$^-$ redox couple. Upon excitation of the sensitizer D to form the

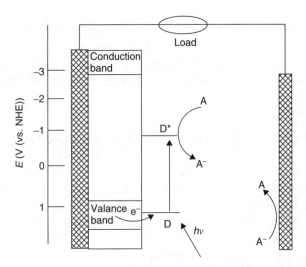

Figure 6.40 Schematic of a photovoltaic cell based on NiO modified with an electron donor and in the presence of the electron relay iodine/iodide

excited state D*, no injection from this excited state into the semiconductor surface is possible. Instead, the excited state may interact with the redox relay in solution. The oxidized ground state, D$^+$, is regenerated by electron *donation* from the valence band in what is in effect a *hole-transfer* process. The injected hole diffuses to the back contact and the reduced acceptor in solution is reoxidized at the counter electrode. This results in a cathodic photocurrent. Thus, in agreement with its p-type behavior, the semiconductor substrate acts as a donor rather than like an electron acceptor, as is the case for TiO$_2$. Based on this principle, photocathodes have been produced based on dyes such as tetrakis(4-carboxyphenyl)porphyrin and erythrosin B, with IPCE values of up to 3 % [19]. A disadvantage of NiO so far has been the small photocurrents and photovoltage. This latter is governed by the difference in potential between the valence band and the redox potential of the solution-based redox couple, which in the case of the iodine redox couple is about 0.1 V. This is in contrast with the higher value of the photovoltage obtained for sensitized TiO$_2$. In order to achieve an improved efficiency, redox relays with less positive redox potentials have to be developed.

6.9 Concluding Remarks

The investigations discussed in this chapter represent clear examples of the advantages that the introduction of a solid substrate has for the design of supramolecular assemblies. The solid-state component is addressable and, by a careful match of the electronic properties of the semiconductor substrate and the molecular components, a strongly interacting assembly is created. Because of the matching electronic properties of substrate and molecular component, photoinduced electron injection into the substrate is fast and a long-lived charge-separated state is obtained. The application of these assemblies as solar cells is a good example of how, by a careful combination of the substrate and the molecular component, real-life devices can be obtained.

It is somewhat surprising that electrochromic devices have also been reported. TiO$_2$ is not electroactive at potentials where the molecular components are. However, electron transport to the underlying ITO surface through an electron-hopping process drives the electrochemical process. A very fast electrochromic device has been developed, further suggesting considerable potential of these assemblies for commercial applications.

The investigation of modified TiO$_2$ surfaces is one of the best studied subjects in the area of surface-bound supramolecular assemblies. There are, however, quite a number of problems which still need to be addressed. The understanding of the recombination reactions is at present very limited. Time-resolved measurements have shown that the recombination reaction is a multi-exponential process. The reason for this is not yet understood, but is most likely related to the nanocrystalline nature of the semiconductor substrate. In order to gain more information about this issue, systematic studies on the effects of surface roughness and particle size are needed.

A variation of the surface structure will also be of great interest in the study of the electron transfer processes of surface-bound molecular dyads. As outlined above, the high porosity of the surfaces allows for through-space contact between the non-covalently bound redox couple of a molecular diad (for example, the osmium center in the studied Ru–Os system) and the semiconductor surface. This limits control of the electron transfer process and flatter surfaces are needed where such a contact is less favorable; this will allow for the development of more sophisticated ways in which the electron transfer can be controlled, and it should be possible to choose whether through-space electron transfer is advantageous or not. At present, surfaces with high surface areas are used, and some wider variation of roughness factors is greatly needed.

References

[1]   Hagfeldt, A. and Grätzel, M. (1995). *Chem. Rev.*, **95**, 49.

[2]   Hagfeldt, A. and Grätzel, M. (2000). *Acc. Chem. Res.*, **33**, 269.

[3]   Bignozzi, C.A., Argazzi, R. and Kleverlaan, C.J. (2000). *Chem. Soc. Rev.*, **29**, 87.

[4]   Brus, L.M. (1984). *Phys. Chem.*, **80**, 4003.

[5]   Bonhote, P., Gogniat, E., Tingry, S., Barbe, C., Vlachopoulos, N., Lenzmann, F., Comte, P. and Grätzel, M. (1998). *J. Phys. Chem., B*, **102**, 1498.

[6]   Hasselmann, G.M. and Meyer, G.J. (1999). *J. Phys. Chem., B*, **103**, 7671.

[7]   Ford, W.E, Wessels, J.M. and Rodgers, M.A.J. (1997). *J. Phys. Chem., B*, **101**, 7435.

[8]   Lees, A.C., Evrard, B., Keyes, T.E., Vos, J.G., Kleverlaan, C.J., Alebbi, M. and Bignozzi, C.A. (1999). *Eur. J. Inorg. Chem.*, 2309.

[9]   Hannappel, T., Burfeindt, B., Storck, W. and Willig, F. (1997). *J. Phys. Chem., B*, **101**, 6799.

[10]   Kelly, C.A., Farzad, F., Thompson, D.W. and Meyer, G.J. (1999). *Langmuir*, **15**, 731.

[11]   Solbrand, A., Henningsson, A., Södergren, S., Lindström, H., Hagfeldt, A. and Lindquist, S.-E. (1999). *J. Phys. Chem., B*, **103**, 1078.

[12]   Farzad, F., Thompson, D.W., Kelly, C.A. and Meyer, G.J. (1999). *J. Am. Chem. Soc.*, **121**, 5577.

[13]   O'Regan, B., Moser, J., Anderson, M. and Grätzel, M. (1990). *J. Phys. Chem.*, **94**, 8720.

[14]   Argazzi, R, Bignozzi, C.A., Heimer, T.A., Castellano, F.N. and Meyer, G.J. (1995). *J. Am. Chem. Soc.*, **117**, 11815.

[15]   Kleverlaan, C.J., Indelli, M.T., Bignozzi, C.A., Pavinin, L., Scandola, F., Hasselman, G.M. and Meyer, G.J. (2000). *J. Am. Chem. Soc.*, **122**, 2840.

[16]   Lees, A.C., Kleverlaan, C.J., Bignozzi, C.A. and Vos, J.G. (2001). *Inorg. Chem.*, **40**, 5343–5349.

[17]   Cummins, D., Boschloo, G., Ryan, M., Corr, D., Rao, S.N. and Fitzmaurice, D. (2000). *J. Phys. Chem., B*, **104**, 11449.

[18]   Beerman, N., Vayssieres, L., Lindquist, S.-E., and Hagfeldt, H. (2000). *J. Electronal. Chem.*, **147**, 2456–2461.

[19]   He, J., Lindström, H., Hagfeldt, A. and Lindquist, S.-E. (1999). *J. Phys. Chem., B*, **103**, 8940.

7 Conclusions and Future Directions

In this chapter, the future of interfacial supramolecular assemblies is considered. In the discussion, important scientific challenges are identified and potential applications in areas such as nanotechnology, biosystems and materials science are considered.

7.1 Conclusions – Where to from Here......?

The aim of this book is to provide the reader with an overview of interfacial supramolecular chemistry. Supramolecular assemblies of the kind considered in this text are truly interfacial, not only because they separate solid- and solution-phase components but also because they represent the 'junctions' where biology, chemistry, physics and engineering meet. In true interfacial supramolecular systems, individual moieties, e.g. the supporting surface and an adsorbed luminophore, interact co-operatively to produce a new function or property. In addition, these two- and three-dimensional structures remain an important step in the evolution of structure from discrete molecules, to interacting assemblies, and finally to solids. In this last chapter, the future of interfacial assemblies will be briefly considered. This discussion will focus on the possibility of integrating such assemblies into practical devices and the identification of the important scientific challenges.

7.2 Molecular Self-Assembly

Much of this book focuses on molecular self-assembly, which represents perhaps the most powerful method currently available for assembling atomically precise materials *en masse*. This approach is reminiscent of biological organisms which are composed of molecular building blocks, such as nucleic acids, proteins and phospholipids, and are equipped to assemble these components into extremely well-organized structures. Self-assembly has always been an integral part of science, but what is new is the idea of looking at these systems as a discrete set of processes that will become an important tool for miniaturization, novel devices and nanotechnology. In this context, molecular self-assembly continues to be widely publicized as a viable approach to the large-scale creation of molecular devices. In this book, we have seen that many elegant 'bottom-up' approaches exist for producing complex structures which are capable of performing a specific function, e.g. electronic switching or solar energy conversion. However, many significant challenges still remain in this field, in particular the interconnection of the micro/nanoscopic and everyday macroscopic worlds. Bottom-up approaches to solving this interconnection problem remain in their infancy while top-down approaches, e.g. addressing a small number of molecules or a single nanoparticle with a scanning tunneling microscope (STM), or by using such an instrument as a nanoplotter, present technical challenges that are likely to prevent their widespread commercial exploitation in the short term.

The following sections consider the likely future developments in specific areas of molecular electronics and considers the role that interfacial supramolecular assemblies may play in fabricating molecular-scale and molecular-based electronic devices.

7.3 Molecular Components and Nanotechnology

As embodied in the famous *Moore's law*, the power of semiconductor integrated circuits currently doubles approximately every 18 months. If this success is going to continue to be realized, engineers must find ways to cram ever more circuits into ever smaller wafers of silicon. However, current chip producing technologies are rapidly approaching a significant barrier. Microchips are manufactured by photolithography, whereby light is shone through a stencil of the circuit pattern. The light travels through lenses, which focus the pattern onto a silicon wafer covered with light-sensitive chemicals. When the wafer is exposed to acid, the desired circuitry emerges from the silicon. The problem is that creating smaller circuits requires shorter wavelengths of light. Current technology uses deep ultra-violet light to create state-of-the-art circuits approximately 100 nanometers wide. Interfacial supramolecular assemblies have two distinct roles to play in eliminating the barrier to further miniaturization of electronic circuits. First, the concept of single-molecule electrical devices has triggered a rising interest in the electrical properties of individual molecules, polymers and nanowires, in particular carbon nanotubes. These remarkable 'wires' with diameters on the molecular scale can behave like ballistic conductors showing quantized conductance. Moreover, they have been used as electrically active components in room-temperature field-effect transistors or chemical sensors. Carbon nanotubes, like other synthetic nanowires or nanoparticles, are promising functional building blocks of future molecular electronic devices. For example, as discussed in Chapter 5, nanoparticles are interesting as 'islands' in single-electron transistors (SETs). A second important approach to miniaturized electronics is to use interfacial supramolecular assemblies (ISAs) to create 'soft' lithographic approaches as a competitive technology to standard electron-beam lithography. In the soft lithography approach, a silicone polymer is replicated on a micro- or nanopatterned silicon wafer. The resulting flexible stamp is inked with a solution, i.e. a sol or emulsion of the chemical species to be printed. The method can be used for feature sizes ranging from a few micrometers down to approximately ten nanometers. Simple adhesion of the inked stamp onto a flat substrate transfers the pattern. Depending on the chemical nature of the ink (e.g. molecules, clusters, gels, etc.), the stamp (e.g. hydrophobic or hydrophilic) and the substrate (e.g. metals or silicon, oxides), the monolayer or multilayers will stay adsorbed in the predefined pattern. The application of conducting and redox-active polymers and of polymeric materials which can produce ordered three-dimensional structures in inks or stamps is worth investigating. Overall, the wide variety of substrate and inks available means that the approach can be used for applications ranging from building nanostructures to miniaturized chemical sensors.

Computer simulations of quantum mechanical electron transport through molecules and the molecular mechanics of manipulation processes are facilitating dramatic advances in STM micropositioning approaches. These studies have provided a better understanding of specific elements of molecular architecture that facilitate two-dimensional stabilization and non-destructive repositioning. In particular, the use of semi-flexible 'legs' with weak absorption characteristics mounted on a rigid 'chassis' has been found to be suitable for two-dimensional assembly operations. Recently, this approach has enabled bi-stable molecular conformations to be used as elements in molecular circuits. Simulations and experiments suggest that tunneling transmission factors can be modulated by factors up to 100 per 0.1 nm mechanical perturbation in molecular shape. Approaches of this kind will lead to new classes of electromechanical amplifiers based on rotation, translation and vertical manipulation of single molecules. It is perhaps important to note the high gain or amplification that can be achieved with systems of this kind, e.g. the electronic conductivity depends proportionately on the degree of molecular twisting. This situation contrasts with many electrochemically based switches where only two states exist, i.e. oxidized and reduced forms, and no amplification is obtained.

Apart from electron-driven and mechanical-switching devices, the development of optical switching devices has been widely considered. This development is driven by the fact that thousands of miles of fiber-optic cable are currently being laid under the streets of cities around the world, providing the superfast telecommunications network of the future. Optical networks use beams of light, rather than electrons, to carry data and in this sense, the 21st Century will be *photonic rather than electronic*. Today, most of these networks use optoelectronic switches to direct network traffic. These switches convert incoming light signals into electronic form, examine their network addressing, and then convert them back to optical signals before forwarding them to the appropriate node on the network. A significant difficulty in this area is that the bandwidth of the network is limited, not by the information-carrying capacity of the fibers themselves, but by the slow switching rates. The next generation of networks use photonic switches which eliminate this conversion step, giving faster performance and higher network capacity. However, this technology currently relies heavily on microscopic mechanical mirrors. Interfacial supramolecular assemblies are likely to be created that can switch optical signals reliably and with high throughput with no moving parts!

7.4 Biosystems

The essential structures of living organisms range in scale from molecular to micron dimensions. An important objective in supramolecular research is to understand how to manipulate matter on this critical length scale. It is reasonable to expect that this level of synthetic capability will revolutionize many aspects of current biotechnology, including medicine, diagnostics and biomaterials. Consequently, increasingly complicated self-organized structures will be built using biocomponents and used to better understand processes within biosystems. These sophisticated structures will lead to advances in sensors, environmental remediation, biocatalysis and biochemical fuel cells. In cell biology, interfacial supramolecular assemblies will advance the understanding of how individual proteins in the cell membrane regulate chemical flow and determine immune response. In genetics, ISAs will provide the test-beds for methods for tagging DNA with single fluorescent molecules or nanoparticles in order to map gene locations along a chromosome.

Prevention of environmentally induced diseases is a major focus of health science research. In particular, the development and validation of alternative models and test systems for detecting environmental pollutants and toxins is becoming an increasingly important objective. Array sensors based on self-assembly methods, state-of-the-art biotechnology, thin-film preparation and processing methods, as well as organic molecular design principles will lead to 'exquisitely' sensitive clinical sensors (a multibillion dollar world market). These sensors will have the ability to identify and measure target molecules (including toxic chemicals and allergens) under real-world conditions with unrivaled efficiency and sensitivity (a few molecules within a 'swimming pool' of sample).

Beyond these traditional areas, biomolecular interfacial supramolecular assemblies are making significant strides in terms of creating optoelectronic circuits that have potential applications in superhigh-resolution video imaging, ultrafast switching, logic devices and solar power generation. For example, self-assembly can be used to isolate and orient naturally occurring leaf proteins onto a gold substrate. The isolated protein centers are naturally occurring photovoltaic and diode structures and can be used to generate electrical current when provided with an electrical contact, e.g. through a film of metal nanoparticles.

7.5 'Smart Plastics'

Self-assembling photonic crystals are another application with enormous potential. Research groups worldwide have built photonic crystals, although their efforts have usually involved laborious and expensive fabrication processes. However, approaches that exploit self-organizing 'smart plastics' are beginning to emerge. For example, it is now possible to create polymers which will self-organize into hollow spheres that then form extended structures – a feat similar to bricks stacking themselves into a wall. The alternating spheres and plastic framework can manipulate light in predictable, precise ways. These self-assembled materials allow the trapping and propagation of light to be controlled, thus allowing high-speed, high-density information storage and communication systems to be created. Potential applications include optical data storage and telecommunications, both of which rely on transmission and detection of specific wavelengths, and eventually holographic data storage. Materials of this kind will stimulate the development of improved light-emitting diodes (LEDs), plastic-based lasers and paints that change color under different light conditions. Additional applications of polymer-based materials may arise from the search for flat display panels for computers televisions and cell phones. Today's most common flat panels, i.e. the liquid crystal displays, are expensive, hard-to-manufacture, and are large energy consumers. These difficulties are the driving force for a new wave of interest in electroluminescent light-emitting devices.

Studies of the electrochemical and conductive behavior of polymers indicate that these properties depend on both the three-dimensional structure of the layer and on the interface between the polymer and the liquid. A particular good example is the observation, discussed in Chapter 4, that the ordering of polyalkylthiophenes with respect to the substrate surface strongly determines the conductivity of the film. This is an important observation and if more general methods could be developed to increase the order in polymer layers, greatly improved electrochemical and conducting properties may be expected through an improved interchain interaction. The integration of block copolymers and/or rod–coil polymers would appear a promising route to increased organization.

In the future, the marked matrix dependence of the electrochemical behavior of redox-active polymers needs to be seriously considered when designing new electrochemical sensors. Furthermore, electrochemical investigations may be extended into the area of biologically important polymers. For example, the investigation of proteins modified with redox-active sites will lead to a better understanding of how protein matrices control electrochemical processes. The observation that the poly(vinyl pyridine)-type materials discussed in Chapters 4 and 5 are efficient electron relays for enzyme-based redox reactions, points to the compatibility of artificial and natural components in this respect. However, it needs to be pointed out that only studies on very well-defined films are likely to succeed. These films need not have a structural organization at the atomic level, but do need to be amenable to structural characterization by using techniques such as electrochemistry, mass-sensitive studies and neutron reflectivity. It is the ability to manipulate their structures by using external variables such as ionic strength or pH that make these systems of particular interest.

7.6 Interfacial Photochemistry at Conducting Surfaces

In contrast to their electrochemical properties, the photophysical and photochemical behaviors of molecular components immobilized onto solid electronically conducting substrates is only relatively poorly understood. Present results indicate that, while electrochemical communication can be maintained between redox-active centers and the electrode over relatively long distances, this is not the case for molecular components in the excited state. The photophysical behavior of monolayers points to a strong interaction with the substrate. On the other hand, polymer films emit and undergo photoinduced ligand exchange processes as observed in solution. However, the role of the electrode surface in deactivating excited states is not fully understood and further investigation on the mechanism and distance-dependence of this deactivation process are likely to provide new insight.

7.7 Modified Semiconductor Surfaces

One of the aims behind the investigation of interfacial supramolecular assemblies is the potential advantage that may be achieved by the interaction between a molecular component and a solid substrate. The best example of this interaction is the development of the TiO_2-based Grätzel-type solar cell. One of the limitations of the present cell design is the presence of acetonitrile as a solvent. This limits the applicability of the cells as solvent evaporation may be expected at high temperatures. The answer may lie in the development of cells based on solid electrolytes, although novel systems will be required since the solid-electrolyte devices produced so far have low photocurrent efficiencies. Another area where advances are needed is in the nature of the surface-based relay, used to reduce the attached dye after photoexcitation. The redox potential of the I^-/I_3^- redox couple is not optimal and leads to considerable losses in cell voltage. So far, no alternative redox couples have emerged.

Modified TiO_2 surfaces have also found application in the design of fast electrochromic devices. The influence of the substrate on the behavior of interfacial assemblies is well illustrated in this book. However, it is important to realize that the electrochromic behavior observed for modified TiO_2 surfaces was not expected. The oxidation and reduction of attached electrochromic dyes are not mediated by the semiconductor itself but by an electron-hopping process, not unlike that observed for redox polymers, where the electrochemical reaction is controlled by the underlying indium–tin oxide (ITO) contact. These developments show that devices based on interfacial assemblies are a realistic target and that further work in this area is worthwhile.

There are also some very important scientific conclusions to be drawn from this work. The attachment of photoactive molecular components to TiO_2 surfaces has shown that the photophysical properties of the resulting interfacial ensemble are significantly different from those observed for the individual components. The semiconductor surface clearly acts as a rectifying one. In order to fully understand the balance between the intramolecular and the interfacial processes, more experiments are needed where the nature and connectivity of the molecular components, as well as the nature of the contacting solution, are systematically varied.

Other opportunities lie in the investigation of substrates based on p-type semiconductors such as NiO. For p-type materials, electron donation from the valence band to the ground state of the attached molecular component is observed. At present, this process is not well understood and further studies are needed to investigate the behavior of interfacial assemblies based on p-type semiconductors.

7.8 Concluding Remarks

Not withstanding the tremendous progress that has been made in the field of molecular self-assembly, many challenges remain and further research is needed simply to improve control of the process. However, the interdisciplinary nature of this field is one of its greatest strengths, hence allowing researchers to tap into rather unusual building blocks. One area of future exploitation may be in the area of 'nanotools' for synthetic manipulation. Biomotors are receiving increased attention for studying molecular-level transport and as potential memory elements in information storage. Such molecular machines exploit the inherent molecular recognition and self-assembly that is so well developed in biology and will probably find uses in information relays for future computing applications.

Another challenge is achieving the self-assembly of multiple materials in a controlled manner. While Chapter 4 describes approaches that yield multilayer structures, much of this book focuses on essentially two-dimensional structures. Many of the existing multilayer structures show interpenetration of the layers, thus making the resulting devices inefficient and limiting the fundamental insights into the optoelectronic properties that can be obtained.

Index